Data Analysis for Experimental Design

Data Analysis for Experimental Design

Richard Gonzalez

THE GUILFORD PRESS
New York London

A Division of Guilford Publications, Inc.
72 Spring Street, New York, NY 10012
www.guilford.com

Printed in the United States of America

This book is printed on acid-free paper.

Last digit is print number: 9 8 7 6 5 4 3 2 1

Library of Congress Cataloging-in-Publication Data available from the publisher.
ISBN 978-1-60623-017-6

About the Author

Richard Gonzalez, PhD, is Professor of Psychology at the University of Michigan. He also holds faculty appointments in the Department of Statistics at the University of Michigan and in the Department of Marketing at the Ross School of Business; is a Research Professor at the Research Center for Group Dynamics, which is housed in the Institute for Social Research, University of Michigan; and has taught statistics courses to social science students at all levels at the University of Washington, the University of Warsaw, the University of Michigan, and Princeton University. He received his PhD in Psychology from Stanford University.

Dr. Gonzalez's research is in the area of judgment and decision making. His empirical and theoretical research deals with how people make decisions. Given that behavioral scientists make decisions from their data, his interest in decision processes automatically led Dr. Gonzalez to the study of statistical inference. His research contributions in data analysis include statistical methods for interdependent data, multidimensional scaling, and structural equations modeling.

Dr. Gonzalez is currently Associate Editor of *American Psychologist*, and is on the editorial boards of *Psychological Methods*, *Psychological Review*, *Psychological Science*, and the *Journal of Experimental Psychology: Learning, Memory, and Cognition.* He is an elected member of the Society of Experimental Social Psychology and of the Society of Multivariate Experimental Psychology.

Preface

To test a hypothesis for significance is relatively easy; to find a significant hypothesis to test is much more difficult.

Allen L. Edwards

The goal of this book is to highlight the relatively simple concepts that drive analysis of variance, rather than to write a comprehensive handbook of every detail related to the analysis of experimental design. You will find material presented from a conceptual standpoint and discussions of computational formulas avoided unless they provide additional intuition. I focus on single degree of freedom tests wherever possible (what we call *comparisons*) and walk the reader through the rationale of each statistical test presented. To keep the discussion simple, I have avoided presenting all the countless "exceptions to the rule" that make most statistics textbooks thick and serve to confuse the beginning reader. Instead, I stick to the most common analysis of variance models (mostly on interval data) and refer you to reference books as needed. Also, I do not worry too much about precise notation, such as using different symbols to distinguish an observation from its theoretical random variable. I prefer to keep subtle notation distinctions to a minimum.

My other goal in writing this book was to show readers how methods and statistics work together. All too often statistics comes along to fix a methodological flaw. A more sophisticated approach is to be mindful of the interplay of method and statistics from the outset. I show how subtle details in methodological design have profound implications for how data should be analyzed. Within chapters you will see discussion of methodological design intermixed with statistical procedure. This will give you a framework for evaluating statistical results against the specifics of the methodological design. I focus on experimental design and relevant statistical procedures, highlighting, where possible, threats to the ability to make causal inference from an experimental design.

To get you started, the first chapter offers an introduction to relevant terminology and an exploration of what motivates experimental research. The second chapter uses an example created by the late Allen L. Edwards—the farmer from Whidbey Island who claimed to have a special gift—to introduce important concepts in statistical testing, such as the comparison of an observed result to that expected by chance. I use a little probability theory in Chapter 2 to highlight the foundation of hypothesis testing in the Fisherian and Neyman–Pearson styles. A little bit of probability goes a long way in developing intuition for the statistical tests that follow. Chapter 3 introduces the normal distribution as a technique to approximate the tedious probability computations performed on the farmer example in Chapter 2. The remaining chapters develop this approximation in different situations that arise in psychological research, such as when the population variances are unknown, one-way between-subjects designs, factorial between-subjects designs, and within-subjects designs. In several sections, I offer the simplifying assumption that there are equal numbers of participants in each condition or that there are no missing data. This allows me to make the points in a simple and clear way. After all, most computer programs will handle the correct computations—what is important is that the human user comprehends what he or she is doing, and I believe that this is best accomplished by keeping things simple and working with special cases. I do, though, point out in several places how to handle unequal samples or missing data (e.g., a multilevel version of within-subjects designs that complements the main presentation in the text). I also avoid detailed discussion of the esoteric features of analysis of variance such as randomized-block designs and Latin squares.

I have elected in this book to focus on a few procedures rather than present an endless encyclopedia of every possible technique and exception to the rule. My reasons for selecting particular procedures over others are explained throughout the text. In short, I wear my presentation choices and opinions "on my sleeve." In some cases an instructor may not agree with my choices. This could make for interesting classroom discussion. The broader pedagogical point is that methodology and statistics are academic disciplines in their own right with their share of debates. Like many, I believe the classroom experience is enhanced by debate and careful considerations of different options rather than merely presenting methods as a set of recipes to follow, as one would find in a cookbook.

Each chapter ends with several exercises and questions. I encourage the reader to try them all. I have found, in both my own learning about statistics and that of students, that it is one thing to read about a statistical test and another to conduct the test, including entering data into a computer program, running the program, and interpreting the output. The former involves book learning; the latter involves skill learning. An analogy to sports may help make the point clearer. It is one thing to *read* about the proper way to swing a tennis racket, kick a soccer ball, or throw a curveball; but to be able to swing, kick, or throw at a sufficient level of proficiency requires *practice*. Most of you are reading this book because you desire to achieve a level of proficiency in data analysis for your research or your work. You need to develop the skill set that includes both knowledge about the "why" and production about the "how." In addition to the exercises and questions, I encourage the reader to follow along with each example in the chapters, including entering data into your favorite computer program and comparing your output to the results presented in this book.

Many of the ideas presented in this book were inspired by discussions with one of my mentors, Allen L. Edwards, who died on July 17, 1994, just months after we agreed to write a new edition of his book *Experimental Design in Psychological Research* (Harper & Row, 1985). It took me a long time after 1994 before I could imagine writing a book on experimental design without him. Readers who are familiar with Allen's "blue book" will find examples and explanations from that book represented in this book that the Allen L. Edwards Living Trust has graciously granted us permission to use. There was a time when almost every graduate student in psychology learned statistics from one of Allen's textbooks. His books had a subtle, but deep, influence on several generations of psychologists. Allen is credited as being one of the people who brought rigor (and statistics) to psychological research. His empirical research gave us important concepts such as social desirability and new methodologies for assessing personality. I hope that this book proves to be a worthy successor to his earlier book and our discussions together.

Several graduate students commented on earlier versions of the chapters. I am grateful for their candid feedback. I also thank the reviewers—Daniel Ashmead from Vanderbilt University, Warren Lacefield from Western Michigan University, and Michael Milburn from the University of Massachusetts, Boston—who provided insightful feedback and greatly improved the quality of the book.

My views on experimental design and analysis were influenced by my mentors and colleagues. In graduate school I particularly benefited from many late-night discussions with Amos Tversky about research, statistics, and mathematical modeling. I have spent many hours discussing the fine points of statistics and research methodology with Phoebe Ellsworth, Fred Feinberg, Dale Griffin, John Miyamoto, Grazyna Wieczorkowska, and George Wu. I have greatly benefited from these discussions, and I suspect that many of their ideas have likely made their way into this book. I want to give special thanks to David Edwards (Allen's son) for his support throughout this project. The following universities made it possible for me to pursue scholarly activity and work on this book: the University of Washington, the University of Michigan, Princeton University, and the University of Warsaw.

Several people helped with the typesetting of the book. Lisa Nichols was instrumental in the early stages of typesetting the book into LaTeX. Sandra Becker worked through the numerous revisions of each chapter, helped with many of the figures, provided numerous comments during the final editing stages, and gave the entire manuscript a thorough reading while she also juggled teaching several courses. The people at Guilford Publications have been wonderful in every possible way throughout the entire process of producing this book.

Last, but not least, I thank my family (Carrie, Christina, and Geoffrey) for putting up with my unconventional work habits while I worked on this project.

Richard Gonzalez

Contents

Data Analysis for Experimental Design

1

The Nature of Research

1.1 Introduction

Everyday observation is one way to initiate research because it may motivate us to formulate hypotheses, to state problems, and to ask questions. But it is through planned and systematic observation that we find answers to these questions. The objective of empirical research is to use observation as a basis for answering our questions. In research, we do not haphazardly make observations of any and all kinds, but we direct our attention to those observations that we believe are relevant to our questions.[1] A good comedian has a keen eye for everyday observation, and our everyday behavior can be quite funny when framed the right way. A good social science researcher must combine a keen eye for observation with rigorous methodology. We want to make sure that our observations are not biased, and we want to minimize the possibility that our observations may lead to incorrect inferences. This

[1]The brief discussion of research design and experimentation in the behavioral sciences given in this chapter does not do justice to the subject. For a more comprehensive treatment of research methods and techniques, see Neale and Liebert (1986); Underwood and Shaughnessy (1975); Judd, Smith, and Kidder (1991); Aronson, Ellsworth, Carlsmith, and Gonzales (1990).

book focuses on one type of research methodology that is used in the social sciences—the experimental method.

Even when observations are made systematically, the raw unclassified observations in their original form may not lend themselves to an obvious interpretation of the questions we have posed. We therefore use techniques that organize and reduce the observations to a more manageable form so that clear answers can emerge from our questions. These techniques involve classifying and operating on the observations to reduce them to a simpler format such as frequencies, proportions, means, variances, correlation coefficients, and other statistical measures. On the basis of these statistical measures, we hope to draw conclusions or inferences that will bear upon the questions of interest.

Consider a simple example of a research question and a ministudy designed to provide an answer. You and your romantic partner are in a disagreement as to what to do Saturday night. You had a long, tiring week and look forward to a relaxing, quiet evening. Your partner, on the other hand, also had an exhausting week but very much wants to go dancing. Your partner decides to "settle this" by flipping a coin and, holding a coin up high, states, "Heads we stay home, tails we go dancing." With anyone else you would probably agree to such a prospect, but in this case you hesitate because you know that your partner makes extra money as a magician entertaining children at birthday parties. You wonder to yourself, "Is that a fair coin?" ... "Is that coin biased, so that tail is much more likely to come up than head?"

In this case you are interested in the question of whether a particular coin is fair. By "fair" we mean that when tossed, the coin is equally likely to land heads or tails. It isn't too difficult to learn how to flip a coin so that one side comes up more frequently than the other side, or to modify the coin so that one side is more likely to land than the other.

Yes, we are well aware that coin flips can lead to instant boredom and daydreaming in a statistics class. Please indulge us because we think that we can make a very useful point.

Let's formalize the question of whether this is a fair coin. An observation will consist of the result of a single toss; that is, we observe the coin flip and record whether the coin lands heads or tails. In an attempt to answer the question of whether the coin is fair, we can flip the coin (in a nonsystematic manner, of course) many times. Assume we make 100 such observations (that is, 100 tosses of the coin), and we observe the sequence HTTHTTT ... H, with each H and T being the record of a single observation as either Head or Tail. But this series of observations presented in that manner does not make it easy for us to determine whether or not the coin is fair. We can reduce the complexity of the observations, however, by counting the number of H's and the number of T's, and these two frequencies will summarize succinctly under some specific conditions the complete set of observations in a manner that can answer our question. A fair coin should have roughly an equal number of H's and T's. Any major departure from such equality would provide evidence against the notion that the coin is unbiased (that is, is evidence favoring the inference that the coin is not fair). We would be suspicious if the coin landed tails 100 times out of 100 tosses. An important problem in research is how to evaluate objectively the evidence provided by a given set of observations. How do we know that a discrepant observation, such as 100 tails out of 100 tosses, is sufficiently discrepant relative to our expectations? How do we formalize our expectations? This book will present several techniques that allow one to examine evidence in a statistically sound way. The methods can be used in a variety of different real-world and laboratory contexts—they do not require artificial coin flips.

The coin toss example, albeit contrived, characterizes fairly well the nature of much research. One or more questions are formulated. Is the coin toss fair so that I can be comfortable that my partner is not trying to manipulate the outcome? Systematic observation is then made of things believed to be relevant to those questions. Having made a series of observations, the observer then reduces the observations to a limited number of measures that provide a summary description of the complete set of observations (such as the number of heads and tails). By means of further operations on the descriptive measures, statistical inferences regarding the questions of interest can be evaluated. That, in a nutshell, is the essence of the statistical side of most empirical research.

A number of additional aspects of research are also illustrated by the coin example. How many observations should be made? Obviously, only one observation would not provide us with adequate information to determine whether or not the coin is biased. If an insufficient number of observations are made, we may not be much better off than with none. On the other hand, if we make more observations than are needed, we will have wasted time and energy that might be more fruitfully spent in other endeavors (such as reading a book or dancing or learning more about research methods). We will later discuss techniques that can help us determine the number of observations needed to answer our research questions.

Determining what observations to make is not always an easy task. In the coin-tossing example, we do not, for example, direct our attention to observing the position of the coin on the floor after each toss, but rather we observe whether the coin falls heads or tails. Nor do we care to observe the color of our partner's shirt or what we had for dinner the previous night. In this example we observe the frequency of H's and T's because we believe these observations are relevant to the question of whether the coin is fair. Our decision to examine the frequency of heads and tails suggests that we did not view a particular sequence of heads and tails (such as HTTHTTT ...) as relevant to the question of the coin's fairness. This may not always be correct, as in the case where the first 50 tosses landed on heads but the second 50 tosses landed on tails. We would view this as a very unlikely event even though the frequencies show an equal number of heads and tails. In other research problems, the relevance of the observations made to the question of interest may not always be so clear-cut. The issue of whether the observations in a given research problem are relevant to the question must always be given serious consideration during the planning of the research project.

The manner in which the coin is tossed is also important. If the coin is tossed in a biased manner such that the tosser imparts a particular spin on the coin so that one side is more likely to land than the other, then these observations would bear not on the nature of the coin—the original problem of interest—but rather on the way the coin was tossed. Such observations would be telling us more about the tosser than the coin. To assess whether the coin itself is fair, we need to make sure that the coin is not tossed in a biased manner.

Just because we make objective observations does not imply that we will always make a correct inference all the time. We may, for example, conclude that the coin is biased when, in fact, it is not. Or we may fail to conclude that the coin is biased when, in fact, it is. These two kinds of errors (the medical community calls these types of errors false positives and false negatives) also require consideration during the planning of research. We will discuss these errors in a later chapter.

Thus, in every research project we need to determine the question of interest, decide what observations are needed, ensure that those observations tell us what we want to know, decide how we will summarize the observations, decide on how to evaluate those summary measures from a statistical standpoint, and decide how we will convert the observations and the summary into a statement relevant to our original research question.

1.2 Observations and Variables

We have emphasized the importance of observations in research. The things that are observed are called **variables**. In the example cited earlier, the variable, the thing observed, was the face of the coin (heads or tails). Any particular observation is called a **value** of the variable. For the face of a coin, there are only two possible values of the variable: heads or tails. The value of a variable indicates the class to which an observation is to be assigned. If a coin lands heads, we consider that observation and all others with the same value of head as belonging to the same class. In order for something we are observing to be considered a variable, there must be at least two possible classes of observations. The classes must also be mutually exclusive; that is, any given observation can be assigned to only one of the available classes. Therefore, to define **variable** more precisely, we say a variable is anything we can observe such that each single observation can be classified into one and only one of a number of mutually exclusive classes. Some variables such as length, weight, and time may have an infinite number of possible values. It is possible for a class to have more than one value, such as when a researcher codes income in terms of interval (for example, $30,000 to $39,999).

1.3 Behavioral Variables

By a **behavioral variable** we mean any variable that refers to some action or response by an organism. At one extreme, we have simple responses such as eye blinks, pressing a key when a light is flashed, marking true or false for an item on a test, and so forth. At the other extreme, we have complex behavior patterns such as those involved in problem solving, aggression, dominance, and leadership.

One commonly used behavioral variable in behavioral research is the time required for some action or response to occur. This variable is called response time. A stimulus is presented to a participant, and the time required for the participant to make a response is measured. For example, a participant is given a puzzle, and the time required to solve the puzzle is measured. Typing skill may be measured by determining how long it takes an individual seated in front of a computer keyboard to type a standard passage.

In other cases, we may hold the time fixed to a particular interval and count the number of responses of a given kind that occur within that fixed period. For example, we may count the number of times an infant looks at Mom during a fixed interval of time. Or we may count the number of problems solved in a fixed interval of time. We may count the number of "aggressive" responses made by a child during a play period or the number of times the child withdraws in response to the aggressive advances of another child.

In still other cases, we may simply count the number of responses of a given type without regard to the time required to make these responses. For example, we may count the number of correct responses made to a fixed number of problems without concern for the time required by the participant to make the responses.

Sometimes behavioral variables are assessed through the use of human raters. Instead of counting the number of responses of a given kind made by a child, we may ask judges to observe the behavior of the child in a specific situation and then rate the degree to which they believe the behavior of the child reflects a variable of interest. For example, observers might rate the overt aggressiveness of children during play with classmates on a five-point scale of aggressive behavior. Rating scales are frequently used in behavioral research to obtain measures of behavioral variables that are not otherwise easily quantified.

Sometimes the "judge" who makes the rating is the participant in the research study. For instance, the researcher may ask participants to rate how proud they are of their own accomplishments on a seven-point rating scale, ranging from "not at all" (1) to "I am extremely proud of my accomplishments" (7). The term "self-report" refers to such cases where the judge and the research participant are the same person—the judge is rating himself or herself.

1.4 Stimulus Variables

Behavior does not occur in a vacuum but occurs in a particular setting or environment. Situation and context matter to many of the research questions that are of interest to behavioral scientists. The general class of things we observe that relate to the environment, situation, or conditions of stimulation we will refer to as **stimulus variables**. The stimulus variables in a behavioral experiment may consist of relatively simple things, such as electric shock, light, sound, or pressure, which may be quantified by measuring the physical intensity of the stimulus.

There are other stimulus variables of interest to the behavioral scientist for which we have no measures corresponding to physical intensity. These may consist of problem-solving situations, intimate situations, social situations, and so forth, and they are relatively more difficult to quantify. Indeed, in much behavioral research, we can only say that the variations that are possible consist of complex combinations of stimuli differing in kind rather than degree. We will refer to the conditions within a given experiment as **treatments**. Thus, the term "treatment" will refer to a particular set of stimulus or experimental conditions.

In an experiment, for example, we may be interested in the behavior of children when they have an "authoritarian" teacher and in the behavior of other children when they have a "democratic" teacher. The behavior of the teacher will vary according to the instructions of the researcher, but we attempt to keep all other aspects of the stimulus situation constant.

In some research problems, treatments are created by having some aspect of the situation either present or absent. Thus in a learning experiment, the performance of undergraduates solving a puzzle may be observed under conditions where alcohol is consumed and under conditions where alcohol is absent. Or one group of participants may study a passage under conditions where they have feedback about their performance; that is, they are required to respond to material in the text and are immediately given information as to whether their response is correct or incorrect. Another group of participants may study the same passage but without feedback; that is, they are not given information as to whether their response is correct. The group for which some aspect or condition of the situation is present (or is given the "active ingredient") is often referred to as the **experimental group** (or the treatment group), and the group for which the aspect or condition of the situation is absent (or is given an "inert ingredient" such as a placebo) is often referred to as the **control group**. Thus in a study to determine whether or not watching violent movies increases the frequency of aggressive behavior, the treatment group might watch a violent movie and the control group might watch the same movie with the violent scenes removed.

1.5 Individual Difference Variables

Individual difference variables arise from ways in which organisms may be classified and from the observations and measurements of physical, physiological, and psychological characteristics of organisms. For example, we may measure the height and weight of an individual. These observations do not correspond to behavioral variables or stimulus variables, but they may be described as **individual difference variables**. They are characteristic ways in which the particular individuals vary. Individuals may be classified according to the color of their hair or eyes. Or they may be classified according to whether they are males or females, or in terms of their age or height, or any other variable on which individuals differ.

Some individual difference variables are inferred from responses. A person's IQ, for example, is determined by observing his or her response to a standardized testing situation. As another example of a response-inferred individual difference variable, consider the case when participants are given a self-report scale measuring anxiety. On the basis of those scores the researcher may classify some participants as "anxious" and other participants as "nonanxious."

1.6 Discrete and Continuous Variables

In a memory experiment a variable of interest may be the number of correctly recalled items made by a participant out of 20 items that were studied. The only possible values for this variable are 0, 1, 2, ..., 20. Such a variable consists of **discrete** numbers. On the other hand, if we measure the time required for a participant to write down the item that was recalled, there are an infinite number of possible values. Such a variable is called **continuous**. Values of continuous variables are never exact (in the sense that there is little certainty about which category the observation falls), because no matter how precisely we measure the variable, there is always uncertainty with respect to the observed or recorded value of the variable. For example, if we measure time in units of 0.001 second, we can never be sure that a recorded value of 1.213 seconds is actually 1.213 followed by an infinite number of zeros, that is, 1.213000 ... seconds, or 1.213 followed by any other pattern of digits. Thus technically all **observed** measurements are discrete, regardless of whether the underlying variable is continuous or discrete.

Because of the uncertainty associated with observed values of continuous variables, we regard such values as representing an interval ranging from one-half unit below to one-half unit above the measurement. For example, if time is measured in units of 0.001 second, then a recorded value of 1.213 seconds would be regarded as representing an interval ranging from 1.2125 to 1.2135 in order to reflect the uncertainty in the resolution of our measurement.

We will take the same position regarding intervals for discrete variables. For example, it will be convenient to regard eight correct responses as representing an interval ranging from 7.5 to 8.5. It will become obvious, in later chapters, that there are practical reasons for regarding measures of discrete variables as representing intervals, just as we regard measures of continuous variables as representing intervals.

1.7 Levels of Measurement

In measurement we assign numbers to the values of a variable. The numbers should be assigned in such a way as to code the information present in the original observations. We do not want the numbers to have more information than what was contained in the observations. For example, it might be convenient to code males with the integer 1 and females with the integer 2. This does not mean that the females scored one point more than the males. The numbers, in this case, serve as names for the two categories of gender. Scales that use numbers as labels for the values of a variable are called **nominal scales**.

When observations carry information about differences in degree, there are several ways of assigning numbers to those observations. If the observations merely reflect order or rank, then the numerical scale that can be associated with those observations is called an **ordinal scale**. An example would be if a research participant is shown six pictures of members of the opposite sex and asked to place the photograph of the person she thinks is most attractive on the far left, the photograph of the second-most attractive person just to the right of the first, and so on. The task of this participant is to rank-order the pictures on the basis of perceived physical attractiveness. One obvious set of numbers that can be assigned to those judgments are the integers 1 to 6; that is, the picture assigned a rank of 1 represents the one judged most attractive, etc.

In the attractiveness example, the original observations do not provide any information about differences between consecutive pictures. In other words, we could not claim that the difference in physical attractiveness between the first and second ranked photograph is the same as the difference between the third and fourth ranked photograph. Even though the assigned numbers might suggest equal intervals, the judgments (that is, the actual observations in this example) do not convey equal intervals because they merely express the rank order. Therefore, in this case the numerical assignment cannot be interpreted to have an "equal-interval" property. This means that any assignment of numbers, as long as the numbers maintain the observed ordering, is allowed. The researcher could assign the integers 5, 20, 22, 50, 100, 2000 to the six photographs, and the critical property of the observations (that is, the rank order) is still preserved by the numerical assignment. An example of a commonly used ordinal scale in science is the Mohr hardness scale, which is used to measure the hardness of minerals (for example, diamond is harder than quartz, quartz is harder than talc).

When observations themselves possess the equal-interval property, then the numerical assignment must preserve that property. This means that if the participant perceives differences between two pairs of stimuli as equivalent, then the numbers assigned to represent those four stimuli must have the comparable property that the difference between the values for one pair should equal the difference for the second pair. Numerical scales that preserve the equal-interval property in the original observations are called **interval scales**. A common interval scale is temperature as measured by the Fahrenheit and Celsius scales. The meaning of a 10° difference in temperature Fahrenheit is the same across the Fahrenheit scale as is the meaning of a 10° difference in temperature Celsius across the Celsius scale.

The equal-interval property refers to how the numbers are used and not the numbers themselves. An aggression researcher who examines the relation between urban violence and temperature may be using temperature as a proxy variable for some other psychological construct (for example, on hot days people may be more likely to go outdoors, so temperature serves as an indirect measure of the likelihood of being outdoors). In that case, the psychological construct does not likely possess the equal-interval property even though the numerical scale that is being used—temperature—does carry that interpretation. The key idea here is that properties such as order or equal intervals must refer to the observations under study, and merely assigning numbers to those observations does not guarantee that the observations satisfy the properties.

Another important property is concatenation. Let's consider length. If a participant perceives that the length of one object is equivalent to putting two smaller objects end to end, then the numbers assigned to the three objects should have this property: the sum of the numbers representing length that are assigned to the two smaller objects should equal the number representing length that is assigned to the large object. Numerical assignments that have this property are called **ratio scales**. They are called ratio scales because all the essential properties of the data are maintained when one multiplies the scale values by a constant (that is, concatenation is preserved). For example, numbers assigned to represent the length of two objects have the same ratio regardless of the scale of measurement (it does not matter whether we use feet, inches, meters, or miles).[2] Another example of a ratio scale is weight. If two lighter objects together weigh as much as a particular heavier object (for example, using a pan balance, we find that two objects on one side balance the third object on the other side), then this necessary property of a ratio scale holds. Again, the property of concatenation should apply to the observations. The objects themselves may be consistent with the properties of the ratio scale, but if the investigator is referring to the perceived weight of those objects, then the situation may be different. One would need to examine whether the perceived weight judgments also satisfy the property of concatenation. It may not hold that the same mechanism that drove the pan balance occurs inside the perceiver's head.

The properties of the observations influence one's choice of the method of analysis and determine how one should interpret the analysis. Clearly, if a random sample of 100 students is drawn from a public university—with males being assigned 1 and females assigned 2—it would not make sense to interpret the average of the gender variable in terms of degree. It would make sense, however, to interpret the proportion of females in the sample.[3]

[2]Multiplying by a constant converts one ratio scale into another. For example, to convert from inches to feet we multiply the number of inches by $\frac{1}{12}$.

[3]The proportion of females in the sample is equivalent to an average of the gender variable when males are coded as 0 and females are coded as 1. Thus, the concern is not with taking an average of nominal variables but in how we interpret the meaning of the average of a nominal scale.

The definition of ordinal, interval and ratio scales is more involved than merely the three properties presented here (maintaining order, preserving equal intervals, and preserving concatenation, respectively). For more information about the important topic of how numbers can and should be assigned to empirical phenomena see Falmagne (1985); Krantz, Luce, Suppes, and Tversky (1971); Mitchell (1990); Narens (2001); Roberts (1979); and Stevens (1946). This flavor of measurement (that is, a concern over how numbers represent empirical phenomena, how numbers are assigned to observations, the uniqueness and meaninfulness of such numerical assignments, etc.) is a major topic of study in mathematical social science.

1.8 Summarizing Observations in Research

Unambiguous, systematic observation is necessary to obtain clear answers to our research questions. In some cases, the questions of interest have to do with the accurate description of a group of participants with respect to one or more variables. An instructor, for example, may be interested in how the intelligence test scores of students in a class are distributed. The systematic observations might consist of the scores of the students on a standardized test of intelligence. These observations may be reduced (classified) so that the instructor has available for each score the number of students who had that score. The instructor may also be interested in the average score for the students in the class and in the range, or spread, of the scores.

In other cases, the questions of interest may concern the degree of association or relationship between two variables. For example, the same instructor may also be interested in determining whether the intelligence test scores of the students are related to or associated with scores on the final examination. Do students who score high on the intelligence test also tend to obtain high scores on the final examination, while those who score low on the intelligence test tend to obtain low scores on the final examination? To answer this research question the instructor can summarize the observation with a scatterplot and a correlation coefficient.

In some situations, it is possible for an investigator to vary quantitatively one variable, usually a stimulus variable, and to study the behavior of research participants under each value of the stimulus variable. For example, we might vary the font size in which a list of words is printed. Research participants are assigned to a given font size, and the words are presented at a constant rate. The observations obtained for each variation in the stimulus conditions might be the number of words correctly recognized by each participant. If the average number of words correctly recognized for each font size is computed, we may then assess the relation between these averages and the font size. We might be interested, for example, in finding out whether the average number of words recognized increases as the font size is increased. A plot representing the mean scores as a function of font size may provide a useful representation.

In the case of the instructor interested in the relation between scores on the intelligence test and scores on the final examination, the instructor is not able to control or manipulate the variables of interest. For example, the instructor may have little control over the intelligence test scores or over the final examination scores of the students. The values of these variables may largely be fixed or determined by each student and cannot be manipulated or changed directly by the instructor. In contrast, in the font size example one of the variables is directly under the control, or manipulation, of the investigator. This variable is the font size and is a stimulus variable. The ability to control or manipulate at least one of the variables is a necessary feature of an **experiment**. Thus the intelligence test example described earlier does not constitute an experiment because the instructor does not manipulate the intelligence of the research participant. A second necessary feature of an experiment is that participants should be **randomly assigned** to the different levels of the treatment.

The variables over which the investigator has control are called the **independent variables**. Independent variables are the variables that the investigator manipulates or varies. As the independent variables are changed or varied, the investigator observes other variables to see whether they are associated with, or related to, the changes introduced by the independent variable. These variables are called the **dependent variables**. In the above example, the independent variable is the font size and the dependent variable is the average number of words correctly recognized for each font size.

In most experiments, interest is directed toward the problem of discovering whether the different treatments result in differences in the observed values of the dependent variable. Sometimes we may be interested in studying the functional relation between a quantitative independent variable and the dependent variable, that is, in determining the nature of the relation between the two variables. In an experiment the researcher is prepared to observe the event, the event can be replicated, the conditions surrounding the occurrence of the event can be manipulated, and participants can be randomly assigned to treatment condition.

As we will see shortly, random assignment is useful because it provides a way to eliminate a class of alternative explanations to a research finding. Not all research problems entail or permit random assignment in a study. We will discuss methods that use statistical procedures to adjust for various types of alternative explanations. We want to emphasize at this early stage that a major concern of research methodology is the reduction of alternative explanations. Random assignment is one tool, but there are many other tools available to the researcher.

1.9 Questions and Problems

1. Explain the following statement: "Naming a variable may suggest the observations to be made, but the observations may define the variable itself."

2. How does an individual difference variable differ from a stimulus variable and a behavioral variable?
3. What is the difference between a control group and an experimental group?
4. What is the difference between a dependent variable and an independent variable?
5. What is the difference between a continuous variable and a discrete variable?
6. Explain the difference between nominal, ordinal, interval, and ratio scales.
7. Explain why it is useful to think about measurement as consisting of a range.

2

Principles of Experimental Design

2.1 The Farmer from Whidbey Island

Some years ago a farmer from Whidbey Island visited the psychology department of the University of Washington. After wandering around the halls and knocking on various doors, he eventually found a professor of psychology at work in his office. The farmer had with him a carved whalebone and told the professor that in his hands the bone was an extremely powerful instrument, capable of detecting the existence of even small quantities of water. The farmer said that several of his neighbors on Whidbey Island had tried unsuccessfully to tap water wells and had called upon him for help. The farmer said that his whalebone was able to detect where the underground water was located. When his neighbors drilled wells at the points he had located with the whalebone, they found water.[1]

[1]Hyman and Vogt (1967) provide an account of water witching in the United States. Interested readers may also visit the website of the American Society of Dowsers (*www.dowsers.org*).

The farmer added that he was unable to explain his peculiar power. His neighbors were unable to use the whalebone in locating water; the whalebone had to be in his hands before it would dip sharply to indicate the presence of water. He was somewhat disturbed by his ability and thought that perhaps the psychologists at the university would be interested in examining him and telling him why it was that he was able to use the bone so effectively while others could not. He himself thought that it had something to do with "magnetism" that emanated from his body. Anyway, he would be willing to demonstrate his ability so that the psychologists could see for themselves. Perhaps then they could explain it to him. The farmer asked for a cup filled with water. When he was given the cup, he placed it on the floor. He then grasped the whalebone and held it stiffly in front of him as he moved slowly about the room. When the apex of the bone passed over the cup of water, his arms trembled slightly and the bone dipped toward the ground. The farmer showed signs of strain and remarked that the force was so powerful he was almost unable to keep the bone in his grip. The psychologist thanked the farmer for his demonstration and said that he would like to test the farmer's ability to locate water under controlled conditions, but that this experiment would require some preparation. Would the farmer agree to return for these tests next week? The farmer agreed and promised to return at the appointed time.

2.2 The Experiment

When the farmer returned to the psychologist's office the next week, he was greeted by the psychologist and taken to one of the laboratory rooms. Spread around the floor of the room were 10 square pieces of plywood about 8 inches $\times$ 8 inches in size. Numbers from 1 to 10 had been marked on the top of each square. The plywood squares were resting on tin cans. The psychologist explained that five of the cans had been filled with water and five had been left empty. He had not used any systematic rule for determining which cans were filled with water and which were left empty, but rather, as he put it, "This was left to chance." The psychologist did not even know which of the cans contained water and which were empty because he had left this task to a laboratory assistant. He was as much in the dark as the farmer, but he hoped that the farmer, with the aid of his whalebone, would soon be able to enlighten him as to which cans contained water. He again emphasized to the farmer that under five of the plywood squares were cans with water, under the five remaining squares were empty cans, and that the arrangement of the empty and filled cans was random.

The psychologist now wanted the farmer to take his whalebone and attempt to select the five cans filled with water. The farmer did not need to make his choice in any particular order; he was merely to select the set of five plywood squares under which he believed cans filled with water would be found.

The **observations** to be made in this experiment consist of the choices the farmer makes regarding the content of the cans. The **outcome** of the experiment is the particular set of choices the farmer makes. We will examine this experiment in some detail. We will pay particular attention to the set of all possible outcomes of the experiment, the question the experimenter hopes to answer by the observations, and the manner in which he proposes to arrive at this answer.

2.3 The Question of Interest

The question that motivated the experimenter to make the observations is not the one of interest to the farmer. The farmer, in his previous conversation, had indicated that he wanted to know why he could divine the presence of water. It is apparent that the farmer implicitly assumes that he can detect the presence of water. The question of interest to the psychologist, on the other hand, is whether the farmer can actually divine the presence of water. It is obvious that to ask why the farmer is successful in divining the presence of water is meaningful only if it can first be determined that he is, in fact, successful in doing so.

The psychologist may reason in this way: Let us assume that the farmer does not possess any particular powers that enable him to locate water with his whalebone; that is, the only factor operating in determining his choice is chance. More specifically, the question that the experimenter wishes to answer is, Can the farmer do any better in his choices than might be expected on the basis of chance? In order to understand what we mean by "expected on the basis of chance" we need to review elementary probability and simulate this experiment.

2.4 Sample Space and Probability

Recall that the farmer's task is to select the five cans he believes contain water from a set of 10. The set of cans selected by the farmer is a **sample** of $n = 5$ observations from a finite population in which the total number of cans is $N = 10$. Lowercase n denotes sample size (the number of cans the farmer believes contain water), and uppercase N denotes population size (the total number of cans). The particular sample selected by the farmer will represent one of the possible outcomes of the experiment. The farmer is told in advance that five cans contain water so in this experiment the farmer knows that $n = 5$.

The set S of all possible outcomes of an experiment is called the **sample space** of the experiment. The set S represents all the ways in which the farmer could state that 5 out of the 10 contain water. The elements of set S are called **events**, and each possible outcome of the experiment must correspond to one and only one event. We will use the letter E to indicate events. Not only do we want to know the number of possible outcomes of the experiment, but we also want to be able, on some reasonable basis, to assign to each event or outcome a number P that gives the probability associated with the event. If the values of P assigned to each event (that is, to each possible outcome) are to be called **probabilities**, then the values must satisfy the following rules:

1. The values of P assigned to each sample must be equal to or greater than 0 and equal to or less than 1; that is, $0 \leq P \leq 1$.

2. If k subsets, $E_1, E_2, E_3, \ldots E_k$, are defined on the sample space, and if the k subsets are mutually exclusive and exhaustive, then we must have the sum

$$P(E_1) + P(E_2) + P(E_3) + \cdots + P(E_k) = 1$$

   That is, when the sample space can be partitioned into nonoverlapping subsets of events, then the sum of the probabilities of each of these events must equal 1.

In the next few sections we will develop the logic underlying how to compute the probability of events in a study. The example of the farmer from Whidbey Island presents a simple illustration of how to compute such probabilities. In later chapters we will develop statistical approximations to such probabilities. Our goal in presenting some of the elementary probability calculations is to enable the reader to learn the foundations of statistical inference and hypothesis testing. The present example, though, illustrates rather clearly the ideas that underlie hypothesis testing.

2.5 Simulation of the Experiment

Recall that we want to simulate the experiment with the farmer so that we can compare the observation made of the farmer to what is expected by chance. One way to calculate "expected on the basis of chance" is to imagine all the possible results that can occur if the farmer simply guesses the content of the cans. We can simulate the idea that the farmer may be guessing by creating a simple model that will provide a concrete way to represent the underlying probability of guessing.

We can place 10 disks in a box to create a random device. We let each disk correspond to an observation without, for the moment, specifying the value of the observation. We identify the disks in the same manner in which the 10 pieces of plywood are identified, that is, by the numbers 1 to 10. We now shake the box thoroughly and then let one disk fall through a small slot in the box. We will assume that this procedure results in a random selection of a disk. By **random selection**, we mean that the probability of any given disk in the box falling through the slot is the same for all disks. Having selected one disk, without replacing it in the box, we again shake the box and let a second disk from the remaining 9 fall through the slot. We again shake the box and let a third disk from the remaining 8 fall through the slot. We continue in this manner until we have a sample of $n = 5$ disks, or observations, from the $N = 10$ disks. We let this sample correspond to one of the sets of 5 choices that could be made by the farmer in the experiment.

If our method of sampling is random, then on each trial each of the disks in the box has an equal probability of being selected. For example, the probability of a particular disk out of the 10 being selected on the first shake of the box is $1/10$; on the second shake after one disk has already fallen out with 9 disks remaining in the box, the probability of a particular disk falling through the slot is $1/9$; and so on, until on the fifth and last shake, the probability of a particular disk being selected is $1/6$, because on the last shake for the fifth disk there are 6 disks remaining in the box.

This method of selecting disks numbered 1 through 10 is random and would be analogous to what the farmer would do if he were merely guessing at which cans contain water. If the farmer really has the ability to detect water, then his performance will be different than chance. By comparing the farmer's actual performance in detecting cans containing water to what we expect under random chance, we can make a decision about whether the farmer has special ability. If the farmer performs better than chance, we find support for the claim that the famer has the ability to detect water. However, if the farmer's performance is indistinguishable from chance, then we don't have much evidence that the farmer possesses special powers to detect water. This is the basic idea underlying what is called **hypothesis testing**. We will discuss hypothesis testing in more detail later in this chapter. We first need to develop some more concepts.

2.6 Permutations

What is the probability that a sample drawn in the manner described in the preceding section will include the observations identified by the numbers 10, 8, 5, 4, and 1? Consider first the probability of obtaining these 5 observations in the order specified. The probability of obtaining the number 10 on the first draw is 1/10. Given that we have drawn 10, the probability of the disk numbered 8 on the second draw is 1/9. Given that we have obtained 10 on the first draw and 8 on the second, the probability of 5 on the third draw is 1/8, and so on. The probability that the sample will be the particular set of five observations drawn in the order specified will be

$$\frac{1}{10} \times \frac{1}{9} \times \frac{1}{8} \times \frac{1}{7} \times \frac{1}{6} = \frac{1}{30{,}240}$$

The denominator of the right-hand side of the above expression is simply the number of permutations of the total number of $N = 10$ objects taken as samples of size $n = 5$ at a time. Permutations refer to the number of different orders in which a set of objects may be arranged. The number of permutations of N objects taken all together is given by

$$_N\mathrm{P}_N = N! \tag{2.1}$$

where $N!$ is called "N factorial" and represents $(N)(N-1)(N-2)\ldots(2)(1)$. That is, to compute the factorial merely multiply all the successive integers from N down to 1.

The number of **permutations** of N objects taken n at a time is given by

$$_N\mathrm{P}_n = \frac{N!}{(N-n)!} \tag{2.2}$$

where $n \leq N$. Note that if $n = N$, then the denominator of Equation 2.2 will be equal to 0!, and by definition 0! is set equal to 1. Thus, in the special case where $n = N$, Equation 2.2 is the same as Equation 2.1. The $_N\mathrm{P}_n$ notation denotes the permutation of N objects taken n at a time.

The farmer selects a subset of n cans from the total number of N cans. In the experiment the number of permutations of $N = 10$ objects taken $n = 5$ at a time is

$$_{10}\mathrm{P}_5 = \frac{10!}{(10-5)!} = 30{,}240$$

This number gives the total number of every possible set of 5 objects arranged in every possible order. That is, any one of the 10 disks may be selected first; this selection may then be followed by any one of the remaining 9; this selection may be followed by any one of the remaining 8; and so on, until 5 have been selected. Thus, there are $10!/5! = 30,240$ different ordered samples, and each of these samples has a probability $1/30,240$ of being selected on the basis of chance alone. Thus, a particular order of $n = 5$ objects has a very small probability of occurring: $1/30,240$.

2.7 Combinations

We have seen that the probability of selecting disks 10, 8, 5, 4, and 1 in that particular order is $1/30,240$. In the experiment with the farmer, however, we are not concerned with the specific order the farmer selects the cans. Instead, we are primarily interested in the number of cans he correctly identifies as containing water. We note that a set of five observations can be arranged in $5! = 120$ different orders. That is, there are 120 orders that can occur when the disks 1, 4, 5, 8, and 10 are selected. Dividing 30,240 by 120, we obtain 252 ways in which five objects can be selected from 10 objects, when the arrangement, or order, is ignored.

Mathematically speaking, the number of **combinations** (ignoring arrangement) of N distinct objects taken n at a time is given by

$$_N\mathrm{C}_n = \frac{_N\mathrm{P}_n}{n!} = \frac{N!}{n!\,(N-n)!} \tag{2.3}$$

or, in the present problem,

$$_{10}\mathrm{C}_5 = \frac{10!}{5!\,(10-5)!} = 252$$

The probability that a sample selected in the manner described will contain the specified 5 observations, the order in which they are drawn being immaterial, will be $120/30,240 = 1/252$.

The probability we have just obtained will be exactly the same for any other specified set of 5 observations. For example, the probability that the sample will contain the observations 10, 4, 3, 2, and 1 is also $1/252$. Thus, as Equation 2.3 shows, there are $10!/(5!5!) = 252$ different **unordered** samples, and each of these samples has a probability of $1/252$ of being selected on the basis of chance alone.

2.8 Probabilities of Possible Outcomes

Now we will use the information we just developed to answer more directly the probability that the farmer correctly selected all five cans containing water if he was merely guessing. We will also compute the probability that the farmer selected four of the five cans if he was merely guessing, the probability that the farmer selected three of the five cans if he was guessing, etc.

For bookkeeping purposes we will assign a value, or label, to each of the 10 disks to indicate whether or not they contain water. A value of W corresponds to a filled or wet can, and a value of D corresponds to an empty or dry can. In the population of 10 cans there are 5 W's and 5 D's. We are interested in the number of W's in each of the 252 possible unordered samples of $n = 5$ observations. There is only 1 out of the 252 samples that can include the 5 observations with values of W (because there is only one way that all 5 cans are all W's), and we can show that the probability of obtaining a sample with 5 W's is 1/252.

What is the probability of obtaining a sample with four W's and one D (that is, the farmer correctly identified four cans but incorrectly identified one empty can as containing water)? Specifically, let the first four observations drawn have the value of W and the last observation have the value of D. The probability of obtaining a W on the first draw is 5/10. If this result occurs, then there will be 4 disks with W's left in the box and 5 disks with D's, and the probability of W on the second draw will be 4/9. If the first two draws are W's, then there will be 3 W's and 5 D's left in the box, and the probability of W on the third draw will be 3/8. If we obtain a W on the third draw, then there will be 2 W's left and 5 D's. The probability of W on the fourth draw will then be 2/7. If this result occurs, then we have one W and 5 D's left in the box, and the probability that the fifth draw will be a D will be 5/6. Therefore, the probability of the sample WWWWD, in the order specified, is

$$\frac{5}{10} \times \frac{4}{9} \times \frac{3}{8} \times \frac{2}{7} \times \frac{5}{6} = \frac{600}{30{,}240} = \frac{5}{252}$$

If we shift the D to any other position in the sequence, we could show, in the same manner, that the probability of this sequence is the same as when D is in the last position. It is clear that, because D can appear in any one of five positions and because the five sequences are mutually exclusive, the probability of drawing a sample with four W's and one D (where the D can occur anywhere in the first five draws) is

$$5 \times \frac{5}{252} = \frac{25}{252}$$

As another example, the probability of obtaining WWWDD in the order specified is

$$\frac{5}{10} \times \frac{4}{9} \times \frac{3}{8} \times \frac{5}{7} \times \frac{4}{6} = \frac{1,200}{30,240} = \frac{10}{252}$$

Thus, the very same methods we have introduced can be used to compute more general cases such as 3 W's and 2 D's in a particular order. This will become useful when we evaluate the performance of the farmer.

2.9 A Sample Space for the Experiment

Now let's use our new knowledge about permutations and combinations to analyze the experiment with the farmer presented at the beginning of the chapter. Let T be the number of cans the farmer correctly identifies as containing water. Thus, T is a variable that can, in the experiment described, take the possible values of 0, 1, 2, 3, 4, or 5. The variable T represents the frequency, or count, of the number of cans the farmer correctly identifies as containing water. If the outcome of interest in the experiment is T, then the sample space for the experiment is $S = \{0, 1, 2, 3, 4, 5\}$. We note that each possible outcome T of the experiment is associated with one and only one sample point. Furthermore, each sample point has associated with it a probability P. The values of P satisfy the probability axioms stated earlier in this chapter.

The best the farmer could possibly do in the present experiment would be to select the sample of five cans that actually contain water. There is only one way this result could happen, and this particular sample would be 1 out of 252 possibilities. If the farmer's selections are being made solely on the basis of chance and if this experiment were repeated an indefinitely large number of times, then we would expect this particular sample to be selected with a theoretical relative frequency of 1/252. In other words, the probability of correctly guessing all 5 cans that contain water, denoted $P(T = 5)$, is $1/252 = 0.004$. This means that only 4 times in 1,000 would the farmer be expected to have all five of his choices correct on the basis of chance alone. Thus if the outcome of the experiment is $T = 5$ (the farmer correctly identified the five cans containing water), then either a relatively improbable event has occurred by chance or else the farmer is not making his selections on the basis of chance (that is, the farmer may be able to detect water with his whalebone). This example illustrates how we use probability to analyze the results of an experiment. The rest of the book uses this basic approach in more complicated situations.

2.10 Testing a Null Hypothesis

We constructed a sample space for this experiment by assuming that the farmer did not have the ability to detect water with his whalebone and that if this was the case, then he is making his selections simply on the basis of chance. This assumption may be regarded as a hypothesis, often referred to as a **null hypothesis**, that the experiment is designed to test. Assuming the null hypothesis to be true, it is possible to determine the probability of each possible outcome of the experiment. Because the outcome $T = 5$ is relatively improbable on the basis of chance, we may conclude that the evidence (the farmer correctly selecting the five cans containing water) supports the conclusion that the farmer has the ability to detect the presence of water.

When testing a null hypothesis, we must make a decision as to how small the probability of a given outcome must be before we will decide to reject the hypothesis. At what point do we find it reasonable to say that the farmer has the ability to detect water (that is, we reject the null hypothesis that the farmer was merely guessing)? When he selects four out of five cans? five out of five cans?

The probability we choose to use in rejecting the null hypothesis is called the **significance level** of the test and is indicated by the Greek letter alpha, α. Researchers typically choose $\alpha = 0.05$ as the significance level of their tests; that is, they reject a null hypothesis whenever the outcome of the experiment under the null hypothesis has a probability equal to or less than 0.05, or 5%. A **statistically significant** result refers to the case when the chance of observing the result when the null hypothesis is true is less than the significance level α.

Some people might regard this value as a relatively lenient standard and might insist that α be smaller than 0.05, say 0.001. In the present experiment, if we use $\alpha = 0.001$ as a standard, it would be impossible to reject the null hypothesis because, even when the farmer perfectly chooses the five cans containing water, the probability of this outcome is 0.004 (as computed earlier in this chapter), a value that is larger than $\alpha = 0.001$. If we choose α to be very small, we decrease the probability of rejecting the null hypothesis. But there is no sense in doing an experiment and testing the significance of the outcome if it is impossible to reject the null hypothesis under any possible observation of that experiment.

No single experiment can establish the absolute proof of the falsity of a null hypothesis, no matter how improbable the outcome of the experiment is under the null hypothesis. Improbable as $T = 5$ is in the present experiment, we have no way of knowing whether this particular outcome occurred by chance or occurred because the farmer has a special ability to detect water. As Fisher (1942, pp. 13–14) pointed out: "In order to assert that a natural phenomenon is experimentally demonstrable we need, not an isolated record, but a reliable method of procedure. In relation to the test of significance, we may say that a phenomenon is experimentally demonstrable when we know how to conduct an experiment which will rarely fail to give us a statistically significant result." We would be more convinced of the farmer's special

Table 2.1: The four combinations resulting from considering one's decision to reject or fail to reject the null hypothesis and whether the null hypothesis is actually true or false. Two cells are correct inferences and two cells are errors.

		Decision	
		Reject	Fail to reject
Null hypothesis	True	Type I error (false positive)	Correct
	False	Correct	Type II error (false negative)

powers if he consistently selected the five cans containing water in repeated experiments.

2.11 Type I and Type II Errors

As we pointed out above, when conducting tests of significance, we will sometimes make decision errors. When the null hypothesis is true (that is, the farmer does not have dowsing ability) but the results of the significance test lead us to decide that the null hypothesis is false, we describe this situation as a **Type I error**. In medical testing an analogous error is called a false positive—the diagnostic test incorrectly signaled the presence of a medical condition such as a tumor. When the null hypothesis is, in fact, false and we decide on the basis of a test of significance not to reject it, we call this a **Type II error**. In medical testing the analogous error is called a false negative. Table 2.1 illustrates these two errors.

The probability of a Type I error is set by α, the significance level of the statistical test. If we always reject a hypothesis when the outcome of the experiment has a probability of 0.05 or less, and if we consistently apply this standard, then in situations when the null hypothesis is actually true we will, on the average, incorrectly reject 5% of the hypotheses we test; that is, we will declare the null hypothesis false 5% of the time when it is in fact true. By choosing a small α, we can decrease the probability of a Type I error, but at the same time we increase the probability of a Type II error, as shown in the following example.

Suppose, in the present experiment, that we choose $\alpha = 0.05$. Then, if the outcome of the experiment is $T = 5$, the null hypothesis would be rejected. However, suppose that the outcome of the experiment is that the farmer correctly identified four of the five cans, $T = 4$. Then the probability of observing four or greater is $P(T \geq 4) = P(T = 4) + P(T = 5) = 0.099 + 0.004 = 0.103$, and in terms of the standard significance level this outcome is not statistically significant because the computed probability exceeds α. Even though this outcome has a relatively small chance of occurring, we do not reject the null hypothesis because the probability is not smaller than the standard we have chosen. If the outcome of the experiment is $T = 4$ and if we decide not to reject the null hypothesis, it is possible that we made a Type II error. The outcome of the experiment may not allow us to reject the null hypothesis. This observation and decision could occur because the farmer's responses are no different than our chance model under the null hypothesis, or it could be that the experiment was not sufficiently sensitive to detect the farmer's ability.

The **power** of a statistical test is defined as

$$1 - P(\text{Type II error})$$

or, equivalently, as the probability of rejecting a null hypothesis when it is false. In general, if we hold α (the probability of a Type I error) constant, we can increase the power of a test by increasing the number of observations in the sample. The relationship between power and the number of observations will be discussed in more detail in subsequent chapters.

It would appear that in the experiment with the farmer the number of observations is so small that the test of significance has relatively little power. The experimenter is somewhat protected against a Type I error (by setting $\alpha = 0.05$) but has little protection against a Type II error. The farmer may be able to do better than chance in locating water, but the experiment is not sufficiently sensitive to test his ability. The experiment could be made more sensitive by increasing the number of observations, that is, by increasing the number of cans that the farmer will dowse for water.

2.12 Experimental Controls

Even if the farmer correctly identified the five cans containing water, the psychologist cannot be completely sure that the farmer has special powers. The psychologist would need to rule out alternative explanations on the basis of experimental controls. What are some examples of alternative explanations?

Without the experimenter knowing about it, the farmer might use the toe of his foot to tap the cans under the board. Because the cans filled with water could be easily distinguished from the empty cans by a subtle foot tap, this trick alone could account for a perfect selection on the part of the farmer. If this trick is the basis of the farmer's selections, then obviously the whalebone has nothing to do with his choices. The farmer might even deny that he is using this cue—the sound of the can when tapped with his foot—if questioned about it. But the psychologist knows that many of our choices and judgments are based on factors of which we are not aware. It is the experimenter's responsibility to rule out such possibilities by proper control.

The psychologist would also want to make sure that the farmer did not tap the tip of the whalebone on the tops of the plywood boards. If the farmer does so, his choice might be determined by the difference in sound of the boards covering the empty and filled cans. He might thus make a perfect selection and the experimenter would reject the null hypothesis of chance. But again the rejection of the hypothesis of chance does not establish the validity of the farmer's claim concerning the influence of the whalebone.

Another possible explanation of a perfect selection might be that the experimenter's assistant had spilled some water on the floor while filling the cans. The water might have been carefully mopped up, but slight cues may have remained. The absence of dust or the cleanliness of the floor under the sections of plywood containing water, as a result of the mopping, might provide cues for the farmer's choices.

If the experimenter, rather than his assistant, had filled the cans, so that the experimenter had knowledge of which cans contained water and which did not, then the experimenter himself might inadvertently give some sign—holding his breath, biting his lips, or some other unconscious gesture—as the farmer moved his whalebone over the sections containing water.

Perhaps the most famous case, described by Pfungst (1911), where a subject was able to respond correctly on the basis of slight cues provided by the presence of an experimenter who knew the correct response, was that of Clever Hans. Clever Hans was a horse. By tapping his foot, he could give the correct answers to arithmetic problems involving addition, subtraction, multiplication, and division. Hans could also spell, read, and do a number of other things that horses ordinarily do not do. After a series of investigations, Pfungst found that Hans was responding on the basis of slight cues unknowingly provided by himself. A very slight and almost imperceptible forward movement of Pfungst's head would start Hans tapping and at another almost imperceptible upward movement of the head, or of the eyebrows, Hans would stop. Initially, Pfungst was completely unaware that he and others who questioned Hans were providing the horse with these cues. It was only after Pfungst happened to notice these almost imperceptible movements on the part of other questioners that he became aware that he too made these slight movements when he questioned Hans. Having discovered the principle by which Hans was obtaining his answers, Pfungst conducted a number of experiments in which another person played the role of questioner and he, Pfungst, played the role of the horse.

Pfungst had questioners concentrate on some number between 1 and 10 or even some larger number. Then, he would begin to tap out the answer using his right hand and would continue to tap until he believed he had perceived a final signal. He tested 25 different individuals of all ages and sexes and differing in nationality and occupation. None of them was aware of the purpose of the experiment. Pfungst found that only in a few isolated cases were the questioners aware of any movement on their part. He also found that, with the exception of two individuals, they all made the same involuntary movements, the most important being a slight upward inclination of the head when Pfungst had tapped the correct number of times.[2]

The farmer's choice might be based on one of the unconscious gestures or reactions of the experimenter without, of course, the experimenter, and perhaps even the farmer, being aware that these cues were the basis of the farmer's choice. Fortunately, the experimenter, in this instance, anticipated this possibility and controlled for it by having his assistant prepare the cans. The experimenter did not have knowledge of which cans contained water, so could not inadvertently communicate cues.

In a well-designed experiment factors that may influence the outcome of the experiment but are not themselves the factors of interest must be controlled if sound conclusions are to be drawn from the experiment. The reader should realize that conclusions are derived from the structure of the experiment and the nature of the controls implemented in the study; conclusions are not drawn solely from the results of a test of significance. This is difficult to remember when one is excited by statistically significant results that appear in the output of a statistics program.

2.13 The Importance of Randomization

An essential notion in evaluating the outcome of the experiment with the farmer is that of **randomness**. The experimenter mentioned that the selection of the five cans filled with water was determined on a random basis. The

[2]Rosenthal (1966, 1967) provides a survey of various ways in which knowledge or expectancies of an experimenter may influence the outcome or results of an experiment.

randomization was done by the assistant; the experimenter (who made the observations of the farmer's choices) was unaware of which cans contained water and which were empty.

Randomization also ensures that the particular probability model used in evaluating the outcome of the experiment is applicable. Suppose, for example, that some slight but perceptible differences existed in the cans such that five of the cans had a slight dent. If the assistant had systematically, but not necessarily consciously, filled either the dented or the undented cans, the farmer may have reacted to this cue and used it as a basis for choice. Randomization offers some assurance that a given characteristic of the cans will not be associated with the presence or absence of water in the cans.

Similarly, when assigning the numbered sections of plywood to the cans, randomization is necessary. We would not want all of the even-numbered or all of the odd-numbered sections of plywood to be assigned to the cans containing water. Randomization at this stage is necessary in order to ensure that there is no systematic association between the characteristics of the pieces of plywood and the numbers on them, on the one hand, and the presence or absence of water in the cans, on the other hand.

If the assistant had made a systematic division of the cans into two sets, he or she may have done so on the assumption that the manner in which the division was made could not in any way influence the selections made by the farmer in the course of the experiment. This assumption may, of course, be true, but it remains an assumption, and it may be difficult to convince others that it is true. The only convincing argument is that of appropriate randomization.

It is essential in experiments designed to compare the performance of two groups of participants under different treatments that the assignment of the participants to the two treatments be made on a random basis. When we randomly assign participants to the treatments, we have some assurance that the two groups will be comparable, on average, with respect to individual difference variables that may be related to performance on the dependent variable. Obviously, if the grouping variable is itself an individual difference variable, then randomization should not occur. For instance, an experimenter cannot randomly assign a participant to either the male condition or the female condition.

2.14 A Variation in Design

We now consider a variation in the experimental procedure. Suppose that the 10 cans are arranged at random into five pairs. One can in each pair is randomly selected to be filled with water. The farmer is told that he will be presented with five pairs of cans, one of which is filled with water and one of which is empty, and that he is to select the can in each pair that he believes contains water. What are the possible outcomes of this experiment?

There are two ways in which the farmer's first choice may be made; and independently of this choice, the second choice may be made in two ways; and independently of this choice, there are two ways in which the third choice may be made; and so on for the 5 choices. Thus, there are a total of $2^5 = 32$ possible outcomes of this experiment.

If the farmer does not have the ability to detect the presence of water with his whalebone, then he should be just as likely to select the empty can as the water-filled can each time he has to make a choice. Let W be the selection of a water-filled can. We denote the probability of detecting a can containing water as $P(W)$. Then $P(W) = 1/2$ in each of the five pairs. Let T be the number of W's in a sample of $n = 5$ choices (that is, the number of times the farmer correctly chooses the water-filled cans out of the five pairs). In this setup the variable T can take the possible values of 0, 1, 2, 3, 4, and 5.

There is only one way in which T can be equal to 5, and in this case the farmer would have to be correct on each of his five choices. Under a simple guessing model, the probability of WWWWW will be equal to $1/2^5 = 1/32$. The probability of WWWWD is $1/2^5 = 1/32$; that is, the probability of the first four choices being correct and the last one wrong. But we note that according to Equation 2.3 there are $5!/(4!1!) = 5$ ways to order four W's and one D, and the probability of each of these is $1/32$. Then, $P(T = 4) = 5/32$. We have $5!/(3!2!) = 10$ ways in which we can obtain $T = 3$ (three W's and two D's), so $P(T = 3) = 10/32$. Similarly, we find that $P(T = 2) = 10/32$, $P(T = 1) = 5/32$, and $P(T = 0) = 1/32$. The values of P assigned to the possible values of T satisfy the axioms for probabilities (see Section 2.4).

With this experimental design, we compute that $P(T = 5)$ is $1/32$ or approximately 0.031, whereas with the previously described experimental design $P(T = 5)$ is approximately 0.004. A major lesson that emerges from these two examples is that a minor modification in the experimental procedure influences the probability of observations under the null hypothesis. Experimenters must be mindful that subtle changes in procedure can impact the likelihood of rejecting the null hypothesis.

Both experimental procedures provide relatively little protection against a Type II error (failure to reject a null hypothesis that is false), and as we have previously stated, one way we could increase the power of the test of significance would be to increase the number of observations. The problem though is that, as we increase the number of observations, the methods described in this chapter for obtaining the probabilities associated with each possible outcome of the experiment become exceedingly laborious. The needed probabilities would require a great deal of computation. In the next chapter, we describe methods that can be used to approximate the probabilities associated with the outcomes. These methods will serve as the basis of the analysis of variance design discussed later in this book.

2.15 Summary

This chapter used the example of the farmer from Whidbey Island to illustrate basic principles of experimental design, such as dealing with alternative explanations through design features. The chapter also showed how to compute models of chance for two different types of experiments that could be conducted in the example with the farmer. These models of chance provide a foundation for computing Type I and Type II errors, for illustrating the role of sample size and power in experimental design, and for pointing out how subtle changes in the nature of a study can have major impacts on the computation of the probabilities underlying inferential procedures.

2.16 Questions and Problems

1. A laboratory rat is placed in a jumping stand and is trained to jump to the smaller of two squares. The right and left positions of the smaller square are randomly alternated so that the experimenter has some confidence that the animal is not merely reacting to position. The experimenter is interested in determining whether the established reaction pattern will generalize to the extent that the animal will react similarly to the smaller of two circles. After the animal has learned to discriminate between the two squares, the rat is given a series of eight trials with two circles. The position of the smaller of the two circles is randomly alternated. We make the assumption that, if generalization of the previous learning is not present, the animal will react to the two circles on the basis of chance. On the other hand, if the animal jumps to the smaller circle with a frequency greater than can be attributed to chance, the chance hypothesis can be ruled out and we infer that generalization has taken place.

    (a) What is the probability of seven or more jumps to the smaller circle in eight trials if the null hypothesis is true?

    (b) If the number of trials is increased to 12, what is the probability of 10 or more jumps to the smaller circle if the null hypothesis is true?

2. It is claimed that an infant stimulated by a loud sound shows a response pattern that is differentiated from the pattern of response present when the infant's movements are restrained. The response to the loud sound is called "fear," and the response to restraint is called "rage." An infant is stimulated four times by loud sound and four times by restraint of movement, and the responses of the infant are videotaped immediately after stimulation. There are a total of eight videotapes, four showing reaction to sound and four showing reaction to restraint. This situation is explained to participants who are to serve as judges. Participants are asked to select the set of four videotapes showing "fear."

   The questions that follow are related to the evaluation of the possible outcomes of the experiment under the null hypothesis that a correct selection is a matter of chance.

   (a) What is the probability of a single participant selecting the set of four correct videotapes from the eight?

   (b) What is the probability of selecting a set of three correct "fear" videotapes and one wrong "rage" videotape?

   (c) What is the probability of selecting a set of two correct "fear" and two wrong "rage" videotapes?

3. An experimenter has a set of four cards, three that are blank and one that has an X printed on it. The cards are shuffled and placed face down in a row. The research participant is to determine the position of the card with the X on it.

   (a) What is the probability that the participant will make a correct selection in a single trial assuming that the participant is selecting by chance?

   (b) If the participant is given four trials, what is the probability that he or she will make precisely three correct choices by chance?

(c) What is the probability that the participant will make precisely one right choice in three trials?

(d) If there are 128 participants who serve in the experiment and each participant is given three trials, then how many participants would be expected to obtain perfect scores by chance alone?

(e) How many of the participants would be expected to obtain scores of two or more correct by chance?

4. A student claims that she can differentiate her own brand of oatmeal from three other popular brands. Outline an experiment that would provide evidence for this claim. What kinds of experimental controls might be necessary in the experiment?

5. Outline an experiment in which one might test a student's claim that he can discriminate beer A from beer B. What experimental controls may be necessary? What role does randomization play in the experimental design? What outcomes of the experiment would be regarded as significant?

6. Suppose we wish to determine whether orange juice can be distinguished from pineapple juice and apple juice when visual and olfactory cues have been experimentally controlled. We block the nasal passages of a male research participant and blindfold him. He is then presented with a set of three test tubes. He is told that one of the test tubes contains pineapple juice, one orange juice, and one apple juice, and that he is to pick the one he thinks contains the orange juice. Fifteen sets of three test tubes are presented to the participant.

   (a) What is the probability that he will make nine or more correct choices if he is responding by chance?

   (b) What are some of the experimental controls that should be considered in planning this experiment?

7. In many experiments the dependent variable is a rating assigned to a person or an object by a judge. For example, judges may be asked to rate the improvement of patients in a mental hospital after they have been treated with a drug or after they have had several months of psychotherapy. In taste laboratories judges may be asked to rate the quality of several foods. Discuss the nature of the experimental controls necessary in such studies. Discuss the role of randomization as a device for concealing from the judges which patients have been treated and which have not and, similarly, which foods have been prepared in one way and which in another way.

8. Under what conditions can we consider a sample of $n = 10$ observations drawn from some defined population to be a random selection from that population?

9. A sample of $n = 3$ is drawn from a set of 10 objects without replacement. The sample is drawn at random. How many different samples can be drawn if order does not count?

10. We have six males and four females. A random sample is drawn without replacement in such a way as to ensure that we have three males and two females in the sample. How many different samples can be drawn if order does not count?

11. We have chocolate bars labeled A, B, C, D, E, and F. We draw a random sample of $n = 3$ bars without replacement.

    (a) How many ordered samples are possible?

    (b) How many different unordered samples are possible?

12. An experimenter rejects a null hypothesis when the null hypothesis is true. What kind of error has been made?

13. An experimenter fails to reject a null hypothesis when it is false and should be rejected. What kind of error has been made?

14. How many ways can 20 people be divided into four groups of 5?

15. Explain the difference between the number of permutations of N things taken n at a time and the number of combinations of N things taken n at a time.

16. What is meant by the power of a statistical test?

17. If two events are described as mutually exclusive, what does this mean?

18. What is meant by the significance level of a test?

19. What is meant by a null hypothesis?

20. What is meant by a sample space for an experiment?

21. Explain what is meant when a researcher says that "the findings were statistically significant."

3

The Standard Normal Distribution: An Amazing Approximation

3.1 Introduction

The experiments described in the preceding chapter involving the farmer from Whidbey Island were limited because they involved few observations, which led to a relatively high probability of a Type II error. Thus, in these experiments we might fail to reject the null hypothesis even when it is true. We suggested that, other things being equal, the probability of a Type II error could be decreased by increasing the number of observations. But as we increase the number of observations, we face the problem of evaluating a larger number of possible outcomes of the experiment, and the methods of the preceding chapter (which give the exact probabilities associated with each possible outcome) become exceedingly laborious. Imagine how time-consuming the computations would be if there were 100 cans in the experiment with the farmer. Fortunately, the probabilities associated with random sampling from a binomial population can be approximated quite satisfactorily by the standard normal distribution, which we will explain in this chapter.

Because we have already obtained the exact probabilities for samples of $n = 5$ observations drawn from two different experiments with the farmer, we will use these same two cases to illustrate the approximation using the normal distribution. With the exact probabilities available as a comparison, we can see how well the approximation works. If they work well for these two specific cases of $n = 5$ observations, we may feel even more confident in applying the approximation to experiments involving a much larger number of observations. The approximation is most useful when there are a large number of observations, which is also when exact techniques from the preceding chapter become too laborious. There is an important result, the central limit theorem, that justifies this approximation mathematically. We now review some elementary results involving means and variances, and then will apply these results to the experiments with the farmer discussed in Chapter 2. The next several sections are somewhat technical, but they will help your understanding of the material that follows. It is important to get a few of the details under your belt in order to understand how the normal approximation works (and when it is likely to fail). In the sections that follow we make an important distinction between a population and a sample.

3.2 Binomial Populations and Binomial Variables

We begin with a random variable X that can take only two values, such as the value of 1 or 0. An example would be whether a research participant answers "yes" (1) or "no" (0) to a survey question. The assignment of 0 or 1 to a particular level is arbitrary.

How might we model this variable mathematically? We will write $X = 1$ to indicate that X takes on the value 1 and $X = 0$ to indicate that X takes on the value 0. Assume that there exists a population of observations and that in this population there are F_1 possible observations with $X = 1$ and F_0 possible observations with $X = 0$ (F denotes frequency of observations). For example, the population may have 5 glasses with water and 5 glasses without water. Then the total number in the population will be $N = F_1 + F_0 = 10$ (that is, the sum of the two frequencies). If an observation is randomly selected from this population, then the probability of choosing a 1 is $P(X = 1) = F_1/N$ (that is, the probability that variable X equals 1 is the number of 1's in the population divided by the total number) and $P(X = 0) = F_0/N$. As expected by the rules of probability, the sum of the two probabilities equals 1, that is,

$$P(X = 1) + P(X = 0) = \frac{F_1}{N} + \frac{F_0}{N} = 1$$

For convenience we let P be the probability that $X = 1$ and Q be the probability that $X = 0$. That is, Q is related to P by this formula: $Q = 1 - P$.

A population in which a variable can take only one of two possible values (for example, yes or no, glass full or glass empty, diabetic or not diabetic) is called a **binomial population**, and the variable is called a **binomial variable**. Table 3.1 shows a binomial population in which $P(X = 1) = 0.5$ and $Q = P(X = 0)$ is 0.5. An example of such a population is the second experiment with the farmer from Chapter 2, where the farmer is given a pair of cans, one containing water and the other not, and is asked to choose the can containing water. In this example, across the five pairs there are 5 cans with water ($F_1 = 5$) and 5 cans without water ($F_0 = 5$), each yielding a probability of 0.5 if the farmer is merely guessing at which can contains water.

Table 3.1: A binomial population with $P(X = 1) = 0.5$ and $P(X = 0) = 0.5$.

(1) i	(2) X_i	(3) P_i	(4) P_iX_i	(5) $P_iX_i^2$
1	1	0.5	0.5	0.5
2	0	0.5	0.0	0.0
	$\sum$	1.0	0.5	0.5

3.3 Mean of a Population

The **mean** is a number that represents the "average" value of a random variable. The mean of a population with a finite number of categories is defined as

$$\mu = \sum_{i=1}^{k} P_i X_i \qquad (3.1)$$

where the right-hand side denotes that we multiply each possible value of variable X by its associated probability and add the respective products. The value k denotes the number of possible values that the random variable can take, that is, the number of categories in the population. The subscript i denotes a counter that cycles through each category of possible values of the random variable. For the special case of a binomial population where X can take only the values of 1 or 0 (that is, two categories, $k = 2$) with corresponding probabilities of P and Q, Equation 3.1 simplifies to

$$\begin{aligned} \mu &= P_1X_1 + P_2X_2 \\ &= (P)(1) + (Q)(0) \\ &= P \end{aligned}$$

This result shows that for a binomial population where the two possible values of random variable X are 0 and 1, the mean μ equals the probability associated with $X = 1$. This is verified in Table 3.1, where $\mu = \sum P_i X_i = 0.5$, and this value is also the probability associated with $X = 1$.

As a concrete example, let X be the random variable for the possible outcomes of the second study with the farmer in Chapter 2. Let's assume that the farmer is merely guessing which can contains water. If we denote a can with water as $X = 1$ and a can without water as $X = 0$, then the mean of the population is 0.5, which is equal to the probability P of a can containing water in the pair.

The mean of the population will play an important role in the normal approximation that we develop in this chapter.

3.4 Variance and Standard Deviation of a Population

The variance is a number that represents a particular way to define variability. Data can be tightly clustered around the mean, or data can have a wide spread around the mean. For instance, the mean height of players on two women's soccer team might be 5 feet 8 inches, but one team has players of very similar heights ranging from 5 feet 6 inches to 5 feet 11 inches (tightly clustered around the mean) and the other team has players whose heights range from 5 feet 4 inches to 6 feet 2 inches (wide distribution around the mean).

The **variance of a population** is an index of the variability of the possible outcomes around the mean μ. The variance is computed as follows: square the difference of each outcome from the mean, weight these squared differences by the probability of that outcome, and add the products. In symbols we have

$$\sigma^2 = \sum_i P_i (X_i - \mu)^2$$

In words, the variance is the average of the squared differences from the mean. The squared differences are weighted by their likelihood (the probability P) of occurring. There is a special case when all the values of X equal the mean: the variance will be zero because each squared difference will be zero. The more the values of X differ from the mean, the greater the variance because the difference $X_i - \mu$ gets larger. Thus the variance is an index of the variability around the mean.

The variance of a binomial population is the special case where

$$\begin{aligned} \sigma^2 &= P(1-P) \\ &= PQ \end{aligned}$$

Thus, the variance of a binomial population is simply the product of $P(X = 1)$ and $P(X = 0)$.

The **standard deviation of a population** is defined as the square root of the population variance, and for the special case of a binomial population, we have

$$\sigma = \sqrt{PQ}$$

For the binomial population shown in Table 3.1, the variance and standard deviation are, respectively,

$$\sigma^2 = (0.5)(0.5) = 0.25$$

and

$$\sigma = \sqrt{(0.5)(0.5)} = 0.5$$

We will discuss the standard deviation later, but for now memorize this fact: the standard deviation is defined as the square root of the variance. The standard deviation will play an important role in the normal approximation that we will use—it will represent the variability in the population.

3.5 The Average of a Sum and the Variance of a Sum

In experiments we usually make more than one observation, so we need to extend the idea of mean and variance to many observations. Suppose we draw a random sample of n observations from a population with mean μ, and we let T be the sum of the n values of X. The concept of mean can be applied to the sum T as follows, where each of the individual random variables X_i have the population mean μ_i:

$$T = X_1 + X_2 + \cdots + X_n$$

and

$$\mu_T = \mu_1 + \mu_2 + \cdots + \mu_n$$

Thus the average of a sum is equal to the sum of the individual averages of each variable X_i that constitute the sum. The average of a sum of several observations will become handy, as we will soon see. When all the observations have the same population mean, then the sum T of the n variables will have mean $\mu_T = n\mu$. In words, the average of the sum T of variables with the same mean is equal to n times the mean of the population, where n denotes the number of observations that make up the sum T.

For the special case of a binomial population with $\mu = P$, we have

$$\mu_T = nP$$

For example, suppose we offer the farmer 10 pairs of cans (one containing water, the other not), as in the second study in Chapter 2. If the farmer doesn't have the ability to detect water and is merely guessing, then for each pair the probability of correctly guessing which of the two cans contains water is 0.5; so, the average number of correct guesses over the 10 pairs is $10 \times 0.5 = 5$, which is the average of the sum of those 10 individual pairs of choices, each having a 0.5 chance of yielding a correct guess.

If the n observations of variable X are independent, then the variance of the sum T will simply be n times the variance of X and can be expressed as

$$\sigma_T^2 = n\sigma^2$$

For the special case of a binomial population, the variance and standard deviation of a sum of individual and independent random variables X_i (each with the same mean and variance), respectively, are given as

$$\sigma_T^2 = nPQ \tag{3.2}$$

and

$$\sigma_T = \sqrt{nPQ} \tag{3.3}$$

3.6 The Average and Variance of Repeated Samples

We define the observed **mean of a sample** of n observations as

$$\bar{X} = \frac{X_1 + X_2 + \cdots + X_n}{n} = \frac{\sum X_i}{n}$$

In words, this is the sum of the observations divided by the number of observations. This is the same as dividing the sum T defined in the preceding section by the sample size n.

Now consider this thought experiment. Imagine repeating the n observations over and over again so that for each sample of size n we compute the sample mean. To be concrete, we take a sample of n observations and compute the sample mean, then we take another sample of n observations and compute the mean of this second sample, then we take a third sample of n observations and compute a third sample mean, etc., repeatedly for a long time. We can ask, "What is the average of these sample means?" (that is, the "mean" of all those sample means).

The expected average of the observed $\bar{X}$ is defined as

$$\mu_{\bar{X}} = \frac{1}{n}\mu_T$$

and because $\mu_T = n\mu$, we have the simple identity

$$\mu_{\bar{X}} = \mu$$

In other words, the average value of the mean of a random sample of n observations is equal to the population mean μ of the individual samples. Thus, if we repeatedly take random samples of n observations, for each of those samples compute the corresponding mean, and then take the average of all the sample means, we would, in the long run, have the population μ. The mean of the sample means approaches the population mean, as we will see later, as the number of n observations increases.

The special case of a random sample from a binomial population is simply

$$\mu_{\bar{X}} = P$$

In much the same way that we can ask about the mean of a sequence of sample means, we can also ask about the variance of a sequence of sample means. That is, what is the variance of a bunch of sample means, each drawn from the same population where each sample is of size n? The **variance of a mean** of a random sample of n independent observations drawn from a given population is

$$\sigma_{\bar{X}}^2 = \frac{\sigma^2}{n} \tag{3.4}$$

The square root of Equation 3.4 is called the **standard error of the mean**. In general, the standard error of the mean is given by

$$\sigma_{\bar{X}} = \frac{\sigma}{\sqrt{n}}$$

Note that the number of observations is in the denominator; so, as n grows large, the ratio on the right-hand side grows small—hence the variability of the sample means $\sigma_{\bar{X}}$ approaches 0. The standard error of the mean, and its sample estimate, will play an important role in statistical tests developed later in this book.

For the special case of a binomial population the variance of the mean is given by this ratio:

$$\sigma_{\bar{X}}^2 = \frac{PQ}{n}$$

and the standard error of the mean of a binomial population is

$$\sigma_{\bar{X}} = \sqrt{\frac{PQ}{n}}$$

In the following sections, we apply the formulas we have just developed to the two experiments with the farmer described in the preceding chapter. We will see that it is possible to determine quickly and easily an approximation to the probabilities of the outcomes of these experiments. Your effort in understanding these technical sections and complicated concepts will soon pay off because we will be able to estimate the probabilities of the farmer's correctly guessing the water cans through some simple approximations. Those approximations will serve as a springboard to a set of useful data analytic techniques that will occupy us for the rest of this book.

3.7 The Second Experiment with the Farmer: μ_T and σ_T

Recall that in the second experiment the farmer was presented with a pair of cans and asked to choose which one in the pair contained water. We can now analyze this situation with the ideas we introduced in this chapter. Let X denote a correct choice on a single trial, that is, the presentation of a pair of cans. We define $X = 1$ if the farmer makes a correct choice (that is, if he selects the can with water) and $X = 0$ if he does not. The farmer was given $n = 5$ pairs, and we let T denote the sum of X over the $n = 5$ trials. Thus, T is the number of correct choices observed in a set of $n = 5$ trials. Each X represents a 1 or 0 depending on whether the trial consisted of a correct choice, and T is the sum over those five 0 or 1 variables. The sum T can range between 0 and 5.

For the moment, let's assume that the farmer does not have the ability to detect the presence of water. This is called the null hypothesis; we'll develop this concept in more detail later in the book. Under this assumption the farmer merely guesses which of the two cans contains water. The probability P of correctly guessing which can contains water is 0.5. Thus, under the null hypothesis this experiment corresponds to drawing a random sample of $n = 5$ independent observations from a binomial population in which $P = 0.5$. Therefore, under the null hypothesis the mean and standard deviation of T are

$$\mu_T = nP = 5\,(0.5) = 2.5$$

and

$$\sigma_T = \sqrt{nPQ} = \sqrt{5\,(0.5)\,(0.5)} = 1.118$$

There is another way that we can compute the mean and standard deviation of T. We can work directly with the probabilities of observing $T = 0$, $T = 1$, etc., that we computed in Chapter 2. Table 3.2 lists the sample space for the experiment and shows the probability associated with each possible outcome of the experiment. The possible values of T under the null hypothesis are given in column (1), and the probabilities P associated with each sample point or value of T are given in column (2). These probabilities are those that we obtained in the preceding chapter. We see that the mean of the sum T is

$$\mu_T = \sum P_i T_i = \frac{80}{32} = 2.5$$

that the variance of the sum T is

$$\sigma_T^2 = \sum P_i T_i^2 - \mu_T^2 = \frac{240}{32} - (2.5)^2 = 1.25$$

and that the standard deviation of the sum T is

$$\sigma_T = \sqrt{1.25} = 1.118$$

The important point is that it is possible to calculate the theoretical values for the population mean of sum T, μ_T, and the population standard deviation of T, $\sigma_T = \sqrt{nPQ}$, by knowing only the values of n and P. In other words, it was not necessary for us to do the tedious calculations involved in finding the probabilities associated with each possible outcome of the experiment, as we did in the preceding chapter (displayed in Table 3.2). With knowledge of n and P the values μ_T and σ_T can be calculated relatively easily. We next need to find a way to convert the mean and standard deviation into information about the probabilities of the underlying events. We will then be able to approximate the probability of a specific observation in terms of the theoretical probabilities under the null hypothesis.

Table 3.2: Probability distribution of the sum T, the number of correct choices, when $n = 5$ independent observations are drawn from a binomial population with $P(X = 1) = 0.5$.

(1) T_i	(2) P_i	(3) P_iT_i	(4) $P_iT_i^2$
5	1/32	5/32	25/32
4	5/32	20/32	80/32
3	10/32	30/32	90/32
2	10/32	20/32	40/32
1	5/32	5/32	5/32
0	1/32	0	0
$\sum$	32/32	80/32	240/32

3.8 Representing Probabilities by Areas

We still need to develop a little more machinery before we can answer the question about the farmer's ability to detect water. We will represent the probabilities of discrete variables graphically as areas of rectangles. Recall that the area of a rectangle is given by the product of the base and the height:

$$\text{area} = \text{base} \times \text{height}$$

We need to represent probabilities in terms of base and height in order to display them as rectangles. For discrete variables that can take only values such as 0, 1, 2, 3, ... the base of each rectangle is equal to 1 (the width of the rectangle is 1 because it is the difference between successive integers—the distance between 0 and 1, between 1 and 2, etc.). The heights of the rectangles, in this instance, are the corresponding probabilities that the discrete value is equal to 0, 1, 2, 3, ... The area for any single rectangle will be

$$\text{area} = 1 \times P = P$$

where P is the corresponding probability. Thus, areas can be used to represent the probabilities that the variable takes the value 0, 1, 2, 3, ...

The complete set of rectangles, constructed in the manner described, is called a **probability histogram**. The total area under the probability histogram must always equal 1, because each rectangle has area equal to P and the sum of all probabilities must be equal to 1. Figure 3.1 shows the probability histogram for the probability distribution of T given in Table 3.2. The probability histogram provides a way to visualize a distribution.

Look at Figure 3.1 closely. Note that the rectangles on the horizontal axis go from 0.5 to 1.5, 1.5 to 2.5, 2.5 to 3.5, etc. In Chapter 1 we discussed the convention of representing discrete numbers as intervals. The width of the rectangles is 1; the heights are equal to the respective probabilities. These intervals will become important in subsequent sections when we approximate probabilities.

3.9 The Standard Normal Distribution

The normal distribution is a theoretical distribution and, because it refers to a continuous variable, can be represented by a curve. The equation of the **standard normal distribution**, sometimes referred to as the **Gaussian distribution**, is

$$y = \frac{1}{2\sqrt{\pi}} e^{\frac{-z^2}{2}} \tag{3.5}$$

where y = the height of the curve at any given point along the baseline; π = 3.1416 (rounded), the ratio of the circumference of a circle to its diameter; e = 2.7183 (rounded), the base of the natural system of logarithms; and $Z = (X - \mu)/\sigma$. The variable Z is the observation minus the population mean μ, and that difference is divided by the population standard deviation σ. The total area under the curve defined by Equation 3.5 is equal to 1.00.

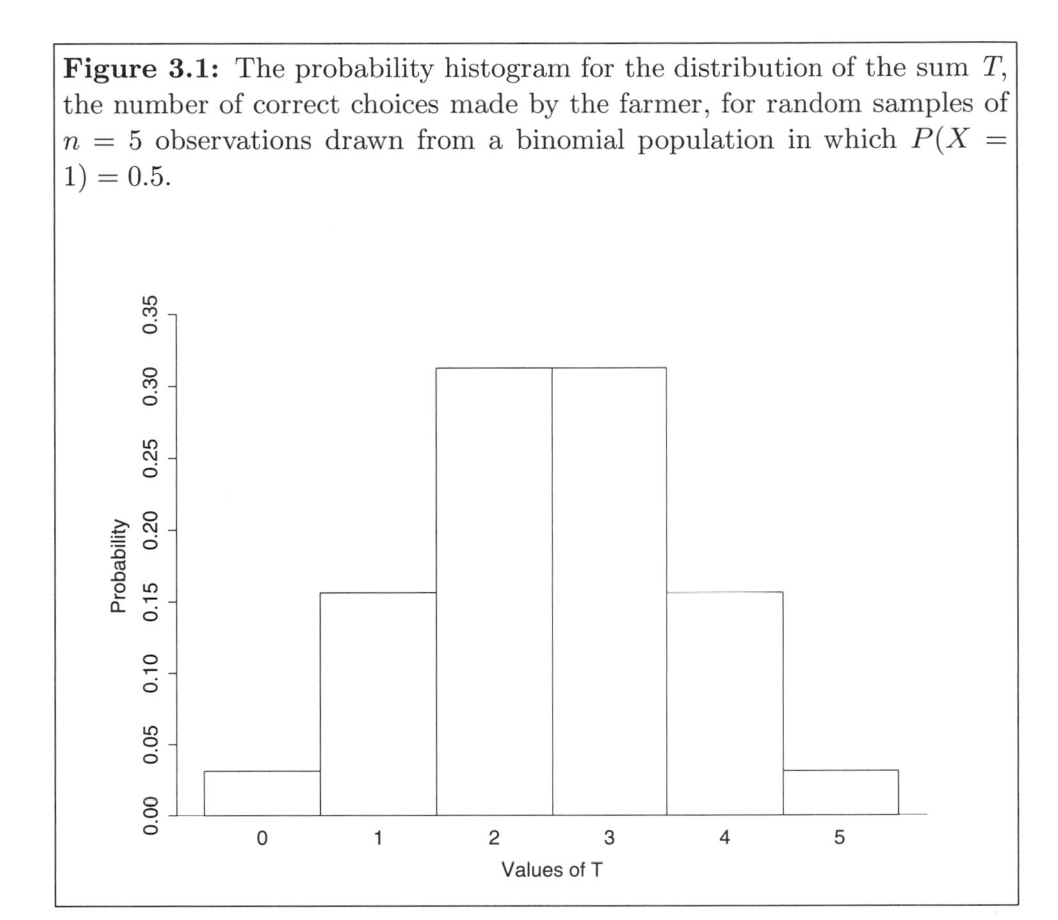

Figure 3.1: The probability histogram for the distribution of the sum T, the number of correct choices made by the farmer, for random samples of $n = 5$ observations drawn from a binomial population in which $P(X = 1) = 0.5$.

Table B.3 in Appendix B is a table of the standard normal distribution. The first column gives values of $Z = (X - \mu)/\sigma$ in units of 0.01. The second column gives the proportion of the total area under the curve between the ordinates computed at $\mu_Z = 0$ and the value of Z given in the first column. The third column gives the proportion of the total area in the larger segment of the curve when an ordinate is computed at Z, and the fourth column gives the proportion of the total area in the smaller segment.

Any normally distributed variable X can be transformed into a standard normal variable Z. A **standard normal variable** is defined by a simple transformation of the variable X

$$Z = \frac{X - \mu}{\sigma}$$

The variable Z is usually referred to as a Z **score**. A Z score is a standardized value because the mean is subtracted from the score (numerator) and the scale is set by the standard deviation (denominator). If we have a sequence of observations from the same population and convert the values of the observations to Z scores using the mean and the standard deviation, then the mean of the set of Z scores will be 0 and the standard deviation of the set of Z scores will be 1.

Converting data into Z scores amounts to what is known as a linear transformation. We can see this by rearranging the Z score formula:

$$\begin{aligned} Z &= \frac{X - \mu}{\sigma} \\ &= \frac{1}{\sigma}X - \frac{\mu}{\sigma} \end{aligned}$$

The slope of this linear transformation is $1/\sigma$, and the intercept is $-\mu/\sigma$. An example of another linear transformation is in converting temperature from degrees Fahrenheit to degrees Celsius. Thus, a linear transformation is merely a change in "scale." Linear transformations do not change the shape of a distribution. If variable X is not normally distributed, neither will a linear transformation of X. Thus, the Z transformation merely rescales the values of X to have a mean of 0 and a standard deviation of 1.

3.10 The Second Experiment with the Farmer: A Normal Distribution Test

We return to the second experiment with the farmer, making use of the concepts introduced in this chapter. Assuming, for the moment, that variable T is a continuous and normally distributed variable with $\mu_T = 2.5$ and $\sigma_T = 1.118$ (as we calculated in Section 3.7), then the standardized score

$$Z = \frac{T - \mu_T}{\sigma_T} \tag{3.6}$$

will be a standard normal variable. Suppose that the farmer correctly chose the can filled with water in each of the five pairs so that the variable $T = 5$. We know, from calculations in the preceding chapter, that the exact probability associated with this outcome is 0.031. This same probability is also approximated by the area of the last rectangle, or the area falling to the right of 4.5, in the probability histogram for T in Figure 3.1.

For discrete variables, the use of the *lower limit* of T in estimating the probability of T greater than or equal to some specified value by means of the standard normal distribution is known as a **correction for discreteness or discontinuity**. For discrete variables, such as T, that can take only the values 0, 1, 2, ... the lower limit will always be given by $T - 0.5$, and the desired probability will be given by the area falling to the right of the corresponding value of Z. To estimate the probability of T less than or equal to some specified value, the correction for discreteness involves the *upper limit* of T, or $T + 0.5$. In this instance, we find the area falling to the left of the corresponding value of Z.

To approximate the probability that $T = 5$ we use the table of the standard normal distribution to the right of $T = 4.5$ (that is, applying the correction for discontinuity). In a normal distribution with $\mu = 2.5$ and $\sigma_T = 1.118$, the area falling to the right of the value 4.5 is given by the area falling to the right of the corresponding Z score

$$Z = \frac{4.5 - 2.5}{1.118} = 1.79 \tag{3.7}$$

in the standard normal distribution.

Using the standard normal distribution table (Table B.3 in Appendix B) we find that 0.037 of the total area falls to the right of $Z = 1.79$. In symbols, $P(Z \geq 1.79) = 0.037$, and this value is also the probability of

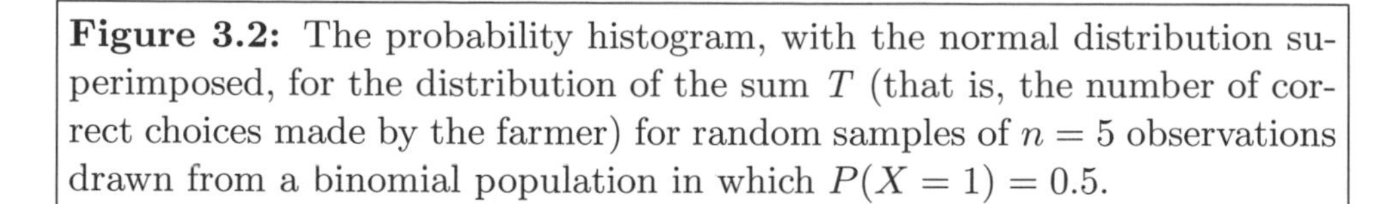

Figure 3.2: The probability histogram, with the normal distribution superimposed, for the distribution of the sum T (that is, the number of correct choices made by the farmer) for random samples of $n = 5$ observations drawn from a binomial population in which $P(X = 1) = 0.5$.

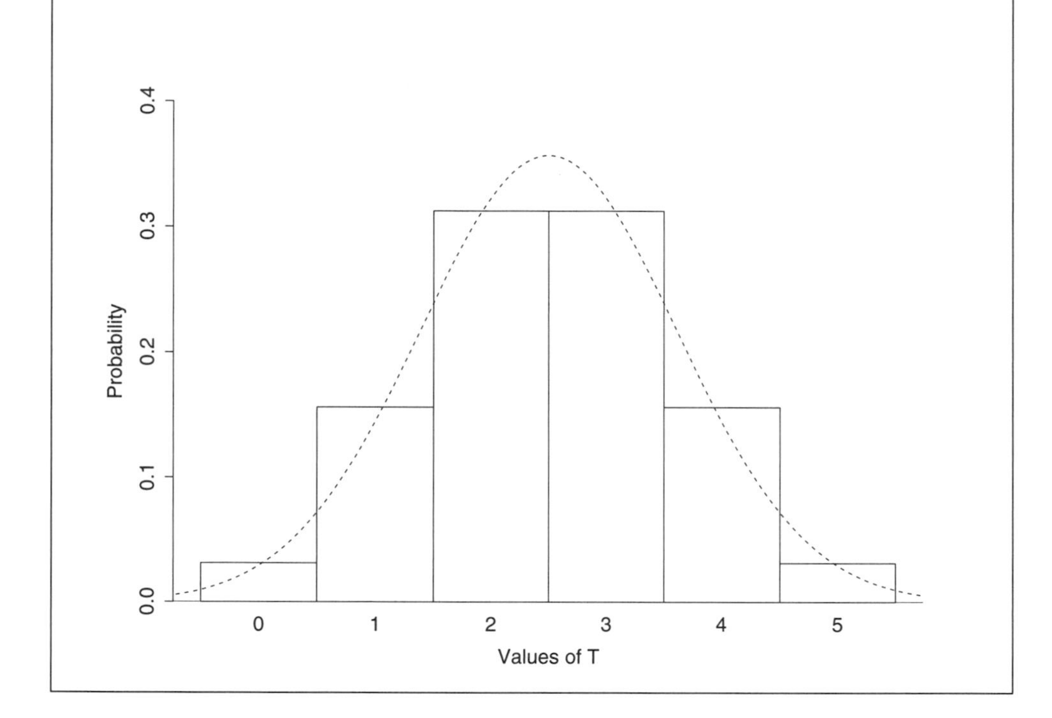

$T \geq 4.5$ if T is continuous and a normally distributed variable with $\mu = 2.5$ and $\sigma_T = 1.118$.

We see that despite variable T being discrete in our experiment and not exactly normally distributed (that is, the rectangles approximate but don't exactly follow a normal curve), the probability of 0.037 we have obtained from the normal distribution approximates quite well the exact probability of 0.031. The approximation is remarkably close to the actual value. Figure 3.2 superimposes the normal distribution on the histogram presented in Figure 3.1. You can see that the continuous bell-shaped curve closely approximates the shape created by the set of rectangles.

In this example if we adopt an $\alpha = 0.05$ of making a Type I error, we see that the computed probability of 0.037 is less than the α criterion. Hence, we reject the null hypothesis that the farmer is guessing which cans contain water. The observed sample of five correct choices is relatively unlikely to occur by guessing. We cannot be sure that the farmer was not guessing—we can merely say that it is unlikely that the farmer guessed correctly on each of the five pairs.

We have used the normal distribution with a relatively small value for n. The approximation method we just developed improves as the sample size n increases. Obviously, the methods of the preceding chapter become more laborious as the sample size n increases, so the benefit of the approximation is easy to appreciate. We do not need to compute the tedious probabilities that we computed in Chapter 2 because they can be approximated through the standard normal distribution. The approximation simply requires these steps:

1. Compute the mean and standard deviation.
2. Compute the Z score corresponding to the observed value (as we did in Equation 3.7).
3. Look up the probability of obtaining that Z score in the normal table.
4. Check whether the probability corresponding to the Z score is less than the α criterion.

We suggest the following heuristic for using the normal distribution test on discrete variables: if both nP and nQ are at least equal to or greater than 5, then for all practical purposes the normal distribution test may be substituted for the exact test. This rule sets a minimum standard, and, of course, we would be better off if both nP and nQ were considerably larger than 5, not only because of the increased accuracy of the normal distribution test but also because of the increased power of the test with a larger sample size n. Obviously, in the example with the farmer we violated our own heuristic because $nP = 2.5$, but we see that the approximation still works well in this example (mostly because the distribution is symmetric because $P = Q = 0.5$, as we will show in a later section). The suggested heuristic is relatively conservative. The normal approximation works quite well, though there are some examples, especially when trying to approximate the extreme tails of a binomial distribution, say 20–30 standard deviations away from the mean, where the approximation can break down (see Jaynes, 2003, ch. 5).

3.11 The First Experiment with the Farmer: A Normal Distribution Test

In the second experiment in Chapter 2, where the farmer was presented with five pairs of cans, the probability of a correct choice remained the same throughout the study. It was equal to 1/2 for each successive pair of cans. The model for this second experiment involved sampling from an infinite population or, what amounts to the same thing, sampling from a finite population with replacement after each draw. If we have a finite population and sample with replacement after each draw, the probabilities associated with the values of X will remain the same on each successive draw. But in the first experiment, where the farmer was to select 5 cans from 10, the probabilities do not remain the same for each of his successive choices. Instead, the probability that the farmer's second choice was correct depended on whether his first choice was correct. For example, if his first choice was correct, then the probability of his second choice being correct was not 1/2 but 4/9. On the other hand, if his first choice was incorrect, then the probability that his second choice would be correct was 5/9. For all subsequent choices, the probability of a correct choice depends on the nature of the previous choices. In our model for this experiment, we said that we would regard the sampling procedure as involving a random sample of $n = 5$ from a finite population of $N = 10$ without replacement after each draw.

With this sampling procedure, we still have the mean of the sum $\mu_T = nP$, but we need to use a finite population correction factor to obtain the correct standard deviation σ_T. Even though the average of the sum T is the same regardless of whether sampling is done with or without replacement, the variance changes depending on the type of sampling. The finite population correction factor is equal to $(N-n)/(N-1)$ where N is the size of the population and n is the size of the sample. The variance of the sum T under sampling without replacement for a binomial variable is given by

$$\sigma_T^2 = \frac{N-n}{N-1} nPQ \tag{3.8}$$

and the standard deviation is

$$\sigma_T = \sqrt{\frac{N-n}{N-1} nPQ}$$

In the case of the first study with the farmer we have the following values: $N = 10$, $n = 5$, and $P = 0.5$. The corrected variance is given by

$$\sigma_T^2 = \frac{10-5}{10-1} (5)(0.5)(0.5) = 0.6944$$

and the corrected standard deviation

$$\sigma_T = \sqrt{0.6944} = 0.833$$

Note that as the size of the population grows infinitely large, then the ratio $(N-n)/(N-1)$ goes to 1. In that case, Equation 3.8 grows closer to the definition of the variance of a sum of variables given in Equation 3.2. In general, if the ratio n/N is less than 1/10, then for all practical purposes we may neglect the correction factor in calculating σ_T.

Table 3.3 lists the sample space for this experiment and organizes the computation into columns. In column (1) we have the possible values of the sum T, and in column (2) we give the probabilities associated with each value of T. These probabilities are the same as those obtained in Chapter 2 using exact methods. We reproduce those computations in this section. Column

Table 3.3: Probability distribution of the sum T, the number of correct choices, when $n = 5$ observations are drawn from a binomial population of $N = 10$ without replacement and $P(X = 1) = 0.5$.

(1) T_i	(2) P_i	(3) P_iT_i	(4) $P_iT_i^2$
5	1/252	5/252	25/252
4	25/252	100/252	400/252
3	100/252	300/252	900/252
2	100/252	200/252	400/252
1	25/252	25/252	25/252
0	1/252	0	0
$\sum$	252/252	630/252	1750/252

(3) gives the values of P_iT_i, and we note that the mean is the same as in the other model involving pairs of cans (that is, the second study in Chapter 2),

$$\mu_T = \sum P_iT_i = \frac{630}{252} = 2.5$$

In column (4) we give the squared values of T multiplied by the corresponding probabilities. The variance of the sum T is given by

$$\sigma_T^2 = \sum P_iT_i^2 - \mu_T^2 = \frac{1750}{252} - (2.5)^2 = 0.6944$$

which is equal to the value we obtained using Equation 3.8. The standard deviation of T is given by

$$\sigma_T = \sqrt{0.6944} = 0.833$$

as before. Thus, the mean and variance computed through the formulas presented in this chapter yield the same results as the mean and variance that result from the exact computations in the preceding chapter.

Figure 3.3 shows the probability histogram for the probability distribution of T given in Table 3.3. In Chapter 2, we found that for this experiment $P(T = 5) = 0.004$. On the histogram the value T = 5 corresponds to the

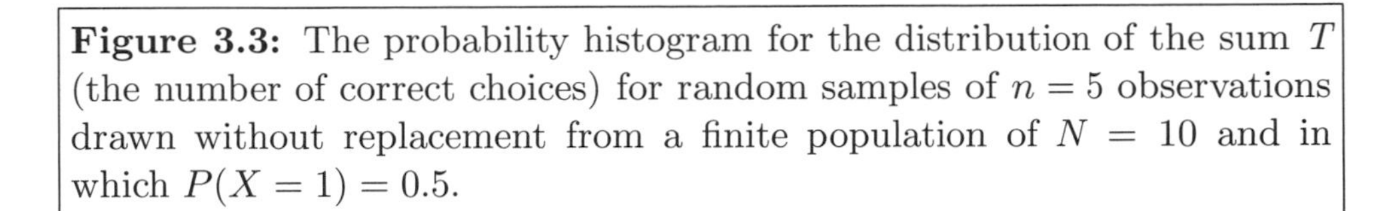

Figure 3.3: The probability histogram for the distribution of the sum T (the number of correct choices) for random samples of $n = 5$ observations drawn without replacement from a finite population of $N = 10$ and in which $P(X = 1) = 0.5$.

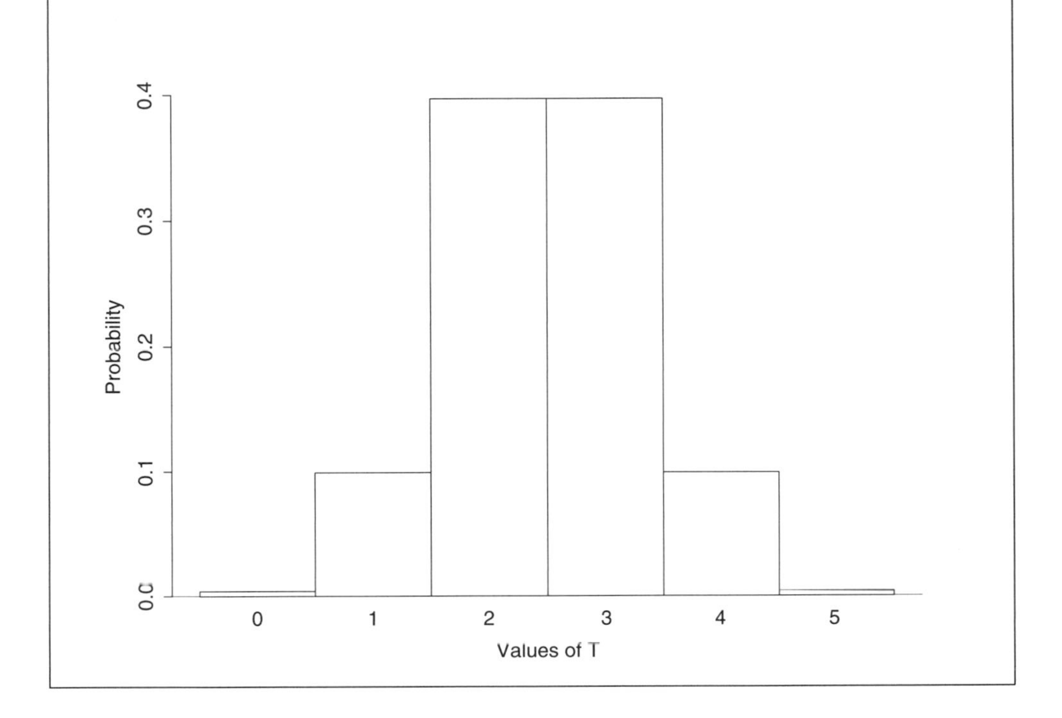

area falling to the right of 4.5. For a normal distribution with $\mu = 2.5$ and $\sigma = 0.833$, the area falling to the right of 4.5 equals the area falling to the right of the Z score

$$Z = \frac{4.5 - 2.5}{0.833} = 2.4$$

in the standard normal distribution. From the table of the standard normal distribution (Table B.3 of Appendix B), we find that 0.008 of the total area falls to the right of $Z = 2.40$. This value is the probability of $T \geq 4.5$ if T is a continuous and normally distributed variable with $\mu = 2.5$ and $\sigma = 0.833$. This result, $P(T \geq 4.5) = 0.008$, is a pretty good approximation to the probability 0.004 that we obtained from the exact test in Chapter 2. Again the probability of observing 5 correct cans under the null hypothesis that the farmer is guessing in this experiment is 0.008, which is less than the Type I criterion of $\alpha = 0.05$, so we reject the null hypothesis.

3.12 Examples of Binomial Models

Many research problems can be evaluated in terms of a model that involves random sampling from a binomial population, either finite or infinite. The dependent variable is one in which there are only two possible responses that a participant can make in responding to some situation or experimental condition.

We might, for example, be interested in determining whether a participant, when presented with fresh or frozen orange juice, can correctly select the fresh juice. Or whether a participant can correctly identify a given brand of cola when presented with two brands. Or whether a participant can distinguish between handwriting specimens of males and females, or correctly judge which of two tones has a higher pitch.

In a psychophysical experiment we may be interested in determining by how much one weight must differ from a standard weight before a participant can detect significantly better than by chance that the weight is heavier than the standard. Or we may wish to find out how salty a solution must be before it can be discriminated significantly better than by chance from a plain solution. There are many examples that involve the binomial distribution.

The methods described in this chapter are general and do not require that $P(X = 1) = 0.5$, as in the two experiments with the farmer. We may have a binomial population in which $P(X = 1)$ is $1/3$, or $1/4$, or any other value. The first experiment with the farmer, for example, might have been modified by having 4 wet cans and 6 dry cans. The task set for the farmer would then be to select a set of 4 cans from the complete set of 10. In this case, P would be 0.4 rather than 0.5, and Q would be 0.6. The approximation using the normal distribution can be computed using those values for P and Q. Similarly, in the second experiment we might have arranged the cans in triplets, rather than in pairs, so that each triplet contained one wet can and two dry cans. In this case we would have $P = 1/3$ rather than $P = 1/2$, and the approximation would carry forth with those values.

Knowing that we have an experiment that involves random sampling from a binomial population with a known or assumed value of P, we can quickly and easily determine μ_T and σ_T. We can then evaluate, by the normal curve, the probability of the sum T equal to or greater than the value obtained in a given experiment.

One reason why the standard normal curve approximations were so good in the two experiments is that in random sampling from a binomial population with $P = Q$ the probability distribution of the sum T is symmetric (see Figure 3.1). However, when $P \neq Q$, then the probability distribution of T will not be symmetric but skewed. If P is larger than Q, the distribution will have a tail to the left; if P is smaller than Q, the distribution will have a tail to the right. The top panel of Figure 3.4 shows the skewness of T for $P - 0.8$ with 10 observations.

As the number of observations increases, the distribution of the sum T becomes more symmetric, even when P is not equal to Q. Note that the rule suggested previously, that both nP and nQ be equal to or greater than 5, provides for the possibility that P may be larger or smaller than Q and that the distribution of T will be skewed when n is also small. For example, if $P = 0.8$, the heuristic suggests a minimum of 25 observations for the normal distribution test to be applied to the evaluation of the experimental outcome. That is, the nonzero portion of the theoretical distribution with $P = 0.8$ and 25 observations is fairly symmetric. The middle panel of Figure 3.4 shows an example satisfying this minimum heuristic. The bottom panel shows the distribution becoming more symmetric with larger sample size ($n = 100$).

Figure 3.4: The probability histogram for the distribution of the sum T (the number of correct choices) for random samples of $n = 10$ (top), $n = 25$ (middle), and $n = 100$ (bottom) observations drawn from a binomial population with $P = 0.8$. As n increases, the binomial distribution is well approximated by the normal distribution. The three distributions are shifted because they have different means due to varying sample size n.

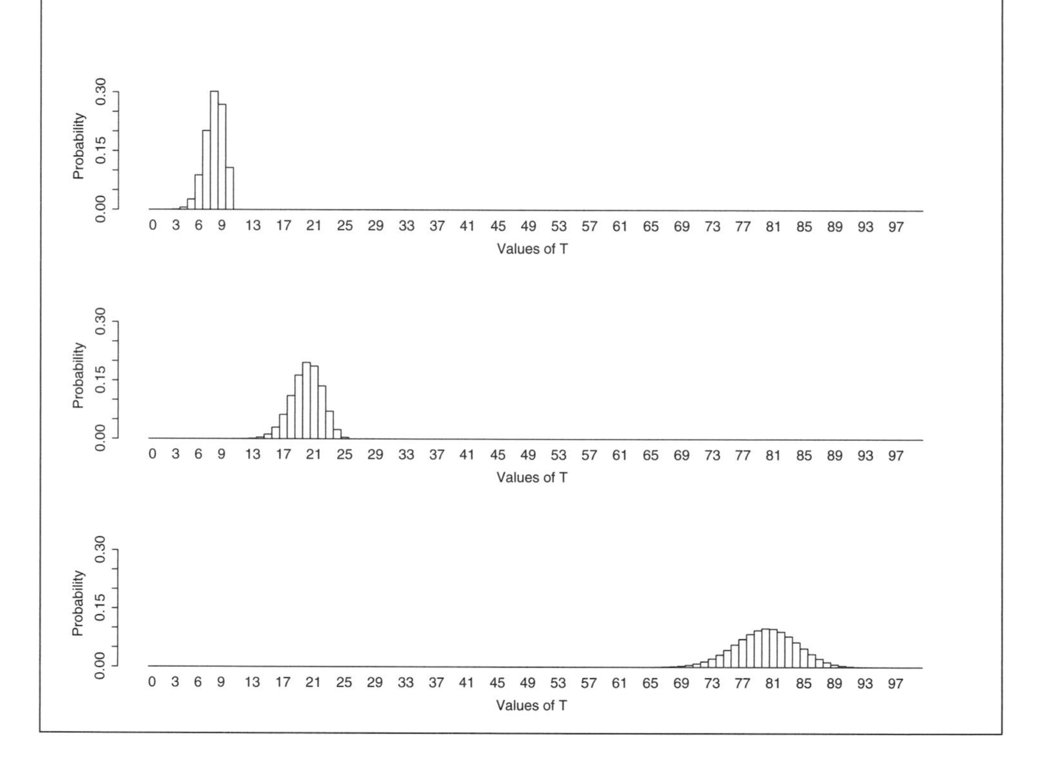

3.13 Populations That Have Several Possible Values

We discussed the distribution of the sum T for a random sample of n independent observations drawn from a binomial population where the individual observation X can take discrete values (that is, $X = 1$ or $X = 0$) with corresponding probabilities of P and Q, where $Q = 1 - P$. But suppose, as is true of some variables in behavioral research, that X can take other possible values such as 2, 3, 4, 5, 6, and so on. The population distribution of X will be given by the corresponding probabilities associated with each possible value of X. We consider three specific examples: (1) a population distribution of X that is rectangular or uniform; (2) a population distribution of X that is U-shaped; and (3) a population distribution of X that is skewed. In all three examples, we assume that the possible values of X are 0, 1, 2, and 3. Such a four-point scale might arise when participants are asked to rate an object or participants are given three questions and their score is the number of questions answered correctly (a number between 0 and 3).

In these examples, we are interested specifically in the distribution of the sum (which we continue to denote as T) of the values of n independent random observations drawn from each of the three populations. We will show that even for a relatively small number of observations the distribution of the sum T for random samples drawn from these nonnormally distributed populations begins to have a common form or shape that approaches a normal distribution.[1]

3.14 The Distribution of the Sum from a Uniform Distribution

Table 3.4 shows the population distribution of a variable X that can take the possible values of 0, 1, 2, and 3, with corresponding probabilities of 0.25, 0.25, 0.25, and 0.25, respectively. The population distribution of X is uniform because each possible value is equally likely. The mean and variance of this population are

$$\mu = \sum P_i X_i = 1.5$$

and

[1] For a more formal development of the central limit theorem see a textbook on mathematical statistics such as Hogg and Craig (1978).

Table 3.4: Probability distribution of X: A uniform distribution.

(1) X_i	(2) P_i	(3) P_iX_i	(4) $P_iX_i^2$
3	0.25	0.75	2.25
2	0.25	0.50	1.00
1	0.25	0.25	0.25
0	0.25	0.00	0.00
$\sum$	1.00	1.50	3.50

$$\sigma^2 = \sum P_iX_i^2 - \mu^2 = 3.5 - (1.5)^2 = 1.25$$

Table 3.5 gives the probability distribution of the sum T for random samples of $n = 1$ to $n = 6$ independent observations drawn from this uniform population. The variable n refers to the number of observations that we make from the population X. When $n = 1$ we make one observation, and the sum T of that single observation can take on the values 0, 1, 2, and 3. When $n = 2$ we make two observations of X, and the sum T of those two observations can take on the values 0, 1, 2, 3, 4, 5, and 6. The argument extends through higher values of n. Figure 3.5 shows the probability histogram for the sum T for random samples when $n = 6$. Note the distribution of the sum T when $n = 6$ is no longer uniform. The distribution of the sum T when $n = 6$ is remarkably close to a normal distribution.

Let us see how well we can approximate the probabilities associated with extreme values of T by using the standard normal distribution. We have a population with mean $\mu = 1.5$ and variance $\sigma^2 = 1.25$. The sum T, based on random samples of $n = 6$ observations from the uniform population X, has a mean and a standard deviation equal to

$$\mu_T = n\mu = (6)(1.5) = 9.0$$

and

Table 3.5: Probability distribution of the sum T of n independent random values of X when X has a uniform distribution and possible values of 0, 1, 2, and 3.

	n					
T	1	2	3	4	5	6
0	0.250	0.062	0.016	0.004	0.001	0.000
1	0.250	0.125	0.047	0.016	0.005	0.002
2	0.250	0.188	0.094	0.039	0.015	0.005
3	0.250	0.250	0.156	0.078	0.034	0.014
4		0.188	0.188	0.121	0.063	0.029
5		0.125	0.188	0.156	0.099	0.053
6		0.062	0.156	0.172	0.132	0.082
7			0.094	0.156	0.151	0.111
8			0.047	0.121	0.151	0.133
9			0.016	0.078	0.132	0.142
10				0.039	0.099	0.133
11				0.016	0.063	0.111
12				0.004	0.034	0.082
13					0.015	0.053
14					0.005	0.029
15					0.001	0.014
16						0.005
17						0.002
18						0.000

$$\sigma_T = \sqrt{n\sigma^2} = \sqrt{(6)\,(1.25)} = 2.74$$

Let's pick a value of 15 for T. To estimate the probability of $T \geq 15$ when $n = 6$, we first convert to the Z score, with a correction for discreteness or discontinuity (that is, $15 - 0.5 = 14.5$),

$$Z = \frac{14.5 - 9.0}{2.74} = 2.01$$

and from the table of the standard normal distribution in Appendix B, we find that $P(Z \geq 2.01) = 0.022$, which is the normal distribution estimate of the exact probability of $T \geq 15$. The exact probability of $T \geq 15$, as

Figure 3.5: The probability distribution of the sum T for random samples of $n = 6$ observations drawn from the uniform population shown in Table 3.5.

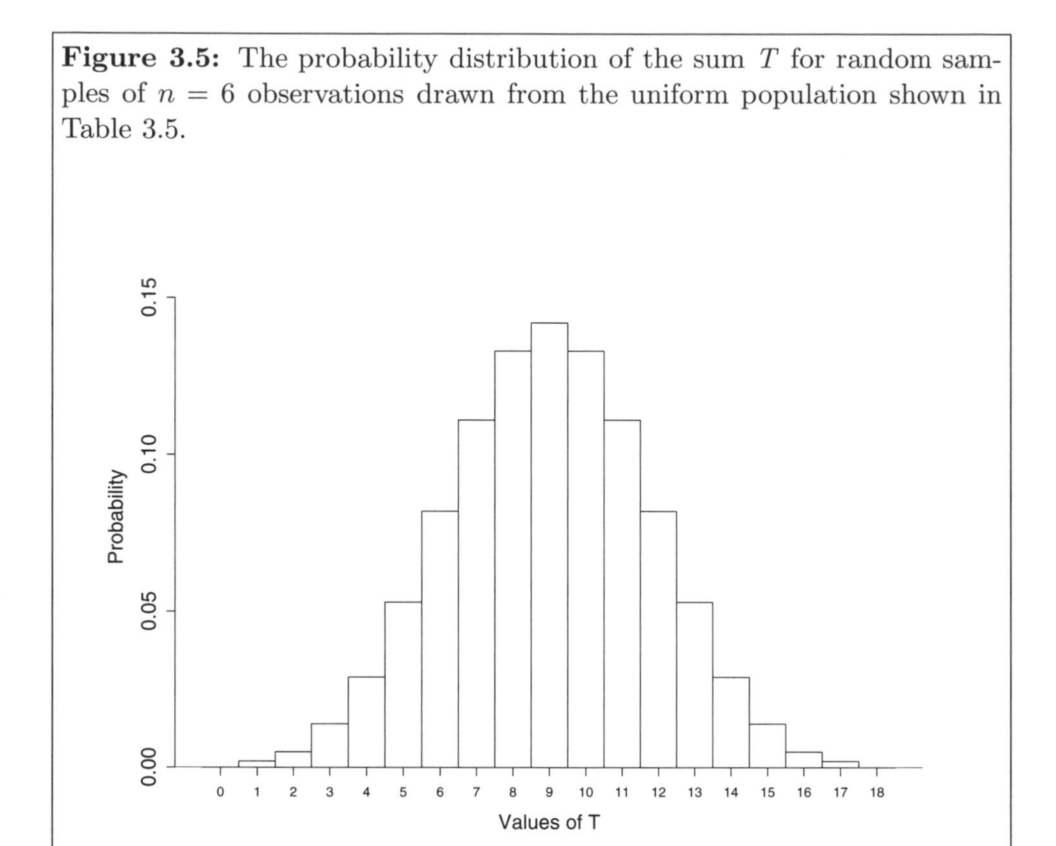

obtained from Table 3.5, is

$$\begin{aligned} P(T \geq 15) &= P(T = 15) + P(T = 16) + P(T = 17) + P(T = 18) \\ &= 0.014 + 0.005 + 0.002 + 0.000 \\ &= 0.021 \end{aligned}$$

The approximation worked remarkably well in this case even though the original distribution was uniform.

As another example, to estimate the probability that the sum $T \leq 3$, we have the Z score with a correction for discreteness or discontinuity

$$Z = \frac{3.5 - 9.0}{2.74} = -2.01$$

and from the table of the standard normal distribution we find that 0.022 of the total area falls to the left of $Z = -2.01$. This value is the normal distribution estimate of the exact probability of $T \leq 3$. The exact probability of $T \leq 3$, as obtained from Table 3.5, is $0.000+0.002+0.005+0.014 = 0.021$. Again, the approximation worked well. Note the symmetry of $P(T \geq 15)$ and $P(T \leq 3)$ both being equal.

We see that for this example, with random samples of only $n = 6$ observations drawn from a uniform population, the standard normal distribution estimates of the exact probabilities associated with extreme values of T are fairly accurate.

3.15 The Distribution of the Sum *T* from a U-Shaped Population

In Table 3.6, we display the population distribution of a variable X that can take the values of 0, 1, 2, and 3, with corresponding probabilities of 0.4, 0.1, 0.1, and 0.4, respectively. In this instance, the distribution of X is U-shaped because the ends ($X = 0$ and $X = 3$) each have a probability of 0.4 and the two middle values ($X = 1$ and $X = 2$) each have a probability of 0.1.

Before we examine the computation, what do you think the shape of the distribution of the sum T when $n = 6$ will look like? For the mean and variance of the population X, we have

$$\mu = \sum P_i X_i = 1.5$$

and

$$\sigma^2 = \sum P_i X_i^2 - \mu^2 = 4.1 - (1.5)^2 = 1.85$$

Table 3.7 gives the probability distribution of the sum T for random samples of $n = 1$ to $n = 6$ independent observations drawn from this U-

Table 3.6: Probability distribution of X: A U-shaped distribution.

(1) X_i	(2) P_i	(3) P_iX_i	(4) $P_iX_i^2$
3	0.4	1.2	3.6
2	0.1	0.2	0.4
1	0.1	0.1	0.1
0	0.4	0.0	0.0
$\sum$	1.0	1.5	4.1

shaped population. In Figure 3.6, we show the corresponding probability histogram for the sum T for random samples when $n = 6$. The distribution of the sum T when $n = 6$ is no longer U-shaped. Note the remarkable effect of taking sums: even though the original distribution of a single observation was U-shaped, the distribution of a sum of six observations begins to approach one that resembles a normal distribution.

Again, let us see how well we can estimate the probabilities associated with extreme values of T in terms of the standard normal distribution. With $n = 6$, the mean and standard deviation of the sum T are

$$\mu_T = n\mu = (6)(1.5) = 9.0$$

and

$$\sigma_T = \sqrt{n\sigma^2} = \sqrt{(6)(1.85)} = 3.33$$

To find $P(T \geq 15)$ using the normal distribution approximation, we make a correction for discontinuity and obtain

$$Z = \frac{14.5 - 9.0}{3.33} = 1.65$$

From the table of the standard normal distribution in Appendix B, we have $P(Z \geq 1.65) = 0.05$, and this value is an estimate of the exact probability

Table 3.7: Probability distribution of the sum T of n independent random values of X, where X has a U-shaped distribution and possible values of 0, 1, 2, and 3, with corresponding probabilities of 0.4, 0.1, 0.1, and 0.4.

			n			
T	1	2	3	4	5	6
0	0.400	0.160	0.064	0.026	0.010	0.004
1	0.100	0.080	0.048	0.026	0.013	0.006
2	0.100	0.090	0.060	0.035	0.019	0.010
3	0.400	0.340	0.217	0.123	0.066	0.034
4		0.090	0.111	0.091	0.062	0.038
5		0.080	0.111	0.101	0.076	0.051
6		0.160	0.217	0.196	0.147	0.099
7			0.060	0.101	0.107	0.090
8			0.048	0.091	0.107	0.099
9			0.064	0.123	0.147	0.139
10				0.035	0.076	0.099
11				0.026	0.062	0.090
12				0.026	0.066	0.099
13					0.019	0.051
14					0.013	0.038
15					0.010	0.034
16						0.010
17						0.006
18						0.004

of $T \geq 15$. The exact probability of $T \geq 15$, as obtained from Table 3.7, is

$$\begin{aligned} P(T \geq 15) &= P(T = 15) + P(T = 16) + P(T = 17) + P(T = 18) \\ &= 0.034 + 0.010 + 0.006 + 0.004 \\ &= 0.054 \end{aligned}$$

Again the approximation is remarkably close to the exact result. To estimate the probability of $T \leq 3$, we have, with a correction for discontinuity,

$$Z = \frac{3.5 - 9.0}{3.33} = -1.65$$

The area falling to the left of $Z = -1.65$ in the standard normal distribution is 0.050, which is the normal distribution approximation of the exact

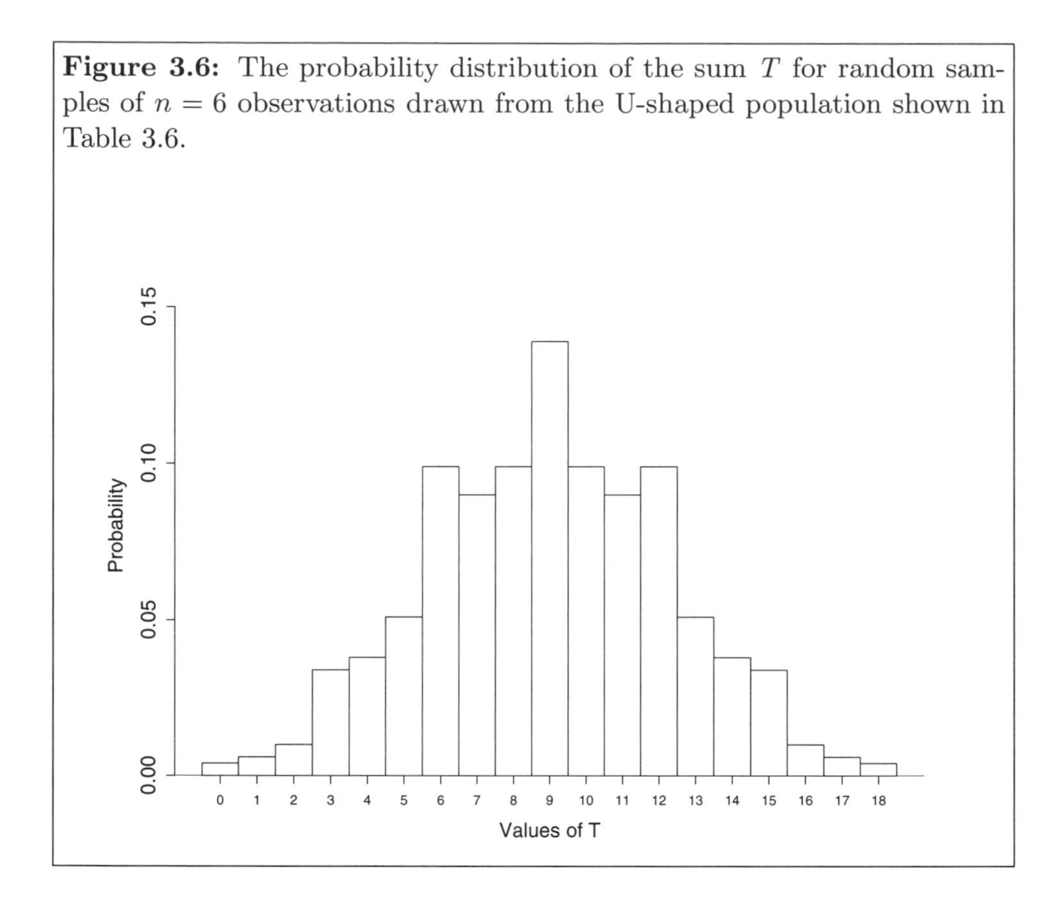

Figure 3.6: The probability distribution of the sum T for random samples of $n = 6$ observations drawn from the U-shaped population shown in Table 3.6.

probability of $T \leq 3$. By comparison, the exact probability of $T \leq 3$, as obtained from Table 3.7, is 0.054. Again, we find that the normal distribution approximation of the exact probabilities associated with extreme values of T for random samples of $n = 6$ observations drawn from the U-shaped population are fairly accurate. As n increases, the approximation gets closer to the exact probability value.

Table 3.8: Probability distribution of X: A skewed distribution.

(1) X_i	(2) P_i	(3) P_iX_i	(4) $P_iX_i^2$
3	0.4	1.2	3.6
2	0.3	0.6	1.2
1	0.2	0.2	0.2
0	0.1	0.0	0.0
$\sum$	1.0	2.0	5.0

3.16 The Distribution of the Sum T from a Skewed Population

We consider a third example. Table 3.8 shows a population where the variable X can take the values of 0, 1, 2, and 3 with corresponding probabilities of 0.1, 0.2, 0.3, and 0.4, respectively. In this instance, the distribution of X is skewed in the sense that low values of X are less likely than higher values. For the mean and variance of the population X, we have

$$\mu = \sum P_iX_i = 2.0$$

and

$$\sigma^2 = \sum P_iX_i^2 - \mu^2 = 5.0 - (2.0)^2 = 1.0$$

Table 3.9 gives the probability distribution of the sum T for random samples of $n = 1$ to $n = 6$ independent observations drawn from this skewed population. Figure 3.7 shows the probability histogram for the distribution of T when $n = 6$. Again, it is obvious that the distribution of the sum T departs markedly from the original skewed population distribution of X and approaches a shape that resembles a normal distribution. Because the distribution of X is skewed, the distribution of the sum T is also slightly skewed. We note, however, that the probabilities associated with the values of T in the left portion, or skewed tail, of the distribution are small and that

Table 3.9: Probability distribution of the sum T of n independent random variables of X when X has a skewed distribution with possible values of 0, 1, 2, and 3 with corresponding probabilities of 0.1, 0.2, 0.3, and 0.4.

			n			
T	1	2	3	4	5	6
0	0.100	0.010	0.001	0.000	0.000	0.000
1	0.200	0.040	0.006	0.001	0.000	0.000
2	0.300	0.100	0.021	0.004	0.001	0.000
3	0.400	0.200	0.056	0.012	0.002	0.000
4		0.250	0.111	0.031	0.007	0.001
5		0.240	0.174	0.065	0.018	0.004
6		0.160	0.219	0.112	0.038	0.010
7			0.204	0.161	0.070	0.023
8			0.144	0.190	0.111	0.044
9			0.064	0.184	0.150	0.073
10				0.138	0.172	0.110
11				0.077	0.167	0.140
12				0.026	0.133	0.158
13					0.083	0.154
14					0.038	0.127
15					0.010	0.087
16						0.047
17						0.018
18						0.004

the distribution of T is more symmetric than is the population distribution of X. With samples of $n > 6$, the distribution of T would be even more symmetric and would approach, as n increases, a normal distribution.

For random samples of $n = 6$, the mean and standard deviation of the sum T are

$$\mu_T = n\mu = (6)(2.0) = 12.0$$

and

$$\sigma_T = \sqrt{n\sigma^2} = \sqrt{(6)(1.0)} = 2.45.$$

Figure 3.7: Probability distribution of the sum T for random samples of $n = 6$ observations drawn from the skewed population shown in Table 3.9.

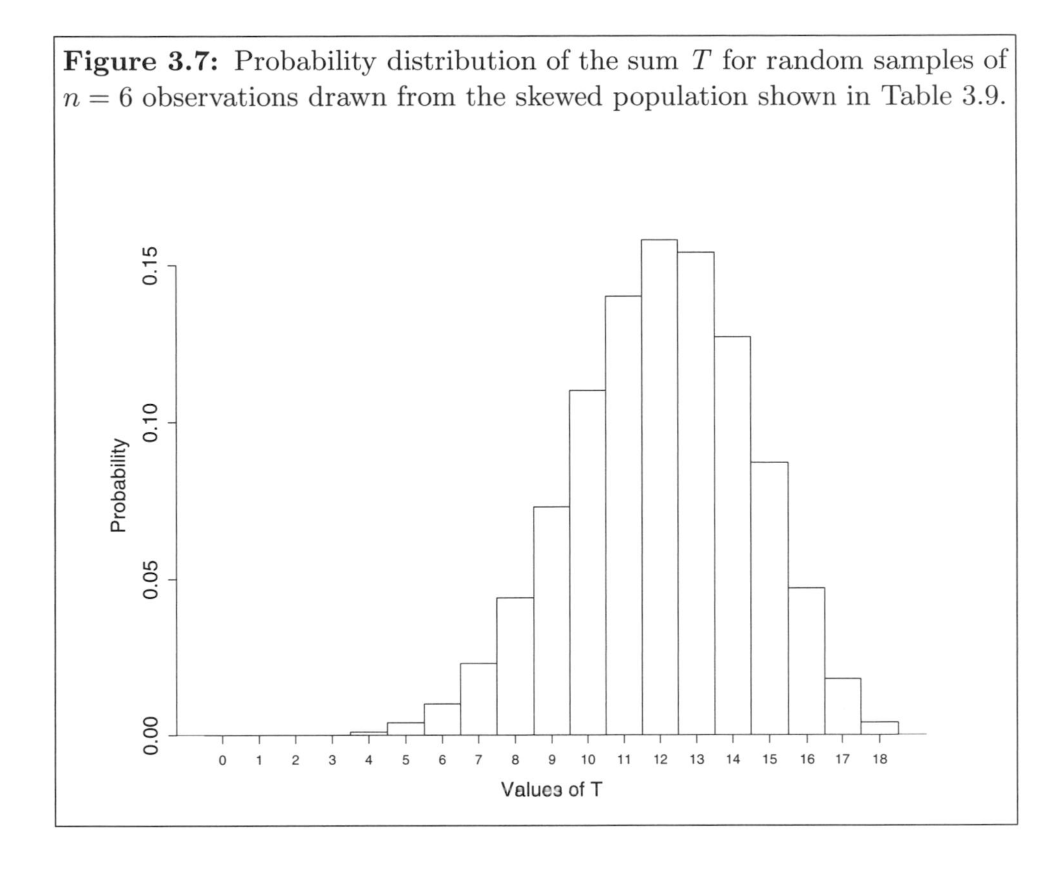

Let's try a different value of T, say 17. We can now use the standard normal distribution to approximate the probability of $T \geq 17$. With a correction for discontinuity, we have

$$Z = \frac{16.5 - 12.0}{2.45} = 1.84$$

From the table of the standard normal distribution in Appendix B, we find that $P(Z \geq 1.84) = 0.033$, and this value is an estimate of the exact probability of $T \geq 17$. The exact probability of $T \geq 17$, as obtained from Table 3.9, is

$$\begin{aligned} P(T \geq 17) &= P(T = 17) + P(T = 18) \\ &= 0.018 + 0.004 = 0.022 \end{aligned}$$

We can see that the skewness in the distribution X makes the approximation of the distribution of the sum T less accurate relative to the previous examples involving the uniform and U-shaped distributions.

For another example with the same skewed distribution X we estimate the probability of $T \leq 7$. The Z score, with a correction for discontinuity, is

$$Z = \frac{7.5 - 12.0}{2.45} = -1.84$$

and the area to the left of $Z = -1.84$ in the standard normal distribution is 0.033. This value is an estimate of the exact probability of $T \leq 7$. The exact probability of $T \leq 7$ is obtained by summing the probabilities associated with all values of $T \leq 7$, as given in Table 3.9, and this sum is 0.038. For this value, the approximation is fairly close to the exact value.

3.17 Summary and Sermon

In this chapter we showed how to approximate the tedious probability computations of the preceding chapter with the normal distribution. The approximation works quite well in many cases, so it can save a lot of time relative to computing the exact probabilities, as we showed in Chapter 2. In this chapter we showed by example that, even with relatively small samples of $n = 6$ drawn from a uniform, a U-shaped, or a skewed population, the probabilities associated with extreme values of the sum T can be approximated fairly well by using the standard normal distribution. For all three populations, the accuracy with which the normal distribution will approximate the exact probabilities associated with the values of the sum T will increase as n increases. Because the three populations we examined represent somewhat radical departures from a normally distributed population, we may be reasonably confident that in experimental work with samples of $n \geq 10$ the distribution of the sum T will be approximately normal in form. There is a mathematical proof, the central limit theorem, that justifies this statement. This result is important because in experiments we are almost always interested in sums (T) or means ($\bar{X}$) of random variables and not in the individual values of the variable. Thus in most experiments, if n is reasonably large, then the mean $\bar{X} = \sum X/n$ will be approximately normally distributed, even though the original population variable X may not be. We will exploit this result—that the sum of individual variables approaches a normal distribution as the sample size gets large—throughout the rest of the book.

Many behavioral scientists incorrectly assume that we use a normal distribution because many phenemona in nature tend to be normally distributed. While it may be true that many phenomena are normallly distributed, this chapter shows that one rationale for the normal distribution in statistics is that the sum of distributions (even those that are not themselves normally distributed) approaches a normal distribution as the sample size increases.

We emphasize that the material in this chapter provides machinery for computing probabilities under a particular set of assumptions. For example, we assume under the null hypothesis that the farmer cannot detect water and that he merely guesses "full" or "empty" with a probability of 1/2 (as in the case of the "second experiment" in Chapter 2). The techniques presented in this chapter can approximate the probability of a particular observation under these assumptions. So, if the farmer correctly "guessed" the five empty cans and the five full cans, we can estimate how likely that observation would be if the farmer had no ability. If the probability is sufficiently low, then we can say that it is unlikely that he does not have the ability to detect water. That is, we can reject the hypothesis that the farmer does not have the ability to detect water whenever we can reject the assumption that the farmer's probability to detect a can of water is .50. However, if the probability of the observation was relatively high under the null hypothesis, then we fail to reject the hypothesis that the farmer cannot detect water (that is, we cannot reject the null hypothesis that the farmer's $P = .50$).

The important point to always keep in mind is that an observation is just that—what you as the experimenter observe. In the case of the experiment with the farmer from Whidbey Island, we observe the number of cans correctly identified as containing water. The fancy computations introduced in this chapter help estimate the probability associated with that observation under a particular set of assumptions. Students frequently overemphasize the computations because they are difficult to learn and may seem strange at first. But the best advice we can provide is to always keep in mind that the important part of the data analysis is describing what you observe. Always highlight the observations or an appropriate summary of the observations. The probabilities associated with those observations allow us to make statistical inferences, and these statistical inferences should not be confused with the observations. The complicated statistics that consume most of the content of this book and books like it are mostly supporting documentation that allows us to say something about the "acceptability" of one's observations. We use the word "acceptability" to highlight that when we deem a result "statistically significant" we do so against an agreed-upon convention such as the criterion of $\alpha = 0.05$ using conventional statistical tests.

3.18 Questions and Problems

1. In a taste discrimination experiment, a participant is presented with two brands of frozen orange juice and one brand of fresh orange juice. The participant's task is to select the fresh orange juice in each presentation of the three samples. The participant is given 15 trials and correctly selects the fresh orange juice in 9 of the 15 trials. Is the hypothesis that the participant is responding by chance (that is, correctly guessing at a rate of 1/3) tenable? Use both the normal distribution approximation from this chapter and the exact probability computation from Chapter 2.

2. A university determined that errors are made on about 6% of the midterms graded by faculty. A new professor grades 500 papers during a given year, and it is found that 40 of these papers are scored incorrectly. We can treat the 500 papers that are scored as consisting of 500 trials of an event for which the probability of making an error on a single trial is 0.06. Has the faculty member made a significantly large number of errors? Explain and justify your answer.

3. Assume we have an binomial population with $P(X = 1) = 1/3$. Let T be the sum of $n = 15$ random observations drawn from this population. Find the mean and variance of T.

4. Assume we have a finite binomial population with $N = 45$ and with $P(X = 1) = 1/3$. A random sample of $n = 15$ observations is drawn from this population without replacement. Let T be the sum of the n observations. Find the mean and variance of T.

5. We have three females and two males, and we draw a random sample of $n - 3$ without replacement. Let T be the number of males in the sample.

   (a) Find the probability that T is equal to 3, 2, 1, and 0. Leave your answers as fractions.

   (b) Is μ_T equal to nP, where n is the sample size and P is the probability of obtaining a male on the first draw?

6. Let T be the number of 5's face up when a fair die is rolled $n = 6$ times. The 6 rolls are assumed to be independent. Find the value of μ_T and σ_T^2.

7. Let $P = 1/2$ be the probability of a correct response on a single trial in a cognitive reasoning experiment. A participant is given $n = 100$ reasoning questions to solve. Let T be the number of correct responses in the set of $n = 100$ responses.

(a) Find the values of μ_T and σ_T.

(b) Use the table of the standard normal distribution (Table B.3 in Appendix B) to find the probability of $T \geq 65$. Be sure to use the correction for discreteness.

8. In a binomial experiment, $P = 1/2$ and $n = 10$.

   (a) Find, using exact methods, $P(T \geq 8)$ and $P(T = 5)$.

   (b) Find, using the table of the standard normal distribution, $P(T \geq 8)$ and $P(T = 5)$. Compare these probabilities with those obtained by the exact method.

9. Suppose X can take the values of 1, 2, 3, 4, and 5, each with a probability of 1/5. Find the value of μ and σ^2.

10. Assume that a variable X is normally distributed with $\mu = 10$ and $\sigma = 2$. A random sample of $n = 4$ is drawn from this population.

    (a) Find the probability of obtaining $T \geq 50$.

    (b) Find the probability of obtaining $T \leq 25$.

11. A random variable X can take the values of 1 or 0 with probabilities of 1/3 and 2/3, respectively. A random sample of $n = 18$ is drawn from this population.

    (a) What is the probability of obtaining $T \geq 10$?

    (b) What is the probability of obtaining $T \leq 2$?

4

Tests for Means from Random Samples

4.1 Transforming a Sample Mean into a Standard Normal Variable

In the preceding chapter we showed that the sum of the values of n observations can be treated as a standard normal variable $Z = (T - \mu_T)/\sigma_T$. Researchers usually work with means, or averages, rather than sums. In this chapter we extend the technique of using the standard normal distribution to cases involving sample means.

The mean is one way to summarize several scores. An example of a mean is your course grade, which is usually computed as a weighted combination of assignments, midterm exams, and final exams, and summarizes your overall performance in the course. Statistics you hear in the news such as "average income" and "average number of children" are also examples of means.

In the preceding chapter we introduced a technique that allowed us to compute probabilities using an approximation based on the standard normal distribution. We had to convert the observation into a Z score in order to use the approximation. The analogous Z transformation for the sample mean is

$$Z = \frac{\bar{X} - \mu}{\sigma_x / \sqrt{n}} \tag{4.1}$$

Equation 4.1 defines a standard normal variable that can be used to evaluate the probabilities associated with $\bar{X}$, the mean of a random sample of n observations. Two primary differences compared to the Z introduced in the preceding chapter are that the numerator uses the observed mean $\bar{X}$ and the denominator includes information about the sample size n. This equation allows us to compute relevant probabilities for the distribution of a mean. We will be able to evaluate results from experiments that use means as the summary statistic. As we saw in the preceding chapter, the normal approximation greatly simplifies the computation of probabilities. We will see that, aside from a change in how the Z score is computed, the procedure is identical to that in Chapter 3 for the special case when we know the population variance.

For the special case of a binomial population, Equation 4.1 becomes

$$Z = \frac{p - P}{\sqrt{\dfrac{PQ}{n}}}$$

For a binomial population, p is the observed proportion, P is the population proportion, and $\sqrt{\frac{PQ}{n}}$ is the population standard deviation of the proportion. For example, we can ask, "What proportion of the cans did the farmer correctly identify as containing water?" In this chapter we will see that the proportion is identical to the mean of a specially defined variable.

4.2 The Variance and Standard Error of the Mean When the Population Variance σ^2 Is Known

We assume that the variable X is continuously and normally distributed in the population. If we take a random sample of n observations from this population, the sample **mean** $\bar{X}$ will be equal to

$$\bar{X} = \frac{\sum_{i=1}^{n} X_i}{n} \tag{4.2}$$

That is, one sums all the observations (the numerator) and divides the sum by the number of observations. For example, if one observes the scores 3, 5, and 7, the mean is $5 = (3 + 5 + 7)/3$. Even when the population is not normally distributed, we assume that the sample size n is sufficiently large that the distribution of the sum T and, consequently, the mean $\bar{X} = T/n$ will be approximately normal. Recall our detailed discussion in Chapter 3 about a sum approaching a normal distribution even when the original distribution is not normally distributed. Dividing the sum T by the number of entries in the sum (what occurs in Equation 4.2) does not change this property. Thus the process for working with sums introduced in Chapter 3 extends to the mean.

The variance of the mean $\bar{X}$ is given by

$$\sigma_{\bar{X}}^2 = \frac{\sigma^2}{n} \tag{4.3}$$

The square root of Equation 4.3 is called the **standard error of the mean** and is denoted

$$\sigma_{\bar{X}} = \frac{\sigma}{\sqrt{n}}$$

The standard error of the mean is related to both the population standard deviation σ (the numerator) and the sample size n (the denominator). As the population standard deviation increases, the standard error of the mean also increases. However, as the sample size increases, the standard error of the mean decreases. We illustrate with some examples. If samples are drawn from a population with standard deviation $\sigma = 10.0$ and if the sample size is $n = 25$, then the standard error of the mean is $\sigma_{\bar{X}} = 2.0$. If $\sigma = 20.0$ and $n = 25$, then $\sigma_{\bar{X}} = 4.0$. To reduce either of these standard errors by 1/2, we would have to quadruple the sample size. With $n = 100$ observations and $\sigma = 10.0$, then $\sigma_{\bar{X}} = 1.0$. If $\sigma = 20.0$ and $n = 100$, then $\sigma_{\bar{X}} = 2.0$.

If the population standard deviation is known, we can use the results presented in Chapter 3 and apply the normal approximation:

$$Z = \frac{\bar{X} - \mu}{\sigma_{\bar{X}}}$$

which is a standard normal variable (equal to Equation 4.1). This approximation could be used to test any null hypothesis regarding μ. In general, however, σ is not known and needs to be estimated. Thus, we need to make a minor modification to this test so that it can handle cases when we do not know the population standard deviation σ. The next section discusses the estimation of the standard error of the mean.

4.3 The Variance and Standard Error of the Mean When Population σ^2 Is Unknown

We extend these results to a more realistic setting—cases where we do not know the population variance, or standard deviation. We will estimate the population standard deviation and then estimate the standard error of the mean. The estimated **variance of a population**, based on a sample of n observations, is

$$s^2 = \frac{\sum_{i=1}^{n} (X_i - \bar{X})^2}{n - 1} \tag{4.4}$$

The variance defined by Equation 4.4 is said to have $n - 1$ (that is, the denominator) degrees of freedom and is an unbiased estimate of the population variance σ^2 for any random sample of n independent observations. Intuitively, the concept of **degrees of freedom** refers to how many independent pieces of information are used to estimate a quantity. We start out with n pieces of information (that is, the original number of observations), but in order to estimate the population variance we also need to estimate the mean $\bar{X}$, which appears in the numerator of s^2. Thus, we lose one "piece" of information when estimating the mean, leaving $N - 1$ "degrees of freedom" to estimate the variance of a population.[1] The estimated **population standard deviation** is defined as

$$s = \sqrt{\frac{\sum_{i=1}^{n} \left(X_i - \bar{X}\right)^2}{n - 1}}$$

Now that we have estimated the population variance from the sample, we can use that term to estimate the variance of the mean. The **variance of a mean** of a random sample of n observations is given by

$$s_{\bar{X}}^2 = \frac{s^2}{n} \qquad (4.5)$$

It is analogous to Equation 4.3 but with the estimated standard deviation inserted into the numerator. Equation 4.5 is an unbiased estimate of $\sigma_{\bar{X}}^2$. The **estimated standard error of a mean** will then be

$$s_{\bar{X}} = \frac{s}{\sqrt{n}}$$

[1]For a more detailed and mathematical treatment of degrees of freedom see the classic paper by Walker (1940). The intuition has to do with how many pieces of information are available in principle and how many pieces of information have been used in the estimation. In the case of the estimation of variance, there are in principle N pieces of information available (one observation for each participant). But in order to compute the sample variance one must also compute the sample mean; so, one piece of information is used and hence "docked" from the total tally, which leads to $N - 1$. The proper description of degrees of freedom relies on linear algebra, and the interested reader is referred to the classic tutorial article by Walker.

4.4 The *t* Distribution and the One-Sample *t* Test

We are now in a position to test whether an observed sample mean deviates significantly from a hypothesized population mean. The difference between a sample mean $\bar{X}$ and a hypothesized population mean μ, divided by the standard error of the mean, is symbolically

$$t = \frac{\bar{X} - \mu}{s_{\bar{X}}} \tag{4.6}$$

with degrees of freedom $n - 1$ (that is, the denominator of s^2). We will define terms shortly, but for now you should recognize the general form of Equation 4.6, which is very similar to the Z score introduced in Chapter 3. The test we will develop compares the value of an observed mean to the expected mean under the null hypothesis; it is called a **one-sample *t* test.** Because we need to use the estimated standard deviation to compute the standard error of the mean, the results from Chapter 3 do not exactly carry over to this situation. The ratio in Equation 4.6 does not have a normal distribution, but instead the ratio follows what is called a t distribution.

The distribution of t depends on the number of degrees of freedom available in the set of n observations used in calculating the estimate of the population variance s^2. Hence, the table of the t distribution is a two-dimensional table that must be entered with both the value of the desired tail area and the number of degrees of freedom. The distribution of t is not approximately normal for small samples (say, less than 30 observations). Its distribution is symmetric, as is the distribution of Z, but beyond a certain point (depending on the number of degrees of freedom available) the curve of t does not approach the baseline as rapidly as does the curve of Z. In other words, t has a longer tail than the normal curve. For example, in order to mark off 2.5% of the total area in the right tail of the standard normal distribution, we will have to go out beyond the value of $Z = 1.96$ that cuts off 2.5% of the total area in the right tail of the normal distribution.

Figure 4.1: Illustration of the longer tail of the t distribution. The upper distribution shows the right 2.5% of the standard normal distribution. The lower distribution shows the right 2.5% of the t distribution with 4 degrees of freedom.

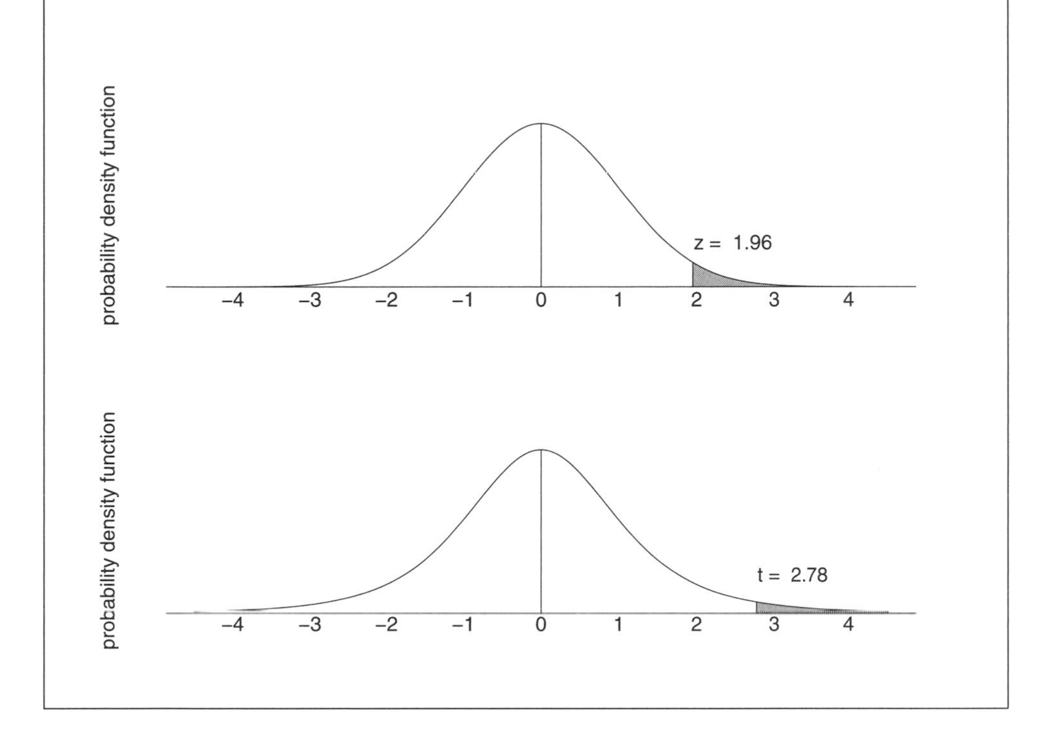

Figure 4.1 shows a normal distribution in the top panel and a t distribution with four degrees of freedom in the bottom panel. Note how the cutoff corresponding to an area of 2.5% for the right tail of the t distribution is shifted to the right relative to that of the normal distribution. Just how far out we shall have to go on the t distribution depends on the number of degrees of freedom available, which is related to the sample size. The greater the degrees of freedom, the more the t distribution resembles the shape of the normal distribution. It is in this sense that the t distribution is a function of sample size.

If you look at the table of the standard normal distribution, Table B.3 in Appendix B, you will find that $P(Z \geq 1.96) = 0.025$; that is, the ordinate

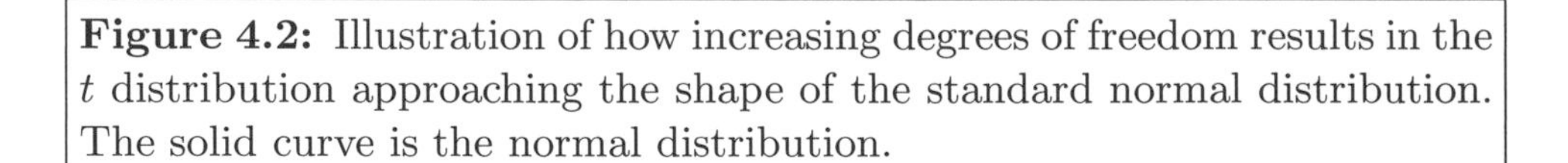

Figure 4.2: Illustration of how increasing degrees of freedom results in the t distribution approaching the shape of the standard normal distribution. The solid curve is the normal distribution.

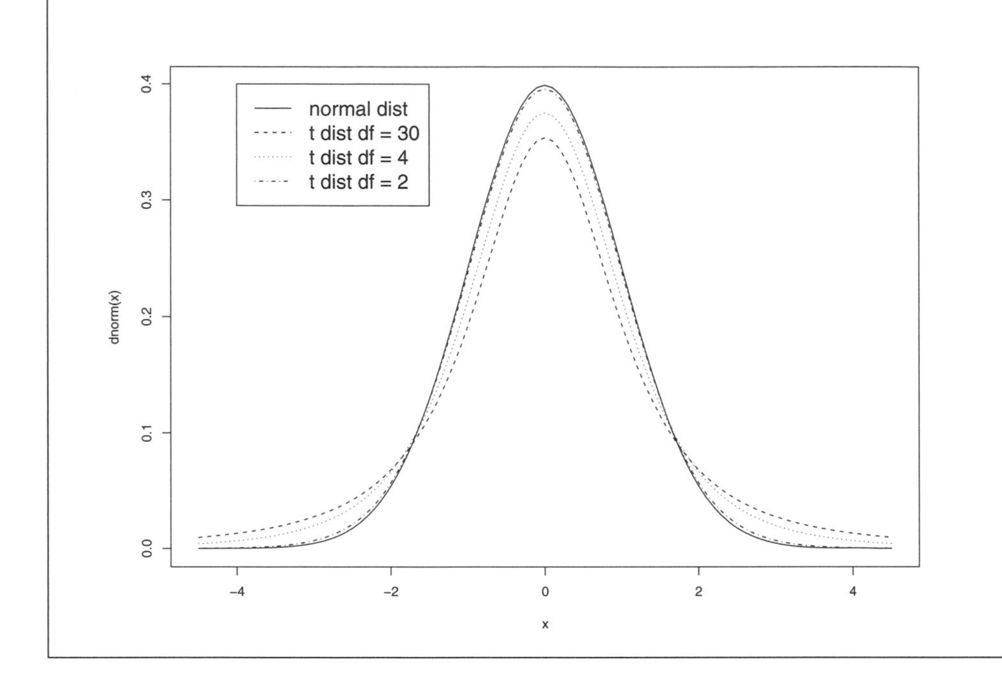

at $Z = 1.96$ will cut off 0.025 of the total area in the right tail of this distribution. Now, examine the table of the t distribution, Table B.1 in Appendix B. As the number of degrees of freedom increases, the value of t that cuts off 0.025 of the total area in the right tail of the t distribution also approaches 1.96. With 30 degrees of freedom (df), we have $P(t \geq 2.042) = 0.025$; with 100 df, $P(t \geq 1.984) = 0.025$; with 300 df, $P(t \geq 1.968) = 0.025$; and with 1000 df, $P(t \geq 1.962) = 0.025$. Figure 4.2 shows how the t distribution approaches the normal distribution as the degrees of freedom increase.

The value of s^2 is itself subject to random variation. As n increases, the accuracy with which s^2 estimates σ^2 improves. For very large values of n, the discrepancy between s^2 and σ^2 may be sufficiently small as to be negligible.

Thus, in the limiting case, with n large, the distribution of t approaches the normal distribution. When samples are as large as about $n = 400$ observations, the values of t are, for all practical purposes, indistinguishable from the values of Z.

4.5 Confidence Interval for a Mean

The standard error of the mean permits construction of confidence intervals around the mean. Confidence intervals provide a probability statement about the value of the mean. Suppose we have a random sample of $n = 49$ observations, with a sample mean $\bar{X} = 62.0$ and an estimated standard deviation of $s = 14.0$. Then, the estimated standard error of the mean is $s_{\bar{X}} = 14.0/\sqrt{49} = 2.0$, and the corresponding t score is

$$t = \frac{62.0 - \mu}{2.0}$$

Even though μ is unknown, we can use the t distribution to construct an interval around the mean. We regard any value of μ such that $(\bar{X} - \mu)/s_{\bar{X}}$ falls within the interval

$$t_{0.025} < \frac{\bar{X} - \mu}{s_{\bar{X}}} < t_{0.975} \tag{4.7}$$

as a reasonable estimate of μ. In Equation 4.7, $t_{0.025}$ is the value of the t distribution that cuts off 0.025 of the total area in the left tail, and $t_{0.975}$ is the value that cuts off $1 - 0.975 = 0.025$ of the total area in the right tail of the t distribution, given $n - 1$ degrees of freedom. With $49 - 1 = 48$ degrees of freedom, these two values of t are -2.01 and 2.01, respectively. Then,

$$-2.01 < \frac{62.0 - \mu}{2.0} < 2.01$$

or

$$(2.0)\,(-2.01) - 62.0 < -\mu < (2.0)\,(2.01) - 62.0$$

Multiplying throughout by -1 and remembering that the direction of an inequality is reversed when all terms are multiplied by the same negative number, we obtain

$$62.0 + (2.0)\,(2.01) > \mu > 62.0 - (2.0)\,(2.01)$$

or

$$66.02 > \mu > 57.98$$

The interval 57.98 to 66.02 that we just computed is called a **confidence interval**, and the limits of the interval are called **confidence limits.** The degree of confidence we have in the statement that μ falls within the confidence interval is called a **confidence coefficient**. In the illustrative example, we have computed a 95% confidence interval because we used the lower 0.025 and upper 0.025 of the t distribution to define the limits of the interval.

Confidence limits are statistics, and like all statistics they are also subject to random variation. If we draw another random sample of $n = 49$ observations from the same population as the first sample, both the sample mean and the sample standard deviation may be different from the values we obtained for the first sample. Therefore, the 95% confidence interval established for the second sample would not necessarily be the same as the one established by the first sample. When we say we are 95% confident that μ falls within the interval, we are expressing our confidence that in repeated samples such an inference concerning μ will be correct 95 times in 100. For any particular sample, the inference will be right or wrong; that is, either the population mean μ falls within the interval computed from the information from the sample data or the population mean μ does not. The confidence interval is a statement about the percentage of times we expect the interval to contain the actual population mean μ.

Note that a **confidence interval implies a test of significance**. In essence, with $\alpha = 0.05$ and with a two-sided test of significance, we would reject, in the example considered, any hypothesis that $\mu \leq 57.98$ or that $\mu \geq 66.02$. Thus, the information one extracts from a hypothesis test is contained in a confidence interval.

It is useful to consider the effects of sample size on the confidence limits. As we just computed, with $n = 49$ and with $s = 14.0$, the 95% confidence limits are 57.98 and 66.02. Increasing sample size n to 100 observations—that is, slightly more than doubling the sample size—will serve to reduce the confidence interval in two ways, assuming that s^2, the estimate of the population variance, remains the same. In the first place, the standard error of the mean will now be $s_{\bar{X}} = 14.0/\sqrt{100} = 1.4$, as compared with the value of 2.0 when the sample consisted of only 49 observations. In the second place, the values of t cutting off 0.025 of the total area in the two tails of the t distribution for 99 *df* are -1.984 and 1.984, rather than the values of -2.01 and 2.01 for 48 *df*. We would then have, as a 95% confidence interval,

$$62.0 + (1.4)\,(1.984) > \mu > 62.0 - (1.4)\,(1.984)$$

or

$$64.78 > \mu > 59.22$$

This 95% confidence interval, based on $n = 100$ observations, has a range of $64.78 - 59.22 = 5.56$, whereas the confidence interval based on $n = 49$ observations had a range of $66.02 - 57.98 = 8.04$. It should be clear that if we want a relatively narrow confidence interval, we will need to have a relatively large number of observations or to find a way to reduce the standard deviation (which can be done by decreasing measurement error). If we interpret a confidence interval as a measure in repeated sampling of the precision of the estimate, then as sample size increases we get more precise estimates because the difference between the confidence limits decreases.

4.6 Standard Error of the Difference between Two Means

The technique of constructing confidence intervals can be extended to the differences between two means. We illustrate with an example. In an experiment on cued recall, two treatments were assigned at random so that 20 participants received Treatment 1 (T_1) and 20 participants received Treatment 2 (T_2). Participants in both groups were presented with a series of paired words such as NURSE–CHAIR and were asked to study each pair. After participants studied all the word pairs, they were presented the first word in the pair (e.g., NURSE) and asked to write down the paired word. Treatment T_1 consisted of giving each participant a mild shock for each incorrect response. In T_2, the participants were not shocked, but instead each incorrect response was followed by the flashing of a red light. The dependent variable X was the number of correct responses made on a delayed memory test 24 hours later. The "cued-recall" scores for the participants in the two groups are given in Table 4.1.

The means for the two treatments are

$$\bar{X}_1 = \frac{220}{20} = 11.0 \quad \text{and} \quad \bar{X}_2 = \frac{160}{20} = 8.0$$

The difference between these two means is $\bar{X}_1 - \bar{X}_2 = 11.0 - 8.0 = 3.0$. If the experiment were repeated under the same conditions, we would not expect to obtain exactly the same values for $\bar{X}_1$ and $\bar{X}_2$ each time. The means of both samples are subject to random variation, and this will also be true of the difference between the means. The distribution of $\bar{X}_1$ will be normally distributed around the population mean μ_1, and $\bar{X}_2$ will be normally distributed around the population mean μ_2. The distribution of the

Table 4.1: Values of X and X^2 for two treatments with $n = 20$ participants assigned at random to each treatment.

	Treatment 1		Treatment 2	
	X_1	X_1^2	X_2	X_2^2
	12	144	4	16
	6	36	12	144
	7	49	9	81
	12	144	9	81
	10	100	14	196
	16	256	9	81
	13	169	11	121
	14	196	0	0
	9	81	9	81
	14	196	11	121
	6	36	1	1
	16	256	8	64
	13	169	10	100
	10	100	8	64
	7	49	6	36
	10	100	8	64
	12	144	9	81
	11	121	9	81
	9	81	10	100
	13	169	3	9
$\sum$	220	2596	160	1522

difference $\bar{X}_1 - \bar{X}_2$ will also be normally distributed around the population mean difference $\mu_1 - \mu_2$. We will use the observed difference $\bar{X}_1 - \bar{X}_2$ to estimate $\mu_1 - \mu_2$.

The **standard error of the difference between the means of two independent random samples** from populations with known standard deviations is given by

$$\sigma_{\bar{X}_1 - \bar{X}_2} = \sqrt{\sigma^2_{\bar{X}_1} + \sigma^2_{\bar{X}_2}} \tag{4.8}$$

But because

$$\sigma^2_{\bar{X}_1} = \frac{\sigma_1^2}{n_1} \quad \text{and} \quad \sigma^2_{\bar{X}_2} = \frac{\sigma_2^2}{n_2}$$

Equation 4.8 can be written as

$$\sigma_{\bar{X}_1-\bar{X}_2} = \sqrt{\frac{\sigma_1^2}{n_1} + \frac{\sigma_2^2}{n_2}}$$

In the present example, σ_1^2 and σ_2^2 are both unknown, but each can be estimated by the sample variance given in Equation 4.4. Then, because we substitute the observed s^2 for the population σ^2, the estimated standard error of the difference between the two means will be

$$s_{\bar{X}_1-\bar{X}_2} = \sqrt{\frac{s_1^2}{n_1} + \frac{s_2^2}{n_2}} \tag{4.9}$$

For the moment, we assume that the population variances are equal, that is, $\sigma_1^2 = \sigma_2^2$, so s_1^2 and s_2^2 are both estimates of the same common population variance σ^2. This is a critical assumption, and it will be discussed in more detail in the next chapter. If we have two or more estimates of a common parameter, the estimates may be combined to provide a single estimate. In the general case of k sample variances, all of which are assumed to estimate the same common population variance σ^2, the single estimate is obtained by the following weighted average (where the weights are the respective degrees of freedom in each group or treatment)

$$s^2 = \frac{(n_1 - 1)\, s_1^2 + (n_2 - 1)\, s_2^2 + \cdots + (n_k - 1)\, s_k^2}{(n_1 - 1) + (n_2 - 1) + \cdots + (n_k - 1)} \tag{4.10}$$

with degrees of freedom equal to $\sum_{j=1}^{k} n_j - k$ (which is the sum of each of the degrees of freedom corresponding to each sample). If all of the n's are equal, then the degrees of freedom term simplifies to $k(n-1)$, where n is the number of observations in each sample and k is the number of samples.

Recall from the definition of the variance that for the kth group we have

$$(n_k - 1)\, s_k^2 = \sum_{i=1}^{n_k} \left(X_{ki} - \bar{X}_k\right)^2 \tag{4.11}$$

or the sum of the squared deviations of the n_k observations in the kth sample from the sample mean. We now substitute the equality in Equation 4.11 into Equation 4.10, and have

$$s^2 = \frac{\sum_{i=1}^{n_1} \left(X_{1i} - \bar{X}_1\right)^2 + \sum_{i=1}^{n_2} \left(X_{2i} - \bar{X}_2\right)^2 + \cdots + \sum_{i=1}^{n_k} \left(X_{ki} - \bar{X}_k\right)^2}{\sum_1^k n_k - k} \tag{4.12}$$

Equation 4.12 is a general equation that will be used throughout this book. The numerator of this general form has a term that is a "sums of squared terms," and the denominator is a term that corresponds to degrees of freedom.

For the present problem, we have two treatment means, so $k = 2$ and Equation 4.12 simplifies to

$$s^2 = \frac{\sum \left(X_1 - \bar{X}_1\right)^2 + \sum \left(X_2 - \bar{X}_2\right)^2}{n_1 + n_2 - 2}$$

with degrees of freedom equal to $n_1 + n_2 - 2$. Substituting the single estimate of the population variance s^2 for the separate estimates s_1^2 and s_2^2 in Equation 4.9, we have

$$s_{\bar{X}_1 - \bar{X}_2} = \sqrt{s^2 \left(\frac{1}{n_1} + \frac{1}{n_2}\right)} \tag{4.13}$$

$$= \sqrt{\frac{\sum_{i=1}^{n_1} \left(X_{1i} - \bar{X}_1\right)^2 + \sum_{i=1}^{n_2} \left(X_{2i} - \bar{X}_2\right)^2}{n_1 + n_2 - 2} \left(\frac{1}{n_1} + \frac{1}{n_2}\right)} \tag{4.14}$$

An equivalent way to express the right-hand side of Equation 4.11 is as follows:

$$\sum_{i=1}^{n_k} \left(X_{ki} - \bar{X}_k\right)^2 = \sum X_k^2 - \frac{\left(\sum X_k\right)^2}{n} \tag{4.15}$$

The right-hand side of Equation 4.15 provides a relatively convenient way to compute the sum of squared deviations from the sample mean. We find that the sum of squared deviations for Treatment 1 is

$$\sum_{i=1}^{n_1} (X_{1i} - \bar{X}_1)^2 = 2596 - \frac{(220)^2}{20} = 176$$

and the sum of squared deviations for Treatment 2 is

$$\sum_{i=1}^{n_2} (X_{2i} - \bar{X}_2)^2 = 1522 - \frac{(160)^2}{20} = 242$$

Note how we used the last row of Table 4.1 to plug numbers into Equation 4.15.

Substituting these values into Equation 4.14 for the standard error of the difference between the two means, we have the final result

$$s_{\bar{X}_1 - \bar{X}_2} = \sqrt{\frac{176 + 242}{20 + 20 - 2}\left(\frac{1}{20} + \frac{1}{20}\right)} = 1.049$$

This section may appear tedious, but the underlying idea is relatively simple. We need to estimate the standard error of the difference between two means in order to build confidence intervals and perform inferential tests. To estimate the standard error of the difference between two means we first assume that both treatments estimate the same population variance. Equation 4.12 shows how to pool the two observed estimates into a single estimate s^2, and Equation 4.13 shows how to use the common estimate of the population variance to estimate the standard error of the difference between two means.

4.7 Confidence Interval for a Difference between Two Means

We can also find confidence limits for the population mean difference $\mu_1 - \mu_2$. Continuing with the cued recall example with $n_1 + n_2 - 2 = 38$ degrees of freedom, let's construct a 99% confidence interval around the difference between the two means. From the table of t (Table B.1) in Appendix B we find that $t = -2.711$ will cut off 0.005 of the total area in the left tail and $t = 2.711$ will cut off 0.005 of the total area in the right tail of the t distribution. Then, the 99% confidence limits will be given by

$$-2.711 < \frac{(11.0 - 8.0) - (\mu_1 - \mu_2)}{1.049} < 2.711$$

or

$$\begin{aligned} 3.0 + (1.049)\,(2.711) > \mu_1 - \mu_2 > 3.0 - (1.049)\,(2.711) \\ 5.84 > \mu_1 - \mu_2 > 0.16 \end{aligned}$$

We can say that under repeated sampling the population mean difference $\mu_1 - \mu_2$ will be within these limits 99% of the time. We remind the reader that "99% confident" means that in repeated sampling 99 out of 100 intervals around the observed difference will contain the true difference between the two population means. We chose to compute a 99% confidence interval in this example to illustrate a different level than the typical 95% interval.

4.8 Test of Significance for a Difference between Two Means: The Two-Sample *t* Test

If our major interest is in determining whether a specified null hypothesis about $\mu_1 - \mu_2$ is to be rejected, then this hypothesis may be tested by the t test. When we do not know the population variances of each sample, we form the t ratio as follows

$$t = \frac{(\bar{X}_1 - \bar{X}_2) - (\mu_1 - \mu_2)}{s_{\bar{X}_1 - \bar{X}_2}}$$

Specifically, if the null hypothesis is that the two population means are equivalent, so that $\mu_1 - \mu_2 = 0$, then under the assumption of equal population variances

$$t = \frac{\bar{X}_1 - \bar{X}_2}{s_{\bar{X}_1 - \bar{X}_2}} \tag{4.16}$$

For the present problem, we have

$$t = \frac{11.0 - 8.0}{1.049} = 2.86$$

with 38 degrees of freedom (that is, $n_1 + n_2 - 2$). With a two-sided Type I error rate $\alpha = 0.01$, the null hypothesis is rejected because the observed value of 2.86 exceeds the critical value of 2.71. In the preceding subsection we computed a 99% confidence interval, which was $5.84 > \mu_1 - \mu_2 > 0.16$. The interval does not include 0, which means that 99% of such intervals will not contain the value 0. This conclusion is the same as that reached from the hypothesis test, which rejected the null hypothesis that the two population means are identical (that is, the difference in population means is zero, which is equivalent to saying that the confidence interval on the difference between the two means does not include zero) at the $\alpha = 0.01$ level.

The reader should place priority on the important attribute of this analysis. The experimenter observed data and reduced those data to two means (11 and 8), each with standard errors. The means provide a summary of what was observed in the two groups. The statistical test merely provides a way to compute a relevant probability that can help evaluate the observed difference between two means against a statistical hypothesis. Likewise, the confidence interval provides a way to evaluate the sampling uncertainty around the observed difference between the means of 3 (that is, $11 - 9$). In this way, the procedure we develop in this chapter is similar to how we used probability in Chapter 2 to evaluate the observations made of the farmer.

Of course, there are many other items that a data analyst should consider in addition to means, standard deviations, t values, and confidence intervals. A couple of issues of concern include whether there were missing data and whether participants were randomly assigned to groups. Such properties could interfere with the ability to make the kinds of inferences that one would expect from experimental designs. For example, if there are more missing data in Treatment 1 than in Treatment 2, the differential pattern of missing data could be attributed to the former treatment having a more severe experience (mild shock) and some participants refusing to provide answers. Such differential attrition in observed data would bias the observed means. If participants were not randomly assigned to the two treatments, it would still be possible to compute the two-sample t test, but the result would need to be interpreted as an association. For example, if the two groups consisted of English and psychology majors, respectively, without random assignment unrelated to one's major, we could only claim that the mean difference observed between the individuals in the two majors was statistically significant but could not claim that the difference was due to their major because there are many reasons people self-select into particular majors and we wouldn't know which of those factors were relevant to the observed mean difference.

It is worth pointing out a subtle feature of the example cued recall study. This study did not compare a group that received an electric shock after each incorrect response to a group that did not receive electric shock. Such a comparison could lead to alternative hypotheses. For instance, if a statistically significant difference between the present versus absent groups was observed, can the difference be attributed to the presence or absence of the electric shock or could it be another component of the study, such as one group receiving a signal after each incorrect response compared to a group that did not receive a signal. Did the electric shock act like a punishment, or did it act like a signal? More generally, there are various ways in which the two groups differed, and these other ways may be the reason that the experimenter observed a difference between the means. The version of the study that was described earlier in this chapter compared a group that received an electric shock after each incorrect response to a group where a red light flashed after each incorrect response. This version equates the two conditions so that each is receiving a signal after an incorrect response, but the nature of the signal differs (electric shock versus red light). We assert that no single experimental design eliminates all possible alternative explanations. The reader may very well think of alternative explanations to the electric shock versus red light study. Multiple studies that test such alternative explanations are needed to round out a program of research. One goal of experimental designs is to create initial designs that minimize the possible alternative explanations.

4.9 Using a Computer Program

The same results can be achieved by using a statistics computer program. The data need to be entered in a particular way to compute a two-sample t test. In most statistics programs these data would be entered in the form of two columns such that one column would contain treatment group codes and the other column the observed data. For example, one column might be a sequence of 1's (denoting that the participant was assigned to treatment group 1) and 2's (denoting that the participant was assigned to treatment group 2), and the second column would contain the observed data corresponding to each participant. A group code is a nominal variable whose categories represent which treatment condition the participant was assigned (recall our discussion of nominal variables in Chapter 1). This format is displayed in Table 4.2. Many statistics programs such as SPSS provide quite a bit of output even for the simplest of analyses. The bare minimum to examine in the output would be the descriptive statistics of the sample, such as the means and standard deviations of each of the two treatment groups, the value of the

Table 4.2: Example showing how to enter the data in Table 4.1 into a statistics program.

Participant number	Group	Data
1	1	12
2	1	6
3	1	7
⋮	⋮	⋮
19	1	9
20	1	13
21	2	4
22	2	12
⋮	⋮	⋮
39	2	10
40	2	3

t test, the *p* value, and the lower and upper bounds of the confidence interval at a specified percentage level, such as 95% or 99%. Table 4.3 presents an example of SPSS syntax and an excerpt from the corresponding output.

4.10 Returning to the Farmer Example in Chapter 2

We can now use the two-sample *t* test to answer new questions regarding the farmer's ability to detect water with a whalebone (recall the example in Chapter 2). Suppose we believe that the ability to detect water can be taught (as opposed to being an innate ability), and we develop a training program to teach people to use a whale bone. To test the effectiveness of this training program we take a sample of 40 farmers and randomly assign them either to participate in the training program or to not participate in the training program. After the training program is complete, we present all 40 participants with a sequence of 10 pairs of cans; within each pair one can contains water and one does not. Each participant can score between 0 and 10, reflecting the number of cans correctly identified as containing water.

Table 4.3: Example illustrating SPSS syntax and output.

SPSS SYNTAX

```
T-TEST
  GROUPS=group(1 2)
  /MISSING=ANALYSIS
  /VARIABLES=data
  /CRITERIA=CIN(.99) .
```

SPSS OUTPUT

Group	N	Mean	Std. deviation	Std. error mean
1.00	20	11.0000	3.04354	.68056
2.00	20	8.0000	3.56887	.79802

					99% Conf. int.	
t	df	p value	Mean difference	St. error diff	Lower	Upper
2.86	38	.007	3.00	1.0488	.156	5.844

Suppose that the 20 participants assigned to the training program correctly detect an average of 8.4 cans of water, whereas the 20 participants not assigned to the training program detect an average of 6.2 cans. We want to compare whether the two observed averages are statistically different from each other. Note how this question differs from the question initially asked in Chapter 2, which involved a single farmer's performance compared to what we expected by chance. Now we are comparing the average number of correctly identified cans from the 20 participants who took the training program to the 20 participants who did not take the training program. That is, the comparison is not of a single farmer against chance but rather between two groups of farmers who differed in the training they received. If the two observed means are statistically equivalent, then we conclude that there isn't sufficient evidence to determine whether the training program improved one's ability to detect water. But, if the mean for the participants in the training program statistically exceeds the mean for the participants who did not take part in the training program, then we have evidence that the training program improves performance (or decreases performance if the means were in reverse order). But how much does the observed difference in the two means have to be before we declare victory and sell our training program to people who want to learn to detect water using a whale bone?

In this example, the difference between the two means was $8.4 - 6.2 = 2.2$. But this is merely a point estimate of the difference between two population means, and we need to take into account the variability around this sample point estimate to determine statistical significance. The estimated standard deviations within each sample are 1.4 and 1.2, respectively, for the two treatment conditions. Using the equations presented earlier in this chapter, we compute the standard error of the difference between the two means to be .4123, which leads to a t value of 5.34. This observed t value exceeds the critical value of 2.02 (two-tailed $\alpha = 0.05$ with 38 degrees of freedom), so we reject the null hypothesis that the population means of the two conditions are equivalent; the 95% confidence interval around the observed difference between the two means has a lower bound of 1.37 and an upper bound of 3.03, thus the interval does not include zero.

4.10.1 Methodological Issues Reiterated

Because we randomly assigned each participant to one of the two conditions, we can attribute a statistically significant difference between the two means to the training program. Unfortunately, training programs usually have many components, and our simple comparison of two means will not allow us to specify which part of the training may be related to the improved ability to detect the presence of water. The difference in performance might be due to something other than the content of the training program, such as the social contact that the training program offered (which was not present in the control condition). It may be useful to compare participants who received the training to a second group of participants who received an inert type of training (rather than no training at all) in order to control for other aspects of the training session such as social interaction.

The statistical test merely tells us whether the difference between the two means was statistically significant; it does not tell us what caused the difference. Further, the random assignment present in this design allows us to rule out some alternative factors. For instance, if instead of randomly assigning the participants to the two conditions, we had instead assigned women to the training condition and men to the other condition, we would have a confound because we would not be able to attribute clearly the reason for the statistically significant difference between the two means—the difference could be due to the training, to the gender composition of the two groups, to some complicated combination of training and gender. Or if we allowed participants to self-select which treatment group to be in (training or no training), then we would not know whether the difference in means could be attributable to the training or to existing differences in the two groups of participants, such as an interest in learning about dousing or their prior belief about dousing in general.

4.11 Effect Size for a Difference between Two Independent Means

The test of significance presented above is influenced, in part, by the sample size. This test of significance is a function of the means, the standard deviation, and the sample sizes. Obviously, if a study has a relatively large number of participants, it will have high power, and so the two sample t test may be statistically significant even though the difference between the means is quite small. Sometimes it is convenient to index the magnitude of the difference between two means in a manner that does not involve sample size.

Some authors have argued for the following index of **effect size**: the difference between the two means divided by the pooled standard deviation (Equation 4.10). In symbols, this effect size index is

$$d = \frac{\bar{X}_1 - \bar{X}_2}{s} \tag{4.17}$$

The only difference between Equation 4.16 (the two-sample t test) and Equation 4.17 (one possible index of effect size) is whether (1) the pooled standard error of the difference between the two means or (2) the pooled standard deviation, respectively, is used in the denominator. The index d can be interpreted as the normalized mean difference ("normalized" because the mean difference is divided by the pooled standard deviation). Thus, the value d can be interpreted as the difference between the means in standard deviation units.

A different index of effect size can be found directly from the value of the t test. This index is called r, and it is computed by the formula

$$r = \sqrt{\frac{t^2}{t^2 + df}}$$

where t is the value computed from Equation 4.16 and df corresponds to the degrees of freedom for that t value (that is, $n_1 + n_2 - 2$, or the sum of the two sample sizes minus two). The value of r can range between 0 and 1, and, as we will see in a later chapter, r is related to the percentage of variability attributable to the treatment effect relative to the all the variability in the observed data, including variability attributable to error. The measure r is easy to generalize to different designs and situations, so we will emphasize r in this book.

In our example with the two groups of 20 farmers the observed t was 5.34 and $df = 38$. The effect size r is equal to $\sqrt{\frac{5.34^2}{5.34^2+38}} = 0.655$.

Even though we sympathize with the fundamental idea underlying the use of effect size measures, we believe that an efficient strategy is to report confidence intervals of the raw effect because the confidence interval separates the raw effect (for example, the difference between the means) from the variability of that raw effect (for example, the standard error of the difference between the means). Statistical tests and measures of effect size currently in use confound these two components because they are created by forming a ratio of the difference and the variability. Simply looking at the definition of the t value (Equation 4.16), which has in the numerator the difference between means and in the denominator the standard error of the difference between means, we see that the ratio takes both of those numbers into a single number. Similarly, the effect size measure d is also a ratio. If the ratio is low, we don't know whether it is because the mean difference (that is, the numerator) was small, the variability was high (that is, the denominator), or some combination of both. A confidence interval highlights these two concepts more clearly by keeping them separate—variability is indexed by the difference between the confidence limits, and the effect is indexed by the difference between means. See Section 4.7 for an example of how to compute a confidence interval. The user of effect size measures should be careful about overinterpreting the meaning of the index. For other concerns involving effect size measures, see Fern and Monroe (1996) and Prentice and Miller (1992). These conceptual issues aside, Bonett (2008) has made progress in developing confidence intervals for effect sizes and their linear combinations.

4.12 The Null Hypothesis and Alternatives

A null hypothesis is usually tested against a class of alternative hypotheses. If the null hypothesis is false, so that one of the alternative hypotheses is true, then we can define the power of a test of significance as

$$\text{Power} = 1 - P\,(\text{Type II error})$$

Because the probability of a Type II error is the probability of failing to reject the null hypothesis when it is false, the power of a test of significance can be said to be the probability of rejecting the null hypothesis when it is false, that is, rejecting it when it should be rejected.

One way to increase the power of a statistical test, that is, to increase the likelihood of correctly rejecting the null hypothesis, is to make the Type I error criterion α large. But we do not like to make α too large because by doing so we increase the probability of a Type I error. If we hold α constant, then we can increase the power of a given test by increasing the number of observations in the sample under consideration. If we hold both α and the number of observations in the sample constant, then we can increase the power of a statistical test against a selected class of alternatives to the null hypothesis by the manner in which we choose the critical region of rejection in the t distribution. That is, if we designate the null hypothesis as H_0 and the alternative to this hypothesis as H_1, then we may be interested in any one of the following three tests:

$$\begin{array}{ll} \text{Test 1} & H_0 : \mu_1 = \mu_2 \text{ with } H_1 : \mu_1 \neq \mu_2 \\ \text{Test 2} & H_0 : \mu_1 \leq \mu_2 \text{ with } H_1 : \mu_1 > \mu_2 \\ \text{Test 3} & H_0 : \mu_1 \geq \mu_2 \text{ with } H_1 : \mu_1 < \mu_2 \end{array}$$

These tests vary in the specific use of equality and inequalities.

Suppose we choose $\alpha = 0.05$. If we conduct Test 1, we reject the null hypothesis if the obtained t falls in either of the two shaded areas of Figure 4.3. With 38 degrees of freedom, the critical values of t, those that would result in the rejection of the null hypothesis, are -2.025 and 2.025. These are the values of t cutting off 0.025 of the total area in each tail of the t distribution. The total area of both sides yields the desired $\alpha = 0.05$ criterion. Because the areas of rejection for Test 1 are in both tails of the t distribution, this test is called a **two-tailed** or **two-sided test**. Test 1 allows rejecting the null hypothesis for either the possibility that $\mu_1 > \mu_2$ or the possibility that $\mu_1 < \mu_2$. In other words, Test 1 is sensitive to the absolute value of the difference between μ_1 and μ_2. Test 1 is the test we should use if we are interested in the absolute magnitude of the difference between μ_1 and μ_2. Most research questions in the social sciences fall into this category. Further, this is the setup that corresponds to the confidence interval because there is both a lower and an upper cutoff.

Figure 4.3: The two-sided test of significance of the null hypothesis $\mu_1 = \mu_2$ against the alternative $\mu_1 \neq \mu_2$. Each of the shaded areas in the two tails of the t distribution is 0.025 of the total area, yielding a total area of 0.05. With $\alpha = 0.05$, the null hypothesis is rejected if the observed value of t falls in either shaded area.

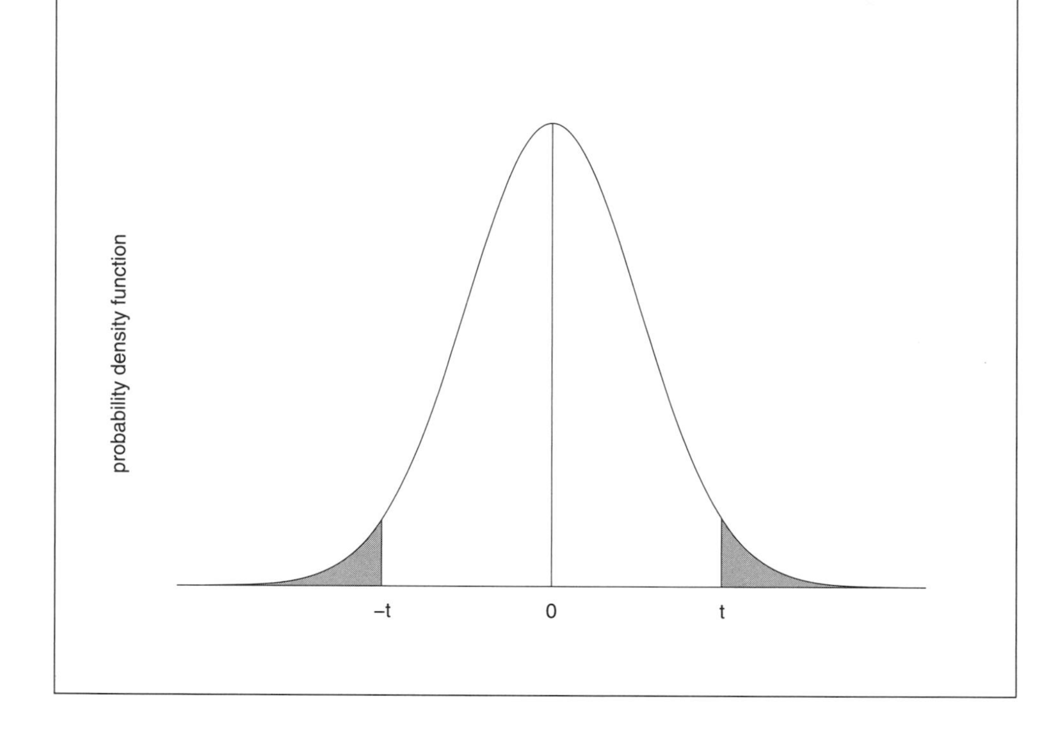

If we perform Test 2, then we reject the null hypothesis only if the obtained value of t falls in the shaded area of Figure 4.4. If $\alpha = 0.05$, then we want the area in the right tail to correspond to 0.05 of the total area of the t distribution. With 38 degrees of freedom, the critical value of t cutting off 0.05 of the total area in the right tail is approximately 1.68. Because the area of rejection is the right tail of the t distribution, Test 2 is referred to as a **right-tailed**, a **one-tailed**, or a **one-sided** test. Test 2 allows rejection of the null hypothesis with the class of alternatives $\mu_1 > \mu_2$. If it is true that $\mu_1 > \mu_2$, then Test 2 will be somewhat more powerful than Test 1 against this class of alternatives, but, unlike Test 1, Test 2 would not allow rejection

Figure 4.4: The one-sided test of significance of the null hypothesis $\mu_1 \leq \mu_2$ against the alternative $\mu_1 > \mu_2$. The shaded area in the right tail of the t distribution is 0.05 of the total area. With $\alpha = 0.05$, the null hypothesis is rejected if the observed value of t falls in the shaded area.

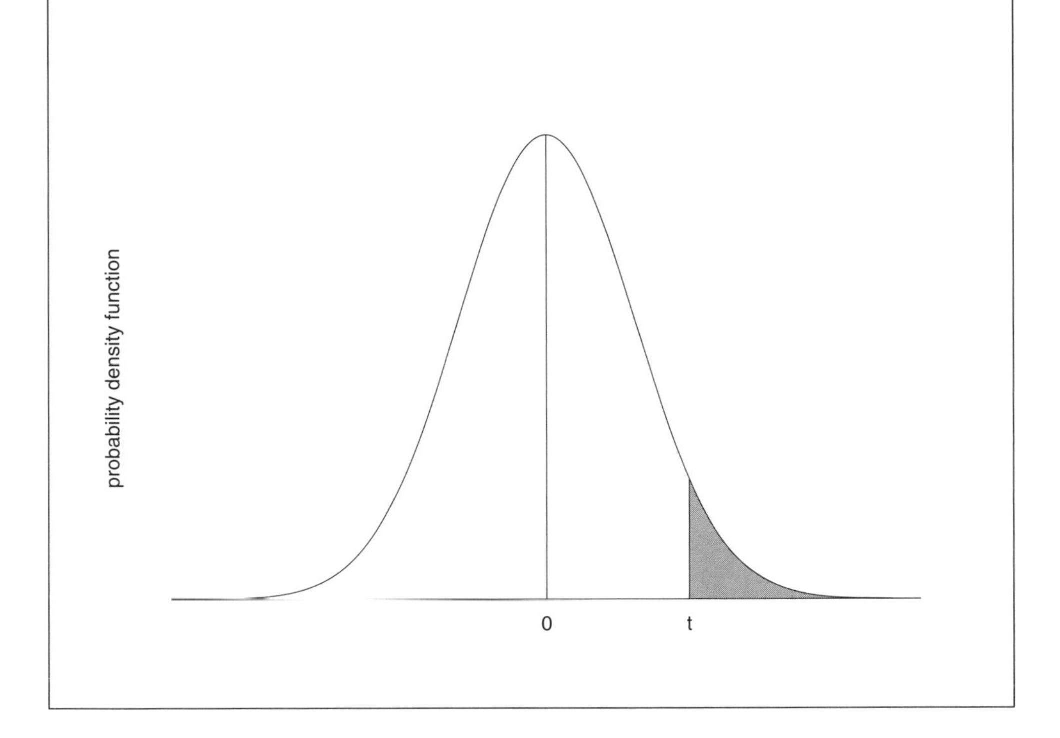

of the null hypothesis with $\mu_1 < \mu_2$. Test 2 should be used only when we have no interest whatsoever in the possibility that $\mu_1 < \mu_2$.

With Test 3, the region of rejection of the null hypothesis is the left tail of the t distribution, as shown in Figure 4.5. If $\alpha = 0.05$ and with 38 degrees of freedom, then for Test 3 the critical value of t is approximately -1.68. Test 3, like Test 2, is a one-tailed, or one-sided test. Test 3 allows rejection of the null hypothesis against the class of alternatives $\mu_1 < \mu_2$. If it is true that $\mu_1 < \mu_2$, then Test 3 will be somewhat more powerful than Test 1 against this class of alternatives but will not allow rejection of the null hypotehsis when $\mu_1 > \mu_2$. Test 3 should be used, therefore, only when we have no interest in the possibility that $\mu_1 > \mu_2$.

Figure 4.5: The one-sided test of significance of the null hypothesis $\mu_1 \geq \mu_2$ against the alternative $\mu_1 < \mu_2$. The shaded area in the left tail of the t distribution is 0.05 of the total area. With $\alpha = 0.05$, the null hypothesis is rejected if the observed value of t falls in the shaded area.

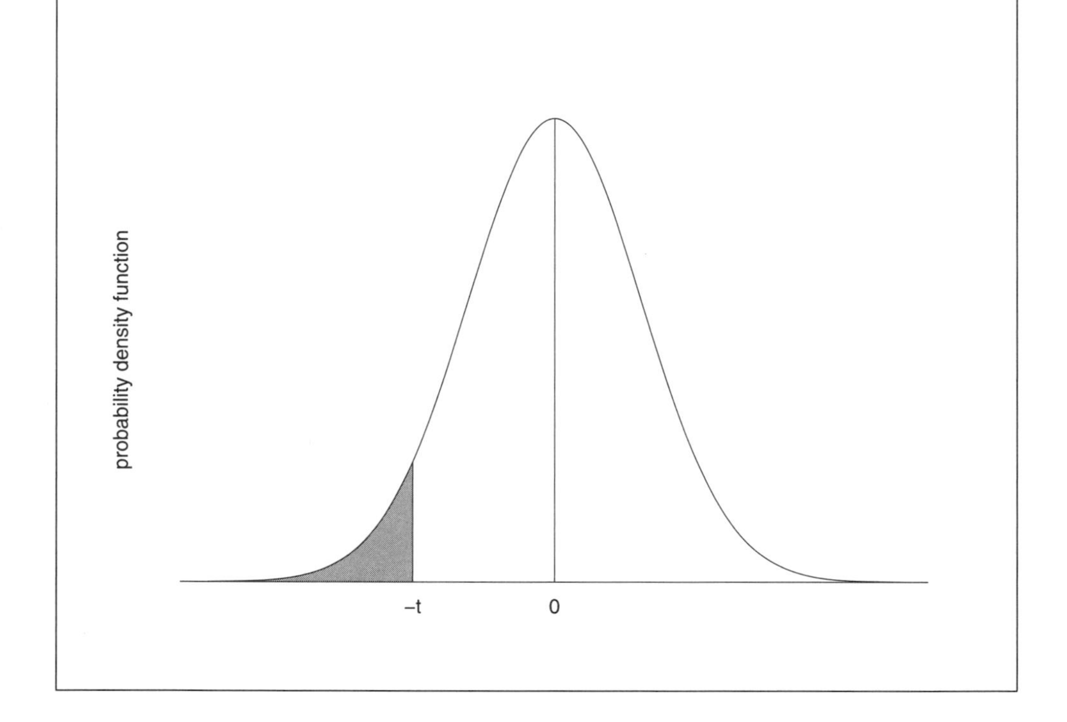

In our opinion, Test 1 is the preferred test in scientific research because it allows for the possibility that either $\mu_1 < \mu_2$ or $\mu_1 > \mu_2$, and, in general, scientists tend to be interested in both possibilities. Even when a scientist makes a prediction about a particular direction, it is desirable to have a test that will allow one to reject a theory when the data are in the opposite direction than the prediction.

If Test 2 or Test 3 is to be made in a given experiment, this decision must be made at the time the experiment is planned and should not be influenced by an examination of the outcome of the experiment. It may sometimes happen that the difference between two means will be declared significant with $\alpha = 0.05$ if a one-sided test is made but nonsignificant with

$\alpha = 0.05$ if a two-sided test is made. To decide, after looking at the outcome of the experiment, that a one-sided test is to be made is unscientific.

We previously discussed some experiments where a one-sided test may be considered appropriate. Consider the case of the farmer from Whidbey Island. The null hypothesis we tested was $P \leq 1/2$ against the alternative $P > 1/2$. When we used the standard normal distribution to evaluate the outcome of this experiment in Chapter 3, the region of rejection was to the right tail of the standard normal distribution; that is, we made a one-sided test corresponding to Test 2. This particular test was made because we were only interested in the possibility that the farmer would do better than chance, and we had no interest whatsoever in the possibility that his performance would be worse than chance.

4.13 The Power of the *t* Test against a Specified Alternative

If the null hypothesis is not rejected, one cannot conclude that the null hypothesis is true. It is possible that the true difference between the population means μ_1 and μ_2 may be quite large, but the test of significance for a given study may have relatively little power to reject the null hypothesis. We saw this in Chapter 2 when five observations (under one of the sampling schemes considered in Chapter 2) did not provide enough power to reject the hypothesis of chance performance.

Establishing a confidence interval for $\mu_1 - \mu_2$ is one way in which we can gain reasonable knowledge about the possible value of $\mu_1 - \mu_2$. For example, assume we have an experiment in which 20 participants are assigned at random to one of two treatments. The observed difference between the two means in the experiment is $\bar{X}_1 - \bar{X}_2 = 2.5$, and the standard error of the difference between the two means is $s_{\bar{X}_1 - \bar{X}_2} = 2.0$. With 18 degrees of freedom, the values of t that would result in the rejection of the null hypothesis with $\alpha = 0.05$ for a two-sided test are $t \geq 2.101$ or $t \leq -2.101$.

The 95% confidence interval for $\mu_1 - \mu_2$ can easily be shown to be -1.702 to 6.702. These limits contain 0, and consequently the null hypothesis $\mu_1 - \mu_2 = 0$ would not be rejected. The 95% confidence interval is relatively wide, reflecting the small sample sizes.

If we fail to reject the null hypothesis $\mu_1 - \mu_2 = 0$ in a given experiment, it is worthwhile to examine the approximate power of the test of significance in terms of some alternative that may be of theoretical or practical interest. Assume, for example, that the alternative hypothesis $\mu_1 - \mu_2 = 3.0$ is, in fact, true in the population. What are the chances that the outcome of the experiment, assuming a pooled standard error of the difference between two means $\sigma_{\bar{X}_1 - \bar{X}_2} = 2.0$, will lead to a rejection of the null hypothesis? What, in other words, is the power of the test of the null hypothesis $\mu_1 - \mu_2 = 0$ against the specific alternative $\mu_1 - \mu_2 = 3.0$ with α set at 0.05 for a two-sided test?

We will reject the null hypothesis if either

$$Z = \frac{\bar{X}_1 - \bar{X}_2}{2.0} \geq 1.96 \quad \text{or} \quad \text{Z} = \frac{\bar{X}_1 - \bar{X}_2}{2.0} \leq -1.96$$

That is, if the observed difference between the two-sample means $\bar{X}_1 - \bar{X}_2 \geq (2.0)(1.96) = 3.92$ or if $\bar{X}_1 - \bar{X}_2 \leq (2.0)(-1.96) = -3.92$, we will reject the null hypothesis. If it is true that the difference in population means is $\mu_1 - \mu_2 = 3.0$, what is the probability of obtaining $\bar{X}_1 - \bar{X}_2 \geq 3.92$? The probability will be given by the area falling to the right of 3.92 in a normal distribution with mean $\mu = 3.0$ and standard deviation $\sigma = 2.0$. It will be easier to convert this to a standard normal variable and compute the area to the right of

$$Z = \frac{3.92 - 3.00}{2.0} = 0.46$$

From the table of the standard normal distribution (Table B.3 in Appendix B), we find that the area falling to the right of $Z = 0.46$ is 0.3228, and this value is an approximate estimate of the power of the test if it is true that

$\mu_1 - \mu_2 = 3.0$. In other words, we have only about 1 chance in 3 of rejecting the null hypothesis under these assumptions. We would find that this value is also the power of the test if the alternative $\mu_1 - \mu_2 = -3.0$ were true. To increase the power of the test for either alternative, $\mu_1 - \mu_2 = 3.0$ or $\mu_1 - \mu_2 = -3.0$, we could increase the number of observations in each of our two treatment groups.

Table 4.4 gives the results of some simple calculations for the power of the significance test for the difference between two means when the population standard deviation is known. If the population σ is known and the two sample sizes are equal, then we can compute the standard error of the difference between two means

$$\sigma_{\bar{X}_1 - \bar{X}_2} = \sqrt{\frac{2\sigma^2}{n}}$$

Suppose, for example, that we have randomly assigned $n = 10$ participants to each of two treatments and that the population standard error is $\sigma_{\bar{X}_1 - \bar{X}_2} = 1.5$. Assume that the null hypothesis is false and that the true difference between the two population means, $\mu_1 - \mu_2$, is equal to 1.5. Then the standardized difference of the two means is given by $\delta_Z = (\mu_1 - \mu_2)/\sigma_{\bar{X}_1 - \bar{X}_2} = 1.00$, and this value is shown in the fifth row of the first column in Table 4.4. Given that $\delta_Z = 1.00$, we want to find the probability of obtaining a value of $\bar{X}_1 - \bar{X}_2 > 0$ that will result in the rejection of the null hypothesis using a two-sided test with $\alpha = 0.05$. For our example, we will reject the null hypothesis if

$$\frac{\bar{X}_1 - \bar{X}_2}{1.5} \geq 1.96$$

or if the observed difference between the sample means $\bar{X}_1 - \bar{X}_2 \geq (1.96)(1.5) = 2.94$. If $\mu_1 - \mu_2 = 1.5$, then the probability of obtaining $\bar{X}_1 - \bar{X}_2 \geq 2.94$ will be given by the area falling to the right of

$$Z = \frac{2.94 - 1.5}{1.5} = 0.96$$

Table 4.4: Probability of rejecting the null hypothesis $\mu_1 = \mu_2$ with a two-sided test, with $\alpha = 0.10$, $\alpha = 0.05$, and $\alpha = 0.01$, respectively, for various values of $\delta_Z = (\mu_1 - \mu_2)/\sigma_{\bar{X}_1-\bar{X}_2} \geq 0$.

	Significance Level (α)		
$\delta_Z = (\mu_1 - \mu_2)/\sigma_{\bar{X}_1-\bar{X}_2}$	0.10	0.05	0.01
0.00	0.05	0.025	0.005
0.25	0.08	0.04	0.01
0.50	0.13	0.07	0.02
0.75	0.18	0.11	0.03
1.00	0.26	0.17	0.06
1.25	0.34	0.24	0.09
1.50	0.44	0.32	0.14
1.75	0.54	0.42	0.20
2.00	0.64	0.52	0.28
2.25	0.73	0.61	0.37
2.50	0.80	0.71	0.47
2.75	0.86	0.76	0.57
3.00	0.91	0.85	0.66
3.25	0.95	0.90	0.75
3.50	0.97	0.94	0.82
3.75	0.98	0.96	0.88
4.00	0.99	0.98	0.92

in the standard normal distribution. From the table of the standard normal distribution, Table B.3 in Appendix B, we find that $P(Z \geq 0.96)$ is approximately 0.17, and this value is the probability shown opposite $\delta_Z = 1.00$ in the column headed 0.05 in Table 4.4. In other words, if the population $\mu_1 - \mu_2 = 1.5$ and if the population standard error $\sigma_{\bar{X}_1-\bar{X}_2} = 1.5$, then the probability of obtaining a positive value of $\bar{X}_1 - \bar{X}_2$ for the difference between the two sample means that will result in rejecting the null hypothesis for a two-sided test with $\alpha = 0.05$ is 0.17. This is not a very high level of statistical power. Most researchers prefer a power of at least 0.80.

One way to increase the probability of rejecting the null hypothesis is to increase the number of observations for each of the two treatments. For example, suppose with $n = 10$ observations in each treatment, we have

$$\sigma_{\bar{X}_1-\bar{X}_2} = \sqrt{\frac{2\sigma^2}{10}} = 1.5$$

But if we increase the sample size to $n = 40$ observations in each treatment, we would have

$$\sigma_{\bar{X}_1-\bar{X}_2} = \sqrt{\frac{2\sigma^2}{(4)(10)}} = \frac{1}{2}(1.5) = 0.75$$

With a population mean difference $\mu_1 - \mu_2 = 1.5$, we would have $\delta_Z = 1.5/0.75 = 2.00$, and the probability of obtaining a positive value of $\bar{X}_1 - \bar{X}_2$ that would result in the rejection of the null hypothesis for a two-sided test with $\alpha = 0.05$ would be 0.52, as shown in the column headed 0.05 opposite the row entry for $\delta_Z = 2.00$.

The probabilities shown in Table 4.4 were calculated by using the table of the standard normal distribution under the assumption that the population standard deviation σ was known. They may be regarded as approximations of the corresponding probabilities based on the t distribution when we have a sample estimate for the unknown population value σ. In general, the probabilities given in Table 4.4 are somewhat larger than those that would be obtained by using the t distribution but become increasingly more accurate as the degrees of freedom on which the sample estimate of σ is based increases. The actual power calculations when the standard deviation σ is estimated requires the use of the noncentral t distribution, which is beyond the scope of this book (see Cohen, 1987 for more information).

4.14 Estimating the Number of Observations Needed in Comparing Two Treatment Means

Sometimes in research settings it is useful to estimate the number of observations needed to achieve a particular level of power. It is instructive to work through the derivation of the equation that permits the estimation of sample size for a given effect size. It would be a waste of time and effort to conduct a study that has very low statistical power. The logic presented in this subsection allows one to determine the required sample size before running the study.

Assume that on the basis of previous research or a pilot study we can estimate the variability in a dependent variable X under a given set of treatments. It will, in fact, simplify the presentation if we assume that the common population variance σ^2 is known. Suppose also that we set $\alpha = 0.05$ and that we have decided upon a two-sided test of significance. For a two-sided test with $\alpha = 0.05$, the critical values of Z are -1.96 and 1.96. We decide that the population mean difference $\mu_1 - \mu_2$ must be equal to or greater than δ or must be equal to or less than $-\delta$, to be of either theoretical or practical interest. Furthermore, suppose we want the probability of a Type II error to be no greater than 0.16 if the true difference between μ_1 and μ_2 is equal to or greater than δ or equal to or less than $-\delta$. We have previously defined the power of a test as $1 - P(\text{Type II error})$. Therefore, we desire the test to have a power of $1 - 0.16 = 0.84$, which is to say that we want the test to have a probability of at least 0.84 of rejecting the null hypothesis if it is true that

$$|\mu_1 - \mu_2| \geq \delta$$

Consider, first, the possibility that $\mu_1 - \mu_2$ is at least δ. Figure 4.6 shows the distribution of $\bar{X}_1 - \bar{X}_2$, when $\mu_1 - \mu_2 = \delta$ (the alternative hypothesis), at the right and the distribution of $\bar{X}_1 - \bar{X}_2$, when $\mu_1 - \mu_2 = 0$ (the null hypothesis), at the left. Let $Z_0 = 1.96$ be the critical value of Z resulting in the false rejection of the null hypothesis when it is true, that is, when $\mu_1 - \mu_2 = 0$. Thus, if the null hypothesis is true, and if $n_1 = n_2 = n$, then we want the value of c in Figure 4.6, expressed as a standard normal variable, to be

$$Z_0 = \frac{c - 0}{\sqrt{2\frac{\sigma^2}{n}}} = 1.96$$

If the null hypothesis is false, that is, $\mu_1 - \mu_2 = \delta$, and if we obtain a difference that falls to the left of c, the null hypothesis will not be rejected and we shall make a Type II error. We want this probability to be no greater than 0.16. Thus when $\mu_1 - \mu_2 = \delta$, we want 0.16 of the total area in the curve at the right in Figure 4.6 to fall to the left of c and 0.84 to fall to the right of c. Let Z_1 be the Z value corresponding to c in this instance. From the table of the standard normal distribution (Table B.3 in Appendix B), we find that $Z_1 = -1.00$. We also have

$$\mathrm{Z}_1 = \frac{c - \delta}{\sqrt{2\frac{\sigma^2}{n}}} = -1.00$$

Using the above two equations, we have

$$c = 1.96\sqrt{\frac{2\sigma^2}{n}} \quad \text{and} \quad c = \delta - \sqrt{\frac{2\sigma^2}{n}}$$

Substituting the former equation into the latter and solving for δ, we have

$$\delta = 1.96\sqrt{\frac{2\sigma^2}{n}} + \sqrt{\frac{2\sigma^2}{n}}$$

and then solving for the sample size n

$$n = \frac{2\sigma^2}{\delta^2}(1.96 + 1.00)^2 \tag{4.18}$$

Figure 4.6: The distribution of $\bar{X}_1 - \bar{X}_2$ when $\mu_1 - \mu_2 = 0$ is shown at the left (labeled "Null") and the distribution of $\bar{X}_1 - \bar{X}_2$ when $\mu_1 - \mu_2 = \delta$ is shown at the right (labeled "Alternative"). The point c is located so as to cut off 0.025 of the total area in the right tail of the distribution at the left (the darker shaded region) and 0.84 of the total area in the right tail of the distribution shown at the right (the lighter shaded region). The value of c, which is in the scale of δ, is then used to determine the necessary sample size.

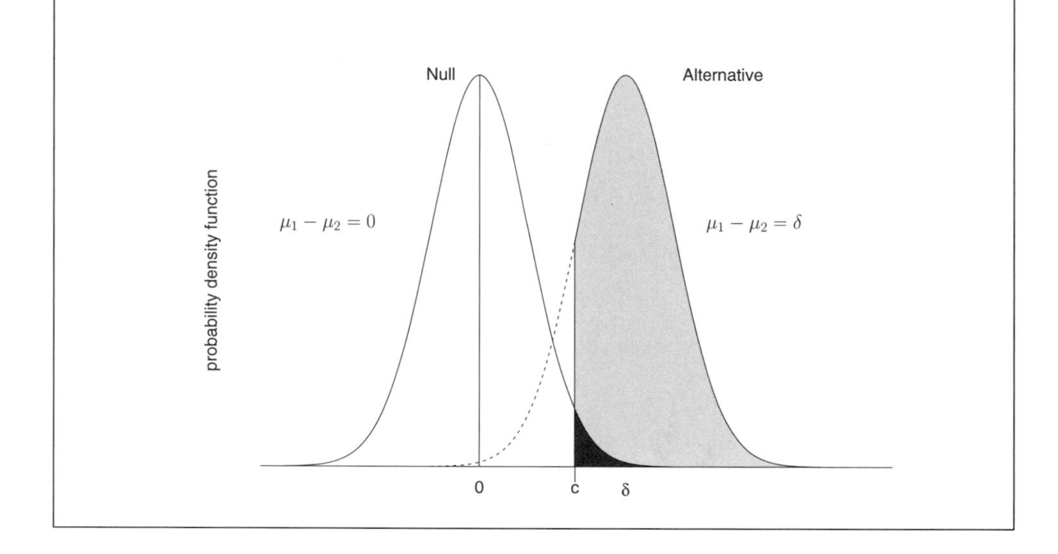

which is the formula that allows us to estimate the sample size required to achieve the desired level of power of a two-tailed test at $\alpha = .05$. Assuming that the population standard deviation is $\sigma = 10.0$ and that $\delta = 5$, we have

$$n = \frac{2\,(10)^2}{(5)^2}\,(2.96)^2 = 70$$

so at least $n = 70$ participants in each treatment group would be needed if the test of significance is to have the properties described. We have considered only the possibility that $\delta = 5$. We would arrive at exactly the same value for the sample size n by considering the possibility that $\delta = -5$.

4.15 Random Assignment of Participants

In the experiment with the farmer described in Chapter 2, we emphasized the necessity of using some random method for determining which cans were to be filled with water and which were to be left empty. If this randomization had not been done, then there might have been systematic differences between the cans filled with water and those left empty, which could, in turn, possibly influence the farmer's selections.

In an experiment designed to compare the difference between the means obtained under two treatments, we want to be able to conclude, in the case of a statistically significant outcome, that the difference is the result of the difference in treatments. In order to do so, we must have some assurance that the difference between the means cannot be accounted for in terms of systematic differences in individual difference variables between the two groups receiving the different treatments. We illustrate with an example. If differences in intellectual ability are related to differences in learning and if we wish to compare the performance of two groups under two different learning conditions, it would not do to have one group composed of individuals with a very high level of intellectual ability and the other of individuals with a very low level. If this situation were the case, then any difference between the mean learning scores of the two groups under the two treatments could be interpreted as being the result of pre-existing systematic differences in intellectual ability between the two treatment groups rather than as a result of differences in the experimental treatments. Under these conditions, there is no logical way that we can make any statement about the explanation of differential effects of the two treatments. Of course, one way around this issue is to conduct a study that randomly assigns high-ability subjects to both experimental conditions and also randomly assigns low-ability subjects to both experimental conditions, thus ensuring that high- and low-ability subjects are in each experimental condition. Techniques for the analysis of such experiments will be discussed in later chapters on factorial design.

In order to conclude that a difference between two means is the result of differences in the treatments, we must be able to rule out other alternative explanations that could account for the difference. If participants were not randomly assigned to the treatments, then an individual difference variable could be an alternative explanation for the difference between two treatment means. Random assignment of the participants, on the other hand, offers assurance that the groups receiving different treatments are not likely to differ systematically with respect to individual difference variables. Thus, under random assignment, we are on stronger footing to attribute a statistically significant difference between two means to the treatments themselves (that is, the experimental manipulation). There is a growing literature surrounding issues of making the strongest possible inferences when random assignment is not possible (see Cook & Campbell, 1979, and Rubin, 2005).

The relation between sample size and power can intuitively be captured with an analogy. Imagine that you are fishing the ocean with a large net that drags from your boat. The fineness of the net's mesh is related to the size of the fish you will catch. If the mesh is coarse so that there are large "holes" in the net, the small fish will go through the net, and you will only catch the fish that are too large to go through the mesh. If the net has a fine mesh, you will catch not only large fish but also the smaller fish that are also caught by the finer net. Sample size, in a sense, dictates the fineness of the "statistical net." A large sample size leads to a more sensitive statistical test, so you can catch smaller effects. A small sample size, much like a coarse fishing net, does not lead to a sensitive test, so only the large effect sizes can be "caught," or deemed statistically significant.

4.16 Attrition in Behavioral Science Experiments

Even though we randomly assign the same number of participants to each of two treatments, it is possible that we may end up with an unequal number of observations in each treatment. Some participants may forget to keep a laboratory appointment, others may be ill at the time they are to be tested, and still others may, for one reason or another, not complete the experimental session.

It is important that the loss of participants not be the result of the treatments themselves. For example, we may be interested in comparing the performance of participants who receive a reward for each correct response in a learning experiment to the performance of participants who receive an electric shock for each incorrect response. If the participants consist of laboratory animals, we can randomly assign the animals to the two treatments and compel them to participate in the experiment. Even though the lab animals assigned to the shock treatment may find the shock unpleasant, they are not in a position to withdraw voluntarily from participation in the experiment. This is not the case with human participants.

It is standard ethical practice in behavioral science experiments involving human participants that they be permitted to withdraw from participating in an experiment at any time they desire to do so. Thus even though we may randomly assign an equal number of participants to a reward and shock treatment, some of the participants assigned to the shock treatment may decide that they do not wish to participate in the experiment. In this instance, it is possible that the participants who do not withdraw and who complete the experiment under the shock condition are no longer a random sample and may not be comparable to the group receiving the reward (most of whom, say, chose to complete the study). If the loss of participants is in any way related to the treatments themselves, we have no way of knowing whether a statistically significant outcome of the experiment is obtained because of differences in the treatments or because the two groups of participants no longer represent random samples from a common population. The way one handles missing data in experimental design is a complicated task and a current topic of research. For an advanced treatment see Little and Rubin (2002). In this book we will assume that missing data are "completely at random," that is, whatever is missing is unrelated to the experimental condition.

4.17 Summary

This chapter made an important transition because it showed how we can use the techniques from previous chapters to address problems involving the comparison of two means. By extending the approximation developed in Chapter 3 for a sum to the comparison of the difference between two means, we now have the machinery to test research questions about different groups or treatments. The use of the two-sample t test presented in this chapter compares the means of two treatments (such as the mean score from a group of participants who received a drug compared to the mean score from a group of participants who received a placebo). We also reviewed the concept of power (originally introduced in Chapter 2) and discussed how to estimate the sample size needed to reach a particular level of power.

4.18 Questions and Problems

1. Compute the 95% confidence interval for a random sample of $n = 16$ observations with $\bar{X} = 22.4$ and $s = 4.3$.

2. The mean score on a standardized test in psychology for a random sample of 200 freshmen college students at University A is 133.8, with the sample standard deviation s equal to 14.7. For a random sample of 140 freshmen at University B, the mean score is 138.4, with s equal to 15.2. Determine whether the two means differ significantly, using a two-tailed test with $\alpha = 0.05$.

3. Forty participants are assigned at random to two treatments, with 20 participants in each treatment. The measures on the dependent variable are given here:

Treatment 1				Treatment 2			
39	41	39	44	36	41	30	39
39	40	39	40	36	39	33	37
37	42	37	43	35	42	36	37
44	38	38	38	34	38	33	31
43	38	41	39	40	32	33	38

Determine whether the two treatment means differ significantly with $\alpha = 0.05$. Compute a 95% confidence interval around the two means. Summarize your results in words.

4. In an experiment, the standard error of the difference between two means was 1.42 with $n = 10$ participants in each treatment group. A repetition of this experiment is planned, and the experimenter wishes to be able to reject the null hypothesis if the absolute difference between the population means is 2.56 or greater. On the basis of the information provided in this question, it is possible to compute the pooled sample variance s^2. In the following questions you may substitute the observed sample variance for the population variance.

   (a) How many participants should the experimenter have in each treatment group if $\alpha = 0.05$ and if the probability of a Type II error is to be no greater than 0.16?

   (b) How many participants should the experimenter have in each treatment group if $\alpha = 0.05$ and if the probability of a Type II error is to be no greater than 0.50?

5. Give a brief interpretation of the meaning of confidence interval.

6. Explain the following statement: A confidence interval implies a test of significance.

7. Discuss briefly the t test of a null hypothesis concerning a difference between two means in relation to the alternatives to the null hypothesis.

(a) Under what conditions should we perform a two-sided test?

(b) Under what conditions should we perform a right-tailed test?

(c) Give an example where each test is appropriate.

8. The null hypothesis to be tested in an experiment is $\mu_1 = \mu_2$. The experimenter sets $\alpha = 0.05$ for a two-sided test of the null hypothesis. Suppose that σ is known and that for the number of participants in the two treatment groups we have $\sigma_{\bar{X}_1 - \bar{X}_2} = 1.5$. If the null hypothesis is false and $\mu_1 - \mu_2 = 6.0$, what is the power of the test of significance?

9. What is meant by the power of a test of significance?

10. We have $n = 10$ participants assigned at random to each of two treatments. The outcome of the experiment is given here:

T1:	4	3	6	7	5	5	5	7	5	5
T2:	2	3	4	2	4	2	5	2	4	4

(a) Test the null hypothesis that $\mu_1 = \mu_2$ with a two-tailed $\alpha = 0.05$.

(b) Find a 95% confidence interval for $\mu_1 - \mu_2$. Does this interval contain 0? What is the relation between this confidence interval and the hypothesis test in part (a)?

11. Assume that for a dependent variable of interest it is known that $\sigma = 5.0$. We are planning an experiment where we want to test the difference between the means of two treatment groups. We plan on making a two-sided test with $\alpha = 0.05$. We have also decided that if the absolute difference between μ_1 and μ_2 is at least 2.5, this difference would be of practical interest.

(a) If we want the power of the test to be at least 0.84, how many subjects would we need in each group?

(b) If we want the power of the test to be at least 0.95, how many subjects would we need in each group?

12. We wish to test the null hypothesis that $\mu_0 = 50.0$ against the alternative that $\mu_1 > 50.0$. We assume a population standard deviation $\sigma = 20.0$ and have $n = 25$ and $\alpha = 0.01$.

    (a) What values of $\bar{X}$ will result in rejection of the null hypothesis?

    (b) What is the power of the test if the alternative $\mu_1 = 60.0$ is true?

    (c) If we increase the sample size to $n = 100$, what values of $\bar{X}$ will result in the rejection of the null hypothesis?

    (d) With the sample size increased to $n = 100$ observations, what is the power of the test against the alternative $\mu_1 = 60.0$?

5

Homogeneity and Normality Assumptions

5.1 Introduction

The preceding chapter reviewed a method for testing the difference between two means based on the critical assumption that the two samples estimate the same population variance. We call this assumption of a common population variance the **equality of variance assumption**. It is also known as the **homogeneity of variance assumption**. Typically, this assumption is a reasonable one, but sometimes the effect of one treatment may increase or decrease variability whereas the effect of the other treatment may not. Such differential variability may make the homogeneity of variance assumption suspect. When that assumption is suspect, then the two-sample t test and the confidence interval around the difference between two means are also suspect.

In this chapter, we describe a test of significance for the difference between two sample variances (that is, the null hypothesis that the two sample variances have the same population variance) and describe a t test that can evaluate the difference between two sample means even when the population variances likely differ. We describe some factors that may lead to a significant difference between two variances. Finally, we review the topics of transformations and boxplots and discuss the assumption of normality.

5.2 Testing Two Variances: The *F* Distribution

It is possible to test the equality of two variances in a manner analogous to how two means are tested for equality. Given two sample variances s_1^2 and s_2^2, based on two independent random samples of n_1 and n_2 observations, respectively, we may test the null hypothesis that the population variances are equal, that is, $\sigma_1^2 = \sigma_2^2$, against the alternative hypothesis that the population variances are not equal, that is, $\sigma_1^2 \neq \sigma_2^2$. The ratio of the two sample variances is distributed in a manner discovered by Fisher (1936), a distribution called F in his honor. The ratio of two sample variances follows an F distribution:

$$F = \frac{s_1^2}{s_2^2} \text{ or } F = \frac{s_2^2}{s_1^2} \tag{5.1}$$

Whether F will be greater than 1.0 or smaller than 1.0 will depend merely on whether the greater variance, s_1^2 or s_2^2, is put in the numerator of the ratio. The tabled values of F, in Table B.2 in Appendix B, are for a one-sided, or one-tailed, test and correspond to the probability that the value of F is greater than 1.0 when the null hypothesis is true. To use the two tables (Tables B.2.1 and B.2.2), we will find the value of F greater than 1.0 in Equation 5.1, which means that by convention we will always put the larger of the two sample variances in the numerator of Equation 5.1.

The F distribution has two types of degrees of freedom, typically called numerator and denominator degrees of freedom. In the application of the F test presented in this chapter, the degrees of freedom term for the numerator is related to the sample size of the sample with the larger variance (that is,

$n_1 - 1$, where n_1 is the sample size in the sample with the larger variance); while the degrees of freedom term for the denominator is related to the sample size of the sample with the smaller variance (that is, $n_2 - 1$).

If the alternative hypothesis is that the two population variances are not equal, $\sigma_1^2 \neq \sigma_2^2$ (as opposed to a predicted direction, or "one-sided" alternative hypothesis), then we need to make a two-sided or two-tailed test; that is, we want to reject the null hypothesis if either $\sigma_1^2 > \sigma_2^2$ or $\sigma_1^2 < \sigma_2^2$. For the two-sided test, with $\alpha = 0.05$, the critical value of F will be the tabled value with a probability of 0.025. Similarly, for a two-sided test with $\alpha = 0.01$, the critical value of F will be the tabled value with a probability of 0.005.

The F test, as defined by Equation 5.1, is often referred to as a test of homogeneity of variance. If a nonsignificant value of F is obtained, the two-sample variances are said to be **homogeneous**; that is, the sample variances are both assumed to be estimates of the same population variance. With a significant value of F, the variances are said to be **heterogeneous**. This test of two independent variances is very sensitive to outliers and violations of the normality assumption.

Keep in mind that, as with any test of statistical significance, the power of the test for two population variances is sensitive to sample size. A large sample may lead to the rejection of a tiny difference between two population variances, whereas a small sample size may fail to detect a very large difference between two variances. When evaluating the result of a statistical test on an assumption, the data analyst needs to be particularly sensitive to the power of the test. It is our belief that it is more important to detect large violations of assumptions rather than statistically significant violations of the assumptions. The criterion of a statistical test to determine the suitability of an assumption has the benefit of providing a clear decision rule; however, it is not obvious that the decision rule is very helpful, especially in the case of large sample sizes and nonnormally distributed data.

5.3 An Example of Testing the Homogeneity of Two Variances

In the experiment on cued recall described in the preceding chapter, we assumed that s_1^2 and s_2^2 did not differ significantly (that is, we assumed a common population value of σ^2). For Treatment 1, the sum of squared deviations from the mean was $\sum_{i=1}^{n_1}(X_{1i} - \bar{X}_1)^2 = 176$, and for Treatment 2, it was $\sum_{i=1}^{n_2}(X_{2i} - \bar{X}_2)^2 = 242$. The sample variances were

$$s_1^2 = \frac{176}{19} = 9.263 \qquad \text{and} \qquad s_2^2 = \frac{242}{19} = 12.737$$

Because s_2^2 is larger than s_1^2, we have the following F test:

$$F = \frac{12.737}{9.263} = 1.375$$

To determine whether the observed $F = 1.375$ is statistically significant, we enter the column of Table B.2.1 with the degrees of freedom corresponding to the numerator of the F ratio and the row with the degrees of freedom corresponding to the denominator. For the obtained $F = 1.375$, we have 19 degrees of freedom for the numerator and 19 degrees of freedom for the denominator. Table B.2.1 has no column corresponding to 19 degrees of freedom, but we can approximate the critical value of F, with $\alpha = 2(0.025) = 0.05$, for 20 and 19 degrees of freedom, as 2.51. The obtained value of $F = 1.375$ is less than this critical value, and with $\alpha = 0.05$ as the criterion the null hypothesis of a common population variance is not rejected.

5.4 Caveats

This particular homogeneity of variance test is discussed here because we want to introduce the F distribution that will be used extensively throughout this book. However, the variance test presented in Equation 5.1 is quite sensitive to departures from normality. Tests with better properties have been developed by Levene (1960) and O'Brien (1981). As a way of checking for differences in variability Levene suggested taking the absolute deviation of each score from its corresponding sample mean and then performing a two-sample t test on those absolute deviations (that is, testing the hypothesis: are the absolute deviations from the mean for one group statistically different from the absolute deviations of the other group?). Absolute deviations are less sensitive to outliers than the variance, which is what makes up the F test, because the variance squares the differences. O'Brien (1981) reviewed several procedures and recommended a variant of Levene's test that is robust to departures from normality.

There is a general problem with the practice of using statistical inference to test assumptions of other statistical tests. As we saw in the preceding chapter, all other things being equal, large sample sizes yield more statistical power. Thus, with a large enough sample a trivial deviation from an assumption will be detected as significant by a statistical test. The logic works the other way too—when the sample sizes are small, tests of inference may fail to reject quite large population differences. Also, it is strange to perform a statistical test on assumptions that are made in order to perform a statistical test, especially when the statistical test for assumptions also makes assumptions that need to be tested. One can fall into a silly series of tests to verify one assumption after another.

We suggest that when examining statistical assumptions researchers use procedures that are relatively insensitive to sample size, distributional properties, and the presence of outliers. The boxplot is a graphical tool that helps one assess the degree to which the variability in two or more groups differs. We now turn to a brief discussion of boxplots and then present a variant of the t test that does not make the homogeneity of variance assumption because it does not use a pooled variance estimate.

5.5 Boxplots

The boxplot (Tukey, 1977) is a useful graphical tool to detect unequal variances as well as other properties of a sample such as outliers. The boxplot is described in most introductory statistics textbooks and will be reviewed here briefly. The three main ingredients of the boxplot are the median (which is the data point corresponding to the 50th percentile in the sample), the data point corresponding to the 25th percentile in the sample, and the data point corresponding to the 75th percentile in the sample. The difference between the 25th and 75th percentiles is called the **interquartile range**. Two additional ingredients of the boxplot yield the whiskers: the data point corresponding to 1.5 times the interquartile range (IQR) above the median and the data point corresponding to 1.5 times the interquartile range below the median. Usually the whiskers are drawn to the nearest data point not exceeding 1.5 times the interquartile range in either direction.

Figure 5.1 shows an example boxplot with two samples. Note that the width of the two boxes differs between the two samples as well as the length of the whiskers. A visual comparison of the two boxplots suggests that these two samples may have different population variances. The sample on the right has more variability than the sample on the left. Boxplots that differ widely in their spread suggest that there may be a problem making the homogeneity of variance assumption, and caution would be needed before proceeding with statistical tests that rely on the equality of population variance assumption.

The boxplot also serves as an indicator of asymmetry in the distribution. Two features to look for to detect asymmetry: (1) whether the median is away from the center of the box and (2) whether one of the whiskers is much longer than the other. The boxplot on the right-hand side of Figure 5.1 is asymmetric, indicating a possible violation from normality. Outliers can be plotted as points that exceed the whiskers (that is, outliers can be defined as observations that exceed 1.5 IQR's from the median).

Figure 5.1: Two boxplots showing the difference in the symmetry and variability of two samples.

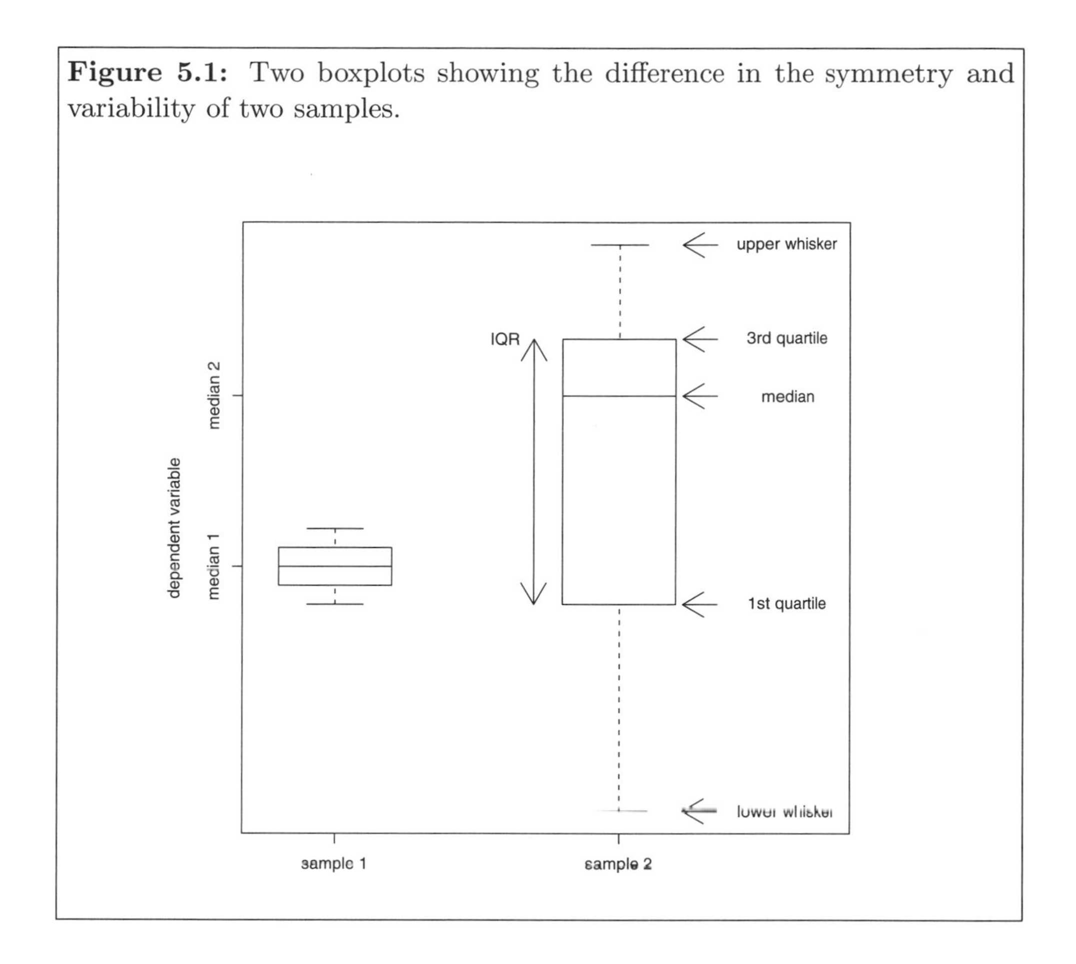

We encourage researchers to use boxplots as a tool to detect unequal variance across treatment conditions, to detect asymmetric distributions, and to detect outliers (data points more extreme than the whiskers). As much as we like boxplots, they aren't perfect. The variability of the median and the IQR is also affected by sample size, and the examination of the properties of the boxplot relies on the subjective perception of the data analyst. Most statistical packages perform boxplots. In the SPSS program the boxplot is part of the EXPLORE command.

We now turn to a version of the two-sample t test that does not make the equality of variance assumption and consequently can be used when the

Table 5.1: Means and variances for two treatments with unequal n's. The observed variances make the equality of population variance assumptions suspect.

Treatment 1	Treatment 2
$\bar{X}_1 = 20.6$	$\bar{X}_2 = 16.0$
$s_1^2 = 28.42$	$s_2^2 = 6.72$
$n_1 = 10$	$n_2 = 20$

boxplots suggest that the variability across the two treatments is substantially different.

5.6 A *t* Test for Two Independent Means When the Population Variances Are Not Equal

Suppose that in an experiment comparing two treatments we observe heterogeneous variances. Consider, for example, the summary data in Table 5.1. Testing for homogeneity of variance, we have

$$F = \frac{28.42}{6.72} = 4.23$$

with 9 and 19 degrees of freedom. With $\alpha = 2(0.025) = 0.05$, the critical two-tailed value of F is 2.88. Because the obtained value of $F = 4.23$ exceeds the critical value, we reject the null hypothesis of a common population variance.

In the preceding chapter, we assumed that s_1^2 and s_2^2 estimated the same population variance, so we combined the separate sample estimates into a single estimate of the common population variance. But in the present example, we reject the null hypothesis of a common population variance $\sigma_1^2 = \sigma_2^2$. Therefore, the sample estimates s_1^2 and s_2^2 cannot be used as estimates of the same population variance, and the formula used in the preceding chapter for the estimated standard error of the difference between the two means

$$s_{\bar{X}_1-\bar{X}_2} = s\sqrt{\left(\frac{1}{n_1}+\frac{1}{n_2}\right)} \tag{5.2}$$

is inappropriate. One solution to this problem is to alter the t test slightly. Two changes are needed: (1) the standard error of the difference between the two means will be estimated in a way that does not pool the two individual sample variances, and (2) the critical t value will be calculated differently to account for the heterogeneity of population variances.

Instead of using a single number, as in Equation 5.2, to find the estimated standard error of the difference between the two means, we will use the separate sample estimates s_1^2 and s_2^2. In this case, we have

$$s_{\bar{X}_1-\bar{X}_2} = \sqrt{\frac{s_1^2}{n_1}+\frac{s_2^2}{n_2}} \tag{5.3}$$

as the estimated standard error of the difference between two means. Using Equation 5.3 for the data in Table 5.1, we have

$$s_{\bar{X}_1-\bar{X}_2} = \sqrt{\frac{28.42}{10}+\frac{6.72}{20}} = 1.783$$

and

$$t = \frac{20.6-16.0}{1.783} = 2.58$$

Because we have used a different procedure for estimating the standard error of the difference between two means, the critical value used to evaluate the observed t must also be altered. The new critical value will be a weighted average (where the squares of the sample standard errors serve as weights) of the critical t's corresponding to each of the two sample sizes n_1 and n_2. To determine whether the observed $t = 2.58$ is statistically significant, we first find, from Table B.1 in Appendix B, the critical values of t_1 for $n_1 - 1 = 9$ degrees of freedom and t_2 for $n_2 - 1 = 19$ degrees of freedom. For a two-sided test, with $\alpha = 0.05$, these two values are $t_1 = 2.262$ and $t_2 = 2.093$. Then, we find the weighted average of the two t values:

$$t' = \frac{t_1 \left(s_1^2/n_1\right) + t_2 \left(s_2^2/n_2\right)}{\left(s_1^2/n_1\right) + \left(s_2^2/n_2\right)} \tag{5.4}$$

The value of t' observed from Equation 5.4 is the critical value against which the observed $t = 2.58$ will be evaluated. Substituting the sample values in Equation 5.4, we have the critical value

$$t' = \frac{2.262\,(28.42/10) + 2.093\,(6.72/20)}{(28.42/10) + (6.72/20)} = 2.24$$

Because the observed $t = 2.58$ exceeds $t' = 2.24$, the null hypothesis that the population mean from one treatment equals the population mean from the second treatment will be rejected. Note that we obtained both a significant value of F in testing the null hypothesis of equal population variances $\sigma_1^2 = \sigma_2^2$ and a significant value of t in testing the null hypothesis of equal population means $\mu_1 = \mu_2$; we conclude that the two treatments resulted in a significant difference in the treatment variances and also in the treatment means. There might be psychological information in knowing that one treatment yielded a different variance than another treatment. We consider some possible explanations in the next sections. Behavioral scientists have mostly ignored differences in variance in their empirical work, focusing mostly on differences in means.

A more complicated procedure for comparing independent samples when the homogeneity of variance assumption is rejected was proposed by Welch (1938, 1947; see also work by Satterthwaite, 1946). It requires more intensive calculations and may lead to noninteger degrees of freedom, making the use of Table B.1 in Appendix B difficult. Several computer programs such as SPSS now include the Welch t test as an option; it is typically labeled as a "separate variance t test" in the output. The interpretation of the Welch t test is just like the standard t test comparing two means. The Welch t test is more general than the standard t test because it does not assume equal population variance. The test does not pool the cell variances and has noninteger degrees of freedom, such that degrees of freedom are essentially penalized in relation to the violation of equality of variances.

5.6.1 SPSS Example

In this section we provide example SPSS output for the material presented so far in this chapter. We use data from Chapter 4 (Table 4.2) that were used to illustrate the two-sample t test. In Figure 5.2 we present the SPSS output for the boxplot, the Levene F test for the equality of two variances, and output for both the standard t test and the t test that does not assume equal variances (that is, the test adjusts degrees of freedom in relation to the degree of violation). The Levene test for testing the equality of variance assumption is different from the simple test presented earlier in this chapter. Recall that Levene's test is more complicated than merely taking the ratio of two variances; it performs a two-sample t test on the absolute deviation of each score from its corresponding sample mean as a way to test whether the two conditions have different variability.

In this example the boxplot suggests that two observations in group 2 may be outliers (the two points at the lower end of the right boxplot). The boxplots show that the spread in the two groups is comparable (that is, the interquartile ranges are similar across the two groups, as are the lengths of the whiskers). The Levene test for equality of variances does not reject the null hypothesis of equal population variances. Because the two variances are comparable, the degrees of freedom in the Welch test are not adjusted very much (the adjusted degrees of freedom are 37.076 instead of the original 38), leading to comparable p values for both the classic test that assumes equal population variances and the test that does not assume equal population variances. Had the equality of variance assumption been violated, the "equal variance not assumed" t test would have yielded much lower degrees of freedom than the classic test that assumes equal variances.

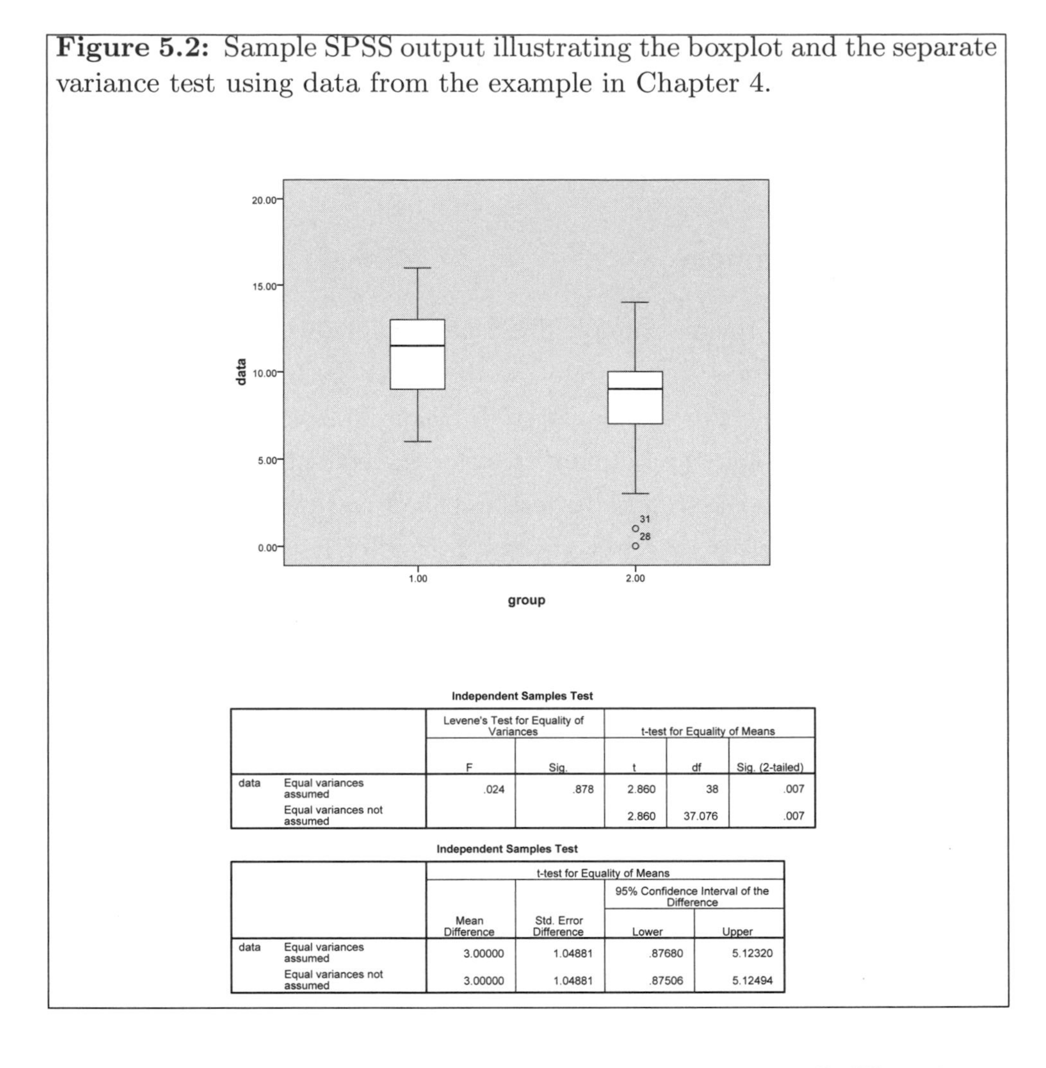

Independent Samples Test

		Levene's Test for Equality of Variances		t-test for Equality of Means		
		F	Sig.	t	df	Sig. (2-tailed)
data	Equal variances assumed	.024	.878	2.860	38	.007
	Equal variances not assumed			2.860	37.076	.007

Independent Samples Test

		t-test for Equality of Means			
				95% Confidence Interval of the Difference	
		Mean Difference	Std. Error Difference	Lower	Upper
data	Equal variances assumed	3.00000	1.04881	.87680	5.12320
	Equal variances not assumed	3.00000	1.04881	.87506	5.12494

Figure 5.2: Sample SPSS output illustrating the boxplot and the separate variance test using data from the example in Chapter 4.

An interesting feature of a statistics package such as SPSS is that it computes the actual p value of the observed data so it is not necessary to use the t table in Appendix B to find the critical value of t to compare with the observed t. The output of the statistics package prints the p value, in this case $p = 0.007$. If the printed p value is less than the critical value of $\alpha = 0.05$, then we reject the null hypothesis. Using either the tabled value of t or the computed value of p will lead to identical conclusions with respect to rejecting or failing to reject the null hypothesis. In this book we will present the method of comparing the observed test statistic such as t against the tabled value of that test statistic and will illustrate the p value comparison to the critical α when presenting output examples from SPSS.

5.7 Nonrandom Assignment of Subjects

In the following subsections we consider conditions under which we may expect the population variances to differ. Indeed, it would be an error for a researcher to always assume equal population variances without checking data to see whether the homogeneity of variance assumption holds. A common culprit of unequal variances is when the participants are not assigned at random to the two groups. If one of the groups initially included participants who are more homogeneous in their performance than the other group, then we might expect the two groups to differ in variability at the conclusion of the experiment. Some variables used in behavioral research such as gender, race, and age do not entail random assignment to conditions. When participants are not randomly assigned to condition (such as when we wish to compare females and males) we not only limit our ability to draw causal inferences from the study, but the nonrandom assignment could also invalidate the underlying assumptions of the standard statistical tests (for example, the two groups may not have a common population variance).

5.8 Treatments That Operate Differentially on Individual Difference Variables

Consider a possible explanation for a difference between two population variances, σ_1^2 and σ_2^2. Suppose we find that s_2^2 is significantly greater than s_1^2. A possible explanation for this result is that a treatment operates differentially with respect to an individual difference variable. For example, if we have divided a group of research participants at random into two groups of n_1 and n_2 participants each, we would expect that these two groups, if tested under identical conditions, would not differ significantly in their means and variances for a particular dependent variable. For example, if we obtained a measure of anxiety about competition prior to the experiment, we would expect the two groups to show only a chance, or random difference, in their means and variances on this scale. This assumption should hold if we randomly assigned participants to the two groups.

Suppose, however, that one treatment operates differentially on participants with high anxiety scores as compared to participants with low anxiety scores. To be specific, let us assume that X is a measure of performance on an intelligence test and that the treatment consists of telling the participants that their performance is to be evaluated in comparison with the performance of other participants. Suppose that high- and low-anxiety participants react differentially to these instructions but we don't know this, and our treatment groups consist of a mixture of participants with low and high anxiety. In a later chapter we will discuss how to deal with this issue as an interaction between treatment and anxiety; however, this will require that we know in advance each participant's anxiety level, which in this present example we do not know. Assume, for example, that highly anxious participants tend to be considerably disturbed by the experimental instructions and, in turn, perform worse on the intelligence test than they would if tested under a control or normal condition. For these participants, we would have an observed score of $X - t_1$, where t_1 represents a constant-treatment effect being subtracted from performance. The idea underlying the difference $X - t_1$ is that the original score is reduced (performance is worse) for participants with high anxiety. Let us also assume that participants with a low degree of anxiety do their best under conditions of competitiveness. For these participants, let t_2 be a constant-treatment effect that increases performance. Then for the low-anxiety participants, we would have $X + t_2$; that is, their performance under the experimental condition would be better than their performance under a control condition.

With the assumptions we have made and an additional assumption that the degree of anxiety does not influence performance under a standard or control condition (that is, under a control condition the mean performance for high- and low-anxiety participants is the same), then in the treatment condition we would have a distribution of measures that extends in both directions from the mean as compared to the corresponding measures for the control condition. Because the X measures for the treatment group are being moved differentially in both directions relative to the control group mean (the high-anxiety participant decreases in performance, but the low-anxiety participant increases in performance relative to their respective controls), we would expect the treatment group to have a greater variance than the control group. If the researcher was unaware of the role of anxiety in this particular example, then he or she would not be able to correct for it and instead would simply observe that the variances in each group differ.

We cannot overemphasize the importance of individual difference variables in accounting for differences in the variability of participants under different experimental conditions. With random assignment, and when the treatment effects are additive, it seems that a plausible explanation for a significant difference in variances is that of the differential operation of a given treatment on an individual difference variable. In many behavioral experiments, it may be of considerable value to obtain measures of one or more relevant individual difference variables. To find that research participants with different values on an individual difference variable react differentially to a given treatment is of perhaps even greater psychological importance than to find that all participants respond to the treatment in the same manner. People differ and if those differences influence the variability of the measurements, then we should be careful about making blanket assumptions about equal variances. Note that random assignment does not help eliminate such differential response to an experimental treatment because this particular source of unequal variances is due to a mixture of different types of people in the same treatment group. Randomization will only guarantee that the relative mixtures of people with high and low anxiety will be comparable across conditions, but if people respond differently to the treatment, or the experimental manipulation, this may be a source of unequal variances.

Underwood (1975) argued that an individual difference approach can be used to test theory. If a theoretical model posits a process, and different groups of people are known to differ in their use of that process, then the theoretical model automatically predicts an individual difference that can be tested. Thus, under such conditions the failure to find an individual difference can be interpreted as evidence that can be used to falsify the theory. Likewise, the presence of such a predicted individual difference is evidence that supports, or is consistent with, the theory. Such a methodology relying on individual difference is commonly used in cognitive neuroscience, where normal participants are compared with patients whose cognitive functioning is generally intact except for one critical difference. This allows a test between two groups of research participants that differ in the use of a very specific psychological process, but the two groups are matched on all other accounts.

5.9 Nonadditivity of a Treatment Effect

Another reason the variances of two treatment groups may differ could be due to the treatment effects not being additive. By additive we mean that if X_1 is the value of a given observation under a control condition (treatment 1), then under the experimental condition (treatment 2) we would have $X_2 = X_1 + t_1$, where t_1 represents a constant-treatment effect. This means that the treatment adds a constant t_1 to the score that would have been obtained in the control condition. If a treatment effect is additive, then it can be shown that the two population variances are equal (that is, $\sigma_2^2 = \sigma_1^2$) because the addition of a constant to all scores (or the subtraction of a constant from all scores) has no influence on the variance. In the preceding section we saw that adding a constant to some scores and subtracting a constant from the remaining scores serves to increase the original variance of that condition.

More generally, when the treatment effect differs across research participants, then there is a possibility of a difference between population variances in the treatment groups.

On the other hand, suppose that the treatment, instead of acting in an additive fashion, acts in a multiplicative fashion. Then, we would have $X_2 = X_1 t_1$, and it can be shown that the variance of X_2 in relation to the variance of X_1 will be $\sigma_2^2 = \sigma_1^2 t_1^2$, because multiplying each value of a variable by a constant serves to multiply the original variance by the square of that constant. Therefore, if a treatment effect is multiplicative, the variance for this treatment group may differ from the variance of a control group or from the variance of another treatment group in which the treatment effect is additive. In cases where one variance is proportional to another, it may be useful to re-express, or transform, the raw data into the log scale, that is, making use of the property that $\log(xy) = \log(x) + \log(y)$. The log conveniently transforms a multiplicative operation between two variables into an additive operation.

5.10 Transformations of Raw Data

When two treatment variances are heterogeneous, it is sometimes possible that a transformation of the original scale of measurement will produce a new scale such that the variances are homogeneous on the transformed scale. We stated that a log transformation may help alleviate the problem of a multiplicative treatment effect. We briefly discuss three additional transformations. The interested reader may consult Hoaglin, Mosteller, and Tukey (1983) for more information on transformations.

The normal distribution has a special property that the mean and variance are independent of each other. Thus, we can test differences in means under the assumption of equal population variances quite naturally. When differences in variances occur, we look at other explanations such as individual difference variables or multiplication effects. However, not all distributions share this independence between the mean and variance—two such distributions are the Poisson and the binomial. In both cases, the variances are functionally related to the mean, so if a treatment is successful at changing the mean, then the treatment will also change the variance, possibly in dramatic ways. Thus, for these distributions a treatment that changes the mean will automatically violate the equality of variance assumption (except in some limited situations of symmetry, as described below). One way to deal with this problem is to transform the data into a scale where the mean and variance are, in effect, independent of each other.

5.10.1 Two Samples from Different Poisson Distributions

Poisson distributions are likely to be obtained when the observations consist of **counts**, such as the number of responses of some kind made in a fixed period of time. A classic example of the use of the Poisson distribution in behavioral science is the number of photons hitting the retina. The Poisson distribution is also relevant in modeling counts of relatively rare events such as the number of emergency room visits made by a class of undergraduate students over a 1-week period. In some cases the shape of the Poisson distribution can be approximated by a normal distribution.

For variables that have a Poisson distribution, the mean and variance are equal, that is, $\mu = \sigma^2$. If we have two treatments such that the dependent variable for Treatment 1 corresponds to one Poisson distribution with $\mu_1 = \sigma_1^2$ and the dependent variable for the other treatment corresponds to another Poisson distribution with $\mu_2 = \sigma_2^2$, and if the two sample means $\bar{X}_1$ and $\bar{X}_2$ differ, we also expect the sample variances s_1^2 and s_2^2 to differ. Thus, because for the Poisson distribution the mean equals the variance, any treatment that influences the mean will also influence the variance, thus creating the potential for unequal variances across treatments. Whether the observed differences in the sample means and sample variances lead to inferences about population means and population variances depends on the statistical power of the inferential test we use.

If the treatment means tend to be proportional to the treatment variances, as occurs when we have two Poisson distributions with different means, then a transformation of the original observations to a new scale may stabilize the variances. For the Poisson distribution, one may try a square root transformation. For example, when the dependent variable is positive, one may transform each value of X by taking $\sqrt{X+0.5}$. The 0.5 additive constant helps deal with frequencies near zero. Freeman and Tukey (1950) have suggested that the variance-stabilizing properties of the square root transformation are improved by using $\sqrt{X}+\sqrt{X+1}$. Given that count data are nonnegative, we won't have the problem of taking square roots of negative numbers, a problem that could arise when one tries to apply the square root transformation to noncount data.

Transformations may seem strange to someone who has never seen them before. It may feel as though the data are being "doctored." It turns out that such "re-expressions" of data (to use the term coined by the famous statistician John Tukey) are common in empirical science. For example, in physics it is common to plot data on "log–log" coordinates in order to simplify the problem to straight lines on the plot rather than more complicated power curves.

We present an example of counts to illustrate the effect of the square root transformation on Poisson data. An investigator observed the number of errors that people made when typing a three-page letter into a computer program. The experimental manipulation involved whether there was or was not time pressure to complete the letter. The left side of Figure 5.3 presents a boxplot of the original data. There are more errors in the group that had time pressure, but more importantly the spread of that group is much wider than the spread of the no time pressure group. The interquartile range of both groups differs, and the length of the whiskers is much longer in the time pressure group than in the no time pressure group. Thus, the assumption of homogeneity of variance is suspect. The right panel of Figure 5.3 shows the same data after applying the square root transformation suggested by Freeman and Tukey. The two boxplots now have spreads that are more

Figure 5.3: Data on counts comparing the number of errors for one group that had time pressure to another group that did not. The left panel presents boxplots of the raw data; the right panel presents boxplots of the same data after applying the square root transformation.

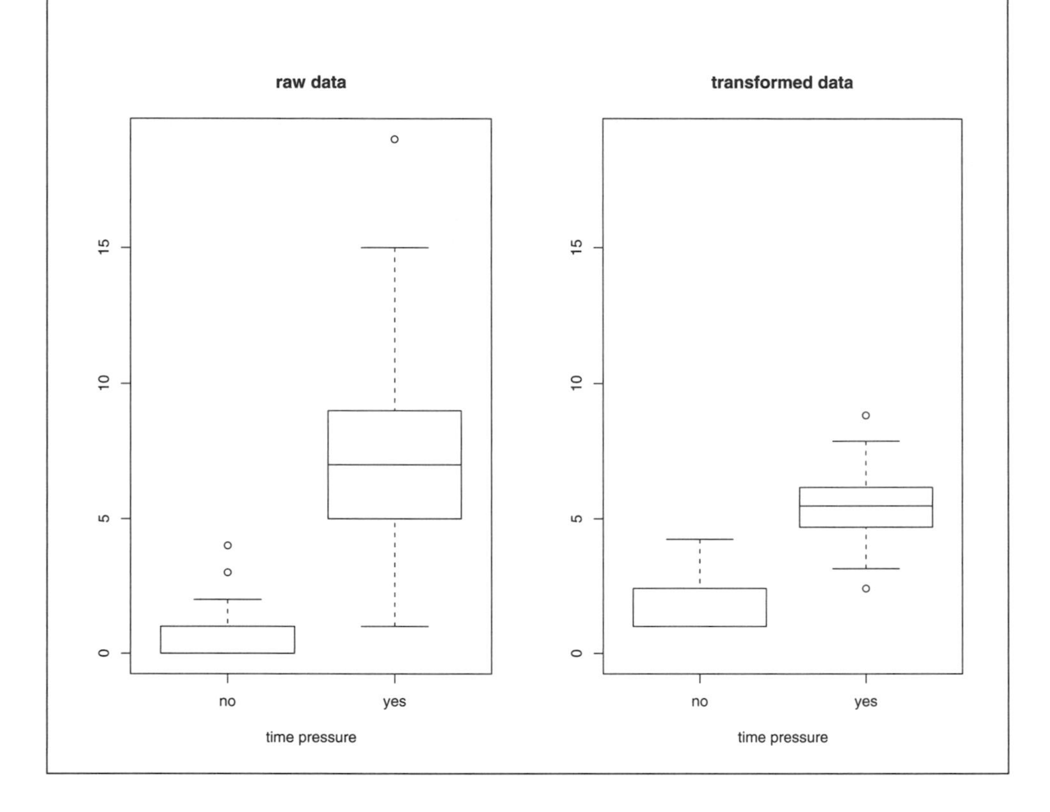

similar, suggesting that in this transformed scale the assumption of equal variances is more plausible. The reason transformations work is that they operate differently at different ends of the scale. For example, the square root transformation takes a score of 1 into a 1 (no effect, or the identity), a score of 25 into a 5, and a score of 100 into a 10. Thus, the higher the number, the more the effect of the transformation. Figure 5.4 illustrates graphically this differential effect of the square root transformation as scores get larger.

Figure 5.4: Illustration of how the square root transformation has more effect on larger positive numbers than on smaller positive numbers.

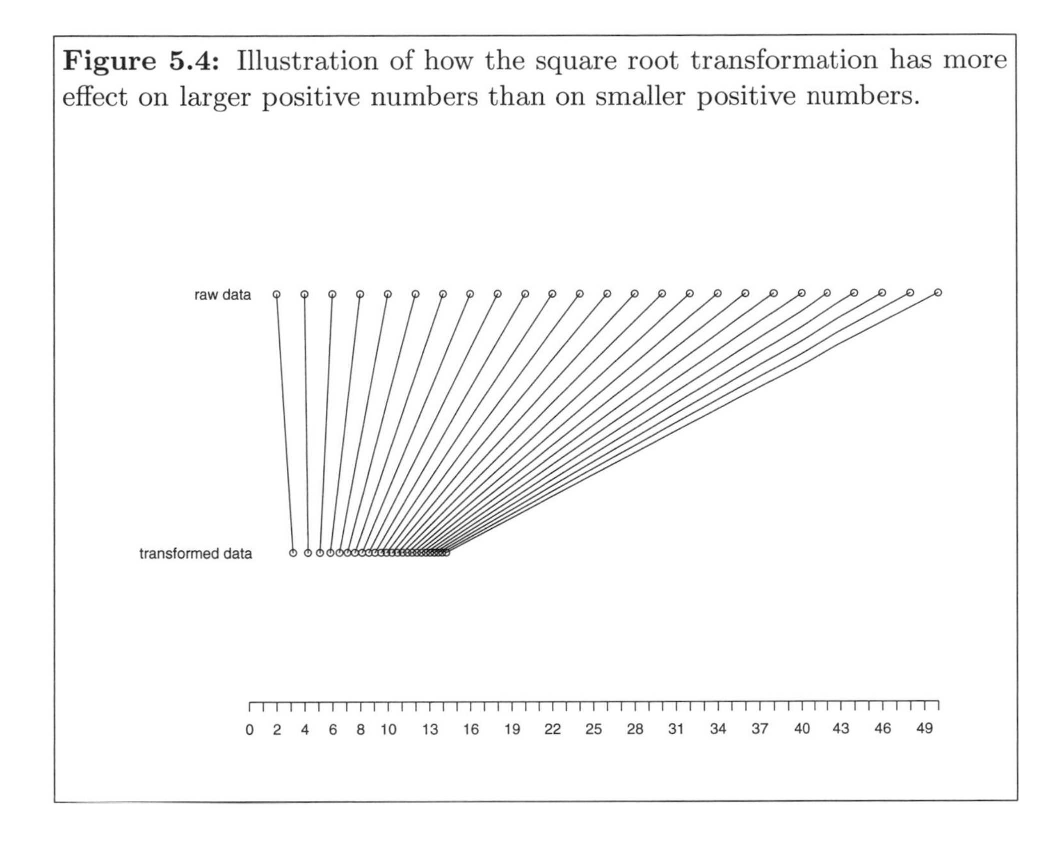

We illustrate the F test comparing the variance of these two groups. The variance for the no time pressure group is 0.498 and the variance for the time pressure group is 7.65. There are 500 observations in each group. Thus, the F test for the null hypothesis that these two groups have the same population variance is

$$\begin{aligned} F &= \frac{7.65}{.498} \\ &= 15.36 \end{aligned}$$

The F critical for a two-tailed test with $\alpha = 0.05$, 499 degrees of freedom in the numerator, and 499 degrees of freedom in the denominator is 1.19. The observed F of 15.36 exceeds the critical value, so we reject the null hypothesis of equal population variances. The statistical test on the variances in the raw scale confirms what is depicted in the left panel of Figure 5.3—we reject the null hypothesis that the population variance is the same in both experimental groups.

Figure 5.5 displays the SPSS syntax for creating a new transformed variable as well as the output for the t tests for both raw and transformed variables. The Levene test shows that the transformed variable had its desired effect because the variances are no longer statistical significant. The t test of the difference between two means was statistically significant in both the raw and transformed versions. Note that the value of the t test is more extreme for the transformed data (that is, leading to a more powerful test). Also, the discrepancy in the degrees of freedom between the t test version that assumes equal variance and the version that does not is much smaller for the transformed variable than it is for the raw data, another indicator that, for this example, the transformed variable better fits the statistical assumptions. Recall that the Welch t test adjusts degrees of freedom relative to the degree the equality of variance assumption is violated, so if the Welch test doesn't adjust the degrees of freedom very much, it is one indication that the assumption is met.

5.10.2 Two Samples from Different Binomial Populations

There is a different transformation for counts based on samples drawn from binomial populations. For example, we may have a binomial population in which the probability of a successful response P_1 is constant from trial to trial. We have a fixed number of n trials, and the dependent variable is the number of successful responses T_1. An example would be the total correct true-or-false questions answered by the research participants or the total number of correct choices by the farmer when given pairs of cans where one member of each pair contains water (as discussed in Chapter 2). Then for participants in this condition we have the expected total number $nP_1 = \mu_{T_1}$ and variance $nP_1Q_1 = \sigma^2_{T_1}$. If for another condition we assume the

Figure 5.5: Sample SPSS syntax and output illustrating a transformation and resulting t tests.

```
compute datasq = sqrt(data) + sqrt(data+1).
execute.

T-TEST
  GROUPS = group(1 2)
  /VARIABLES = data datasq
  /CRITERIA = CI(.95).
```

Independent Samples Test

		Levene's Test for Equality of Variances		t-test for Equality of Means						
									95% Confidence Interval of the Difference	
		F	Sig.	t	df	Sig. (2-tailed)	Mean Difference	Std. Error Difference	Lower	Upper
data	Equal variances assumed	429.281	.000	-54.146	998	.000	-6.91000	.12762	-7.16043	-6.65957
	Equal variances not assumed			-54.146	563.693	.000	-6.91000	.12762	-7.16066	-6.65934
datasq	Equal variances assumed	2.573	.109	-67.383	998	.000	-3.93024	.05833	-4.04470	-3.81578
	Equal variances not assumed			-67.383	947.437	.000	-3.93024	.05833	-4.04471	-3.81578

probability of a successful response P_2 is also constant from trial to trial and if we have the same number of trials as for the first treatment, then for participants tested under the second treatment, we have mean $nP_2 = \mu_{T_2}$ and variance $nP_2Q_2 = \sigma^2_{T_2}$. Obviously, the only way in which $s^2_{T_1}$ can be equal to $s^2_{T_2}$ is if $P_1 = P_2$ or if $P_1 = Q_2$ (the case when $P_1 = Q_2$ is when the two conditions are symmetric around 0.5 such as $P_1 = 0.2$ and $Q_2 = 0.8$). But, for example, if one treatment has a different probability than the other, such as $P_1 = 0.5$ and $P_2 = 0.9$ with $n = 100$ trials, then the population mean and variance for each treatment are $\mu_{T_1} = 50$ and $\mu_{T_2} = 90$ with $\sigma^2_{T_1} = 25$ and $\sigma^2_{T_2} = 9$. In this example, the two population variances are not equal $\sigma^2_{T_1} \neq \sigma^2_{T_2}$ because the two population proportions are not equal and the binomial has a relation between the mean and variance.

Instead of using T as the dependent variable, we could just as well use the observed proportion p $= T/n$, where n is the number of trials. The transformation that stabilizes the variance of a proportion is the arcsin of the square root of the proportion, which is expressed as

Figure 5.6: Illustration of the arcsin transformation for proportions.

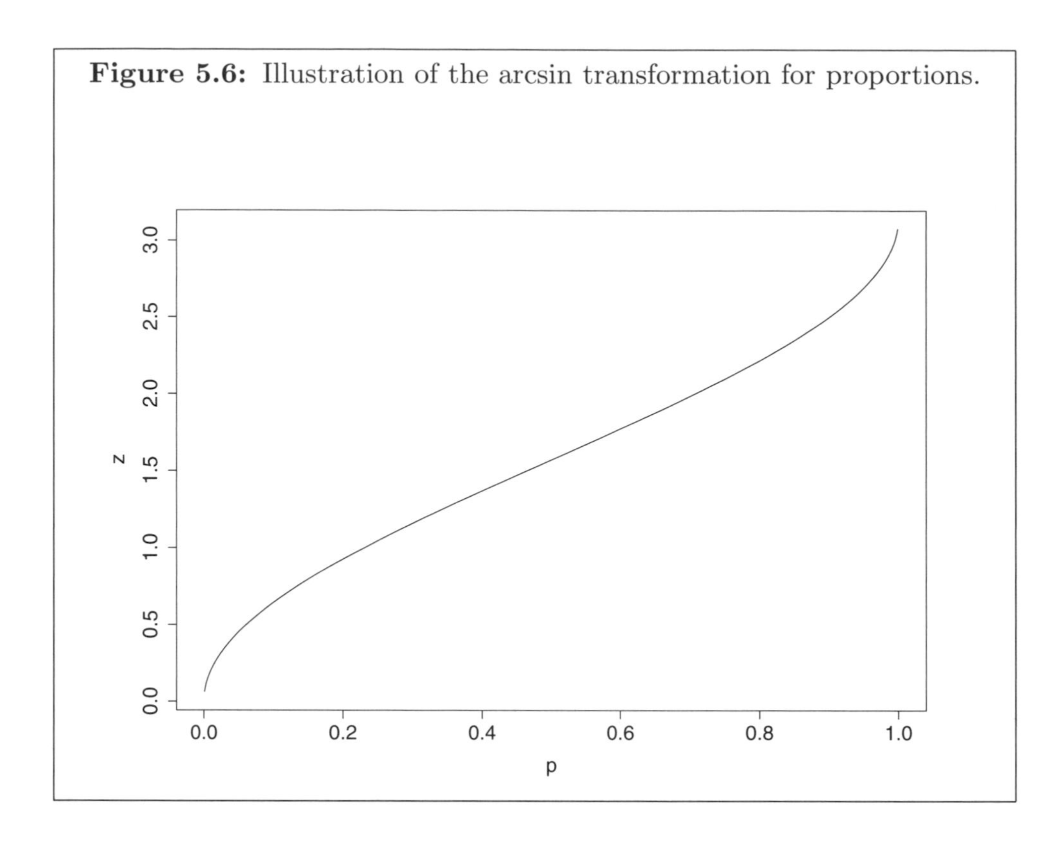

$$Z = 2\sin^{-1}\sqrt{\mathrm{p}}$$

This transformation is illustrated in Figure 5.6 as a function of the proportion p.

5.10.3 Heterogeneity of Variance: Time as a Dependent Variable

A reciprocal transformation $1/X$ may be useful in experiments where time is the dependent variable. For example, the reciprocal transformation may be useful in reaction-time experiments or in studies of problem-solving ability where the time taken to solve the problem is the dependent variable.

When time is a dependent variable, it may happen that for some of the participants in one treatment the time measure is quite long. For example, a few participants may take an excessively long period of time in arriving at a solution. The presence of a few extreme measurements will increase the variance for this treatment, and the variance of this treatment may be significantly greater than the variance for the other treatments. Transforming the time measures to $1/X$ may serve to make the variances more homogeneous on the transformed scale. Ratcliff (1993) presents a detailed account of dealing with outliers in reaction time data.

5.10.4 When Means Are Proportional to Standard Deviations: The Power Transformation

The **power transformation** X^b is a useful transformation in many statistical problems. Different values of the exponent b yield standard transformations. For example, when $b = 1/2$, then the power transformation is the square root transformation; when $b = -1$, then the power transformation is the reciprocal transformation; when $b = 0$, then we define the power transformation as the log. Thus, the power transformation provides a useful way to organize several transformations into one "family." We need a procedure for finding b for a data set that provides the best correction for the violation of the statistical assumption.

When the means and standard deviations are proportional to each other, there is a relatively simple procedure for finding a variance-stabilizing power transformation. Compute the log of the standard deviation for each treatment, and compute the log of the mean for each treatment. Plot the log standard deviation against the log mean for each treatment group, compute the regression line between these two measures, and record the slope of that regression line. The exponent b will be the quantity (1 − slope of the regression line). For example, if the slope of the regression line is 0.5, then the exponent of the power transformation will be $1 - 0.5 = 0.5$, which is equivalent to the square root transformation. This procedure was developed by Box and Cox (1964) and has come to be known as a **spread and level plot**, Some statistics programs such as SPSS implement this plot in a slightly different way: the log of the interquartile range (instead of the standard deviation) is plotted against the log of the median (instead of the mean) in order to make the procedure robust to outliers and skewness. This variant was originally suggested by Tukey (1977), as an attempt to present a plot that would be less sensitive to outliers and asymmetric distributions.

We present an example of the spread and level plot as defined in SPSS using the count data presented earlier in Section 5.10.1. We see that the slope of the regression line is 0.667 for a suggested power transformation of 0.333. Rounding off to the nearest 0.5, we have a suggested exponent for the power transformation of 0.5, which corresponds to the square root transformation. The spread and level plot from SPSS is displayed in Figure 5.7. The spread and level plot provides the analyst with a useful first pass at a potential transformation. The procedure works best when the plot of the variability against central tendency measures is linear. Of course, when there are only two conditions as in the examples in this chapter, there will always be a straight line. The spread and level plot becomes more interesting when there are more than two conditions in the study.

5.10.5 Transformation of Scale: General Considerations

Even though transformations may be helpful in situations where the variances are heterogeneous, transformations come at a cost. The researcher must realize that transformations change the scale and consequently may change the interpretation of the hypothesis being tested. Steiger (1980) made this point for the special case of binomial proportions, but the argument generalizes to any nonlinear transformation. Another side effect of transformations is that other aspects of the data may be altered when a transformation is performed. For instance, a transformation intended to sta-

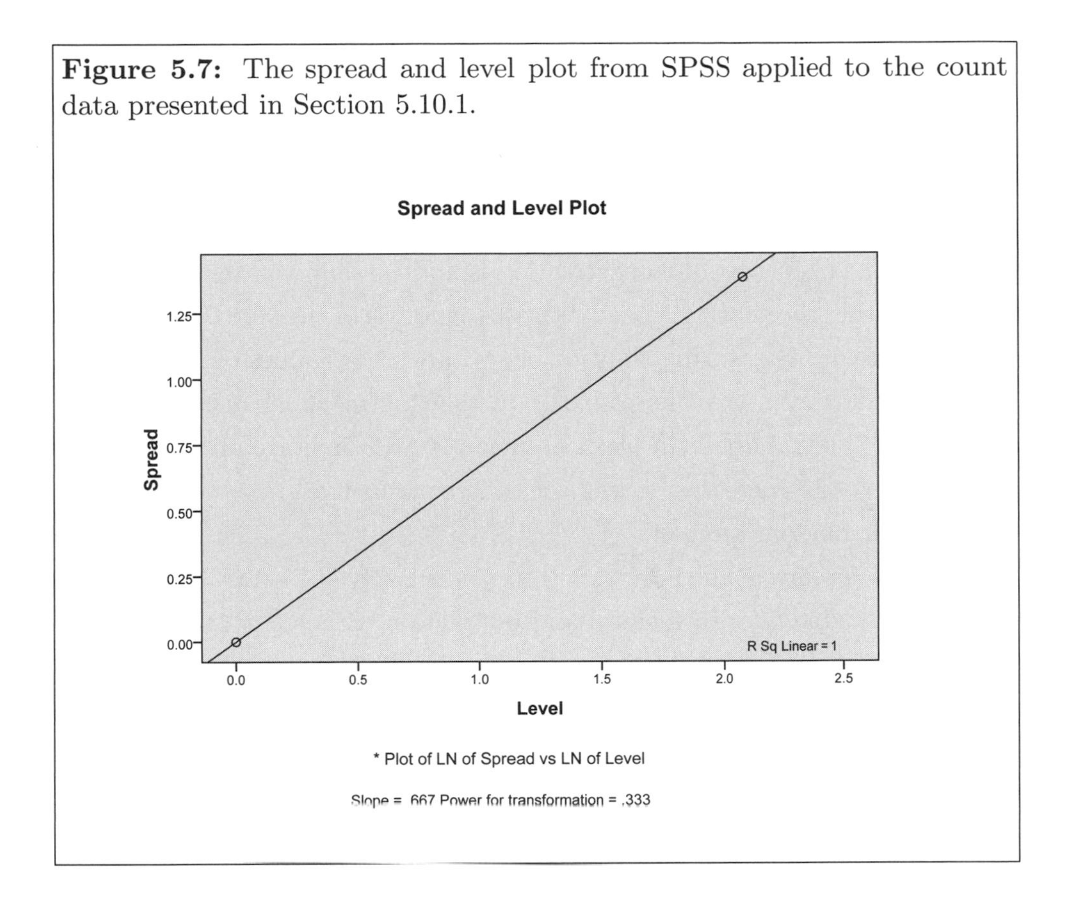

Figure 5.7: The spread and level plot from SPSS applied to the count data presented in Section 5.10.1.

bilize variance may have the negative side effect of changing the shape of the distribution (for example, a variable that appears to be normally distributed in the original raw scale may appear to have a different distribution on the transformed scale), so a transformation designed to "fix" the heterogeneity of variance problem may backfire by creating an asymmetric distribution in the transformed scale. Box and Cox (1964) were aware of this problem, and they developed their transformation as a simple way to address normality and equality of variance. Data analysis can be seen as a juggling act that requires proper balance between meeting the statistical assumptions of the test one would like to use and staying true to the meaningfulness of the scale of measurement. For a general discussion of transformations we suggest the edited volume by Hoaglin et al. (1983).

5.11 Normality

Although we have assumed that X is a normally distributed variable, there is considerable evidence suggesting that the t test for the difference between two means is relatively insensitive to departures from normality in the distribution of X. The important consideration is not the shape of the distribution of X but rather the shape or form of the distribution of the sum T of n values of X, or equivalently, the distribution of the mean of n values of X. In experimental work, the interest is primarily in the treatment means. As we saw in Chapter 3 for random samples, the mean $\bar{X}$ will be normally distributed as the sample size increases, as will the differences between the means of two independent random samples.

This is an important point worth repeating. By the central limit theorem for a variable X with mean μ and population variance σ^2, regardless of the shape or form of the distribution of X, the distribution of the mean $\bar{X}$ for random samples approaches that of a normal distribution as n increases. How large n must be before the distribution of $\bar{X}$ approaches that of a normal distribution depends on the original distribution of X. We do not get into the technical details of when the central limit theorem applies.

In Chapter 3, we considered three different distributions of X. One distribution was U-shaped, another was uniform, and the third was skewed. We showed that even with a relatively small number of observations the distribution of the sum T, and consequently the mean $\bar{X}$, for random samples from these three populations no longer resembled the population distribution. It has been our experience that whenever the distributions in each condition are symmetric, then the t test between two independent means is usually robust even when sample sizes are relatively small. However, Wilcox (1995) has argued that in some cases quite minor violations of normality can have a detrimental effect on standard significance tests.

Still, the t test is a relatively robust test in most situations. A test of significance is described as "robust" if it is mostly insensitive to small violations of its assumptions. The t test is relatively insensitive to nonnormality of distribution and heterogeneity of variance, provided that (1) the sample sizes are equal, that is, $n_1 = n_2$, (2) we have a sufficient number of observations for each treatment, and (3) for skewed populations we make a two-sided test of significance. With $n_1 = n_2 \geq 25$, we can be reasonably confident that a statistically significant value of t obtained from the t test offers adequate assurance that the population means are not equal, $\mu_1 \neq \mu_2$, despite the fact that the variable X may not be normally distributed, provided that the departures from normality are of the same kind for both treatment populations.

Under the conditions described above, there is considerable evidence to show that the t test is a robust test with respect to Type I errors. It is also important that a test be robust with respect to Type II errors; that is, if the null hypothesis is false, then we would hope that the power of the test when assumptions are violated would be approximately the same as when the assumptions are met. Some limited investigations into the robustness of the t test with respect to Type II errors have been made under conditions of nonnormality and heterogeneity of variance. The evidence, although limited, indicates that the t test is also robust with respect to Type II errors (Donaldson, 1968; Tiku, 1971).

Researchers concerned about the distributions of their data might consider the graphical aid of the normal probability plot that is now implemented in many statistical packages. This plot pits the observed data against what would be expected under a normal distribution (or any other distribution of choice). We will not make use of the normal probability plot in this book; for details of this plot see Cleveland (1993).

5.12 Summary

This chapter reviewed the important topic of statistical assumptions. Too frequently researchers do not pay adequate attention to whether or not their data satisfy the assumptions of the statistical tests they use. One must realize that the theory underlying statistical tests requires particular assumptions to be met for the theory to work appropriately. The computation of confidence intervals and p values depends on those statistical assumptions holding.

A major limitation of using statistical tests to check assumptions of other statistical tests is that, as with any inferential test, power increases with sample size. Thus, with a large enough sample size it is possible to reject relatively small, even trivial, deviations from the assumptions. The flip side of this argument also holds: with a small enough sample size it is possible to fail to reject very gross violations of the assumptions.

Our recommendation is that researchers use diagnostic plots such as the boxplot to assess the statistical assumptions. These plots will help answer important questions. Are there outliers? Do the two groups show comparable variability? Are the data in each treatment group symmetric about the treatment median? We also recommend that researchers use statistical tests that are robust to minor violations of the homogeneity of variance assumption such as the separate variance t test presented in this chapter.

5.13 Questions and Problems

1. An experiment was conducted where children were randomly assigned to watch one of two music videos. The videos differed in the amount of violence that was depicted. Children were then observed in play, and aggressive behavior was measured (higher scores correspond to more aggressive behavior). The following scores were obtained for the low- and high-violence groups.

Low-violence music video		High-violence music video			
11	15	4	15	10	10
11	10	4	3	7	12
10	8	8	13	6	14
12	10	9	9	1	8
8	8	12	9	5	5

(a) Test the null hypothesis that $\sigma_1^2 = \sigma_2^2$ with $\alpha = 0.05$.

(b) Test the null hypothesis that $\mu_1 = \mu_2$ with $\alpha = 0.05$, using the separate variance t test presented in Section 5.6.

(c) Summarize your answer to both parts (a) and (b) in a short paragraph describing the tests you performed, the questions you tested, and the statistical conclusion you reached.

2. Another investigator replicated the same study in Question 1 and obtained these data:

Low-violence music video				High-violence music video			
12	10	16	11	15	4	10	15
8	13	10	10	12	4	7	3
12	11	11	9	10	8	6	13
11	9	10	9	5	9	1	9
11	11	9	7	8	12	5	9

(a) Test the null hypothesis that $\sigma_1^2 = \sigma_2^2$ with $\alpha = 0.05$.

(b) Test the null hypothesis that $\mu_1 = \mu_2$ with $\alpha = 0.05$.

(c) Summarize your answer to both parts (a) and (b) in a short paragraph, describing the tests you performed, the questions you tested, and the statistical conclusion you reached.

3. Explain in your own words how differential processes can interact with individual differences to produce different observed variances across treatments. Give an example.

4. Explain the features to look for in a boxplot that signal a violation of the assumption of homogeneity of population variance and a violation of symmetry.

5. Discuss the benefits and drawbacks of transformations.

6

The Analysis of Variance: One Between-Subjects Factor

6.1 Introduction

We have already discussed the application of the t test to research questions involving the difference between the means of two independent random samples. We now generalize this method to test the significance of the differences between two or more means. The technique is known as the **analysis of variance**.

The early development of the analysis of variance as a powerful tool in experimental and research work was largely the accomplishment of Ronald A. Fisher and his associates. Fisher (1934, p. 52) had this to say about the analysis of variance:

> We were together learning how to use the analysis of variance, and perhaps it is worthwhile stating an impression that I have formed—that the analysis of variance, which may perhaps be called a statistical method, because that term is a very ambiguous one—is not a mathematical theorem, but rather a convenient

> method of arranging the arithmetic. Just as in arithmetical textbooks—if we can recall their contents—we were given rules for arranging how to find the greatest common measure, and how to work out a sum in practice, and were drilled in the arrangement and order in which we were to put the figures down, so with the analysis of variance; its one claim to attention lies in its convenience. It is convenient in two ways: (1) because it brings to the eyes and to the mind a summary of a mass of statistical data in which the logical content of the whole is readily appreciated. Probably everyone who has used it has found that comparisons which they have not previously thought of may obtrude themselves, because there they are, necessary items in the analysis. (2) Apart from aiding the logical process, it is convenient in facilitating and reducing to a common form all the tests of significance which we may want to apply. I do insist that its claim to attention rests essentially on its convenience.

Fisher's own view about the analysis of variance should put this chapter, and the rest of the book for that matter, in perspective. We will be learning techniques that help organize data in a convenient way. The techniques will facilitate interpretations that are made from the data; interpretation is one of the ultimate goals of data analysis. The testing of statistical significance is merely a step that helps compare an observed result against a standard that has been accepted by the scientific community. The preceding two chapters showed the two-sample t test where one compares the observed t against the critical, tabled value of t, which serves as the standard accepted by the scientific community.

To make the following discussion concrete, we illustrate the analysis of variance for a **one-way between-subjects design** in which 40 participants have been assigned at random to one of $k = 5$ treatments with $n = 8$ participants for each treatment. For instance, there could be five different conditions that we want to assess the ability to detect water, so we randomize 40 individuals into one of the five conditions. A condition could be a particular type of training program, or the use of different tools such as a whale bone, a dog bone, a broom stick, a steel rod, and a shovel. The goal of such a study is to compare the means of the five conditions in a single study. This generalizes what we presented in the preceding two chapters, which focused on studies involving only two conditions. We can now handle the statistical comparison of any number of treatment means.

We say "one-way" because there is one factor that has been manipulated (as opposed to many factors, which we will cover in later chapters) and "between-subjects" because each of the five treatments has a different group of participants; in this example, the 40 participants are randomly assigned into the five conditions, so there are 8 participants in each condition.

The relatively simple formulas presented in this chapter require an equal number of participants per treatment, that is, equal sample sizes across conditions. We present the special case of equal sample sizes for clarity and will discuss complications due to unequal sample sizes in a later chapter—the formulas vary slightly when there are unequal sample sizes across the conditions. The formulas presented here are definitional, and the reader is encouraged to study the formulas carefully because they illuminate the underlying concepts.

6.2 Notation for a One-Way Between-Subjects Design

In Table 6.1, we introduce notation for the observations in a one-way between-subjects design with 5 treatments and 8 observations for each treatment. We will find this notation convenient in the analysis of variance. Each observation in the table is identified by two subscripts, the first corresponding to a particular treatment and the second to a particular observation for the treatment. For example, X_{32} is the second observation for treatment 3. We let X_{kn} be a general symbol for any observation. In words, we say the nth observation in the kth treatment. In the example depicted in Table 6.1, the

Table 6.1: Data arrangement for a one-way between-subjects design with $k = 5$ treatments and $n = 8$ observations for each treatment.

Treatments				
1	2	3	4	5
X_{11}	X_{21}	X_{31}	X_{41}	X_{51}
X_{12}	X_{22}	X_{32}	X_{42}	X_{52}
X_{13}	X_{23}	X_{33}	X_{43}	X_{53}
X_{14}	X_{24}	X_{34}	X_{44}	X_{54}
X_{15}	X_{25}	X_{35}	X_{45}	X_{55}
X_{16}	X_{26}	X_{36}	X_{46}	X_{56}
X_{17}	X_{27}	X_{37}	X_{47}	X_{57}
X_{18}	X_{28}	X_{38}	X_{48}	X_{58}
$\bar{X}_{1.}$	$\bar{X}_{2.}$	$\bar{X}_{3.}$	$\bar{X}_{4.}$	$\bar{X}_{5.}$

first subscript can take any value from 1 to 5, because there are 5 treatments, and the second subscript can take any value from 1 to 8, because there are 8 observations for each treatment. The reader should be careful when comparing formulas across different textbooks because some authors use the notation X_{ij} to refer to the jth participant in the ith treatment (as we do), but other authors reverse the meaning of the row and column and instead use X_{ij} to refer to the ith participant in the jth treatment. The choice is merely a matter of preference, but as always readers need to be mindful of notation.

The sum of all kn ($5 \times 8 = 40$) observations will be represented by $X_{..}$, where the dots that replace the subscripts kn indicate that we have summed all kn values of X_{kn}. Similarly, we let $\bar{X}_{..}$ be the mean of all kn observations. The five treatment sums are represented by $X_{1.}$, $X_{2.}$, $X_{3.}$, $X_{4.}$, and $X_{5.}$, where the dot that has replaced the subscript n denotes that we have summed over the n observations in that particular treatment. Then, $X_{k.}$ will be a general symbol for any given treatment sum, and $\bar{X}_{k.}$ will be a general symbol for a particular treatment mean. In some cases, we will want to refer to two different treatments without necessarily specifying which two. In this instance, we will use $X_{i.}$ and $X_{j.}$ for the two sums in treatments i and j, and $\bar{X}_{i.}$ and $\bar{X}_{j.}$ for the two treatment means.

6.3 Sums of Squares for the One-Way Between-Subjects Design

We now write the following identity:

$$X_{kn} - \bar{X}_{..} = (X_{kn} - \bar{X}_{k.}) + (\bar{X}_{k.} - \bar{X}_{..}) \tag{6.1}$$

Let us make sure that you understand precisely, in terms of our notation, what Equation 6.1 means. The difference to the left of the equals sign says that the deviation of any observation (X_{kn}) from the mean of all observations in the study ($\bar{X}_{..}$) can be expressed as the sum of two parts: the deviation of that same observation (X_{kn}) from the mean of the treatment group ($\bar{X}_{k.}$) to which X_{kn} belongs plus the deviation of that same group mean ($\bar{X}_{k.}$) from the overall mean ($\bar{X}_{..}$). For example, for X_{11}, the first observation in treatment 1, we have the equality

$$X_{11} - \bar{X}_{..} = (X_{11} - \bar{X}_{1.}) + (\bar{X}_{1.} - \bar{X}_{..})$$

This may seem like a silly move because all we have done is add and subtract $\bar{X}_{1.}$ on the right-hand side, but it turns out that it is useful conceptually. If we square both sides of Equation 6.1 and sum over the n observations for the kth treatment, then for any given treatment we have a decomposition of the sum of square deviations from the grand mean:

$$\begin{aligned}\sum_{i=1}^{n} (X_{ki} - \bar{X}_{..})^2 = \sum_{i=1}^{n} (X_{ki} - \bar{X}_{k.})^2 + 2 (\bar{X}_{k.} - \bar{X}_{..}) \sum_{i=1}^{n} (X_{ki} - \bar{X}_{k.}) \\ + n (\bar{X}_{k.} - \bar{X}_{..})^2\end{aligned} \tag{6.2}$$

We can simplify this complicated expression because one term equals 0; that is,

$$\sum_{i=1}^{n} (X_{ki} - \bar{X}_{k.}) = 0$$

It is always the case that the sum of the deviations of n observations from their corresponding treatment mean will equal zero. Therefore, Equation 6.2 simplifies to

$$\sum_{i=1}^{n} (X_{ki} - \bar{X}_{..})^2 = \sum_{i=1}^{n} (X_{ki} - \bar{X}_{k.})^2 + n\,(\bar{X}_{k.} - \bar{X}_{..})^2 \qquad (6.3)$$

There is an Equation 6.3 that corresponds to each treatment (namely, every value of the first subscript). We then sum over the k treatment groups to get an important result:

$$\sum_{j=1}^{k}\sum_{i=1}^{n} (X_{ji} - \bar{X}_{..})^2 = \sum_{j=1}^{k}\sum_{i=1}^{n} (X_{ji} - \bar{X}_{j.})^2 + n\sum_{j=1}^{k} (\bar{X}_{j.} - \bar{X}_{..})^2 \qquad (6.4)$$

or, in words, total sum of squares = within-treatment sum of squares + treatment sum of squares.

The term on the left-hand side of Equation 6.4 measures the variation of the kn observations about the mean of all kn observations and is called the **total sum of squares**. The term "sum of squares" is actually a shortened phrase. The full phrase is "sum of squared deviations from the mean." As Equation 6.4 shows, the total sum of squares can be separated into two parts, corresponding to the two terms on the right of Equation 6.4. The first term on the right measures the variation of the n observations in each treatment group about the mean for that particular treatment group, and this sum of squares is called the **within-treatment sum of squares**. The second term on the right measures the variation of the treatment means about the mean of all kn observations weighted by n, the number of observations in each treatment group; it is called the **treatment sum of squares**.

We have shown that whenever we have k groups with n observations per cell it is always possible to decompose the total sum of squares into two parts: the within-treatment sum of squares and the treatment sum of squares. This decomposition leads to the comparison of two estimates of

Figure 6.1: Heuristic pie chart illustrating that the total sum of squares is a sum of two components: the treatment sum of squares and the within-treatment sum of squares. SS is an initialization for sum of squares.

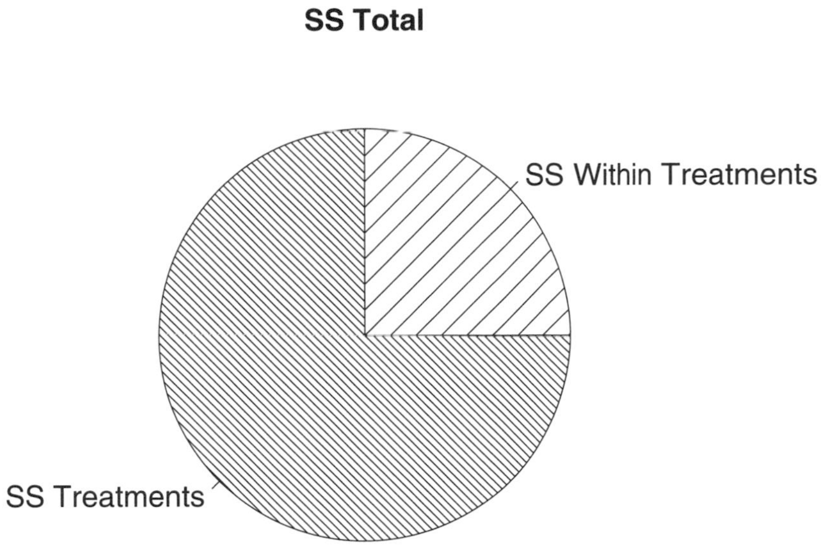

SS Total

SS Within Treatments

SS Treatments

variability: an estimate of the variability of the observed scores within each treatment and an estimate of the variability of the treatment means.

It is helpful to illustrate this decomposition with a pie chart (see Figure 6.1). The entire pie chart represents the total sum of squares. The two pieces that constitute the pie are the within-treatment sum of squares and the treatment sum of squares.

The decomposition of total sum of squares suggests a simple measure that indexes the variability across treatments. Letting SS denote sum of squares, the simple index

$$R^2 = \frac{SS_{treatment}}{SS_{total}}$$

gives the proportion of total variability that is accounted for by knowledge of the variability across the treatment means. That is, R^2 provides the proportion of total variability that can be attributed to variability in the treatment means (see Figure 6.1). The R^2 index is widely used in behavioral research even though it is well known that this index is slightly biased (see Hays, 1988).

The degrees of freedom decompose in a similar way as the sum of squares. The degrees of freedom associated with the total sum of squares will be $kn - 1$. The within-treatment sum of squares will have $k(n - 1)$ degrees of freedom, and the treatment sum of squares will have $k - 1$ degrees of freedom. Note that, in a manner analogous to the sum of squares, the degrees of freedom within treatments plus the degrees of freedom for treatment equals the total degrees of freedom. That is, $k(n - 1) + (k - 1) = kn - 1$.

The idea of decomposing the total sum of squares is the fundamental idea of the analysis of variance. The different types of designs reviewed in this book differ in their specific decomposition, but they all have the general form of a within-treatment sum of squares on the one hand and treatment sum of squares on the other hand.

6.4 One-Way Between-Subjects Design: An Example

We illustrate these concepts with a concrete numerical example. In Table 6.2, we give the values of a dependent variable for a one-way between-subjects design with $k = 5$ treatments and with $n = 8$ participants assigned at random to each treatment. For instance, students could be assigned to one of five possible exercise programs. The dependent variable could be some

Table 6.2: Randomized group design with five treatments and eight subjects randomly assigned to each treatment.

	Treatments					
	1	2	3	4	5	
	16	16	2	5	7	
	18	7	10	8	11	
	5	10	9	8	12	
	12	4	13	11	9	
	11	7	11	1	14	
	12	23	9	9	19	
	23	12	13	5	16	
	19	13	9	9	24	
$\overline{X}$	14.5	11.5	9.5	7	14	(overall mean = 11.3)

measure of overall fitness after a 4-week period. The researcher might be interested in finding whether the five exercise programs produced different mean scores on the dependent variable, thus examining whether one treatment program produced better fitness results than another. In this case, the researcher does not have a hypothesis as to which exercise program is predicted to be better. Rather, the researcher is interested in exploring whether the five exercise programs produced different means on the fitness measure. A different researcher may have a more sophisticated hypothesis that the two exercise programs involving heavy cardiovascular activity produce better fitness than the three exercise programs in the study that do not involve heavy cardiovascular activity. In this case, the researcher is making a particular prediction for the pattern of means over the five exercise programs. Below we present the analysis for answering the general question "Is there a difference among the five treatment means?"; in subsequent chapters we will present techniques for testing specific hypotheses over the treatment means.

By examining the treatment means presented in Table 6.2, we see that exercise programs 1 and 5 produced relatively high fitness scores (means of 14.5 and 14, respectively) and exercise program 4 produced the worst result. Figure 6.2 presents a boxplot of the five treatments, and Figure 6.3 presents the means along with $\pm$ one standard error.

Figure 6.2: Boxplot representing data in the five treatments give in Table 6.2.

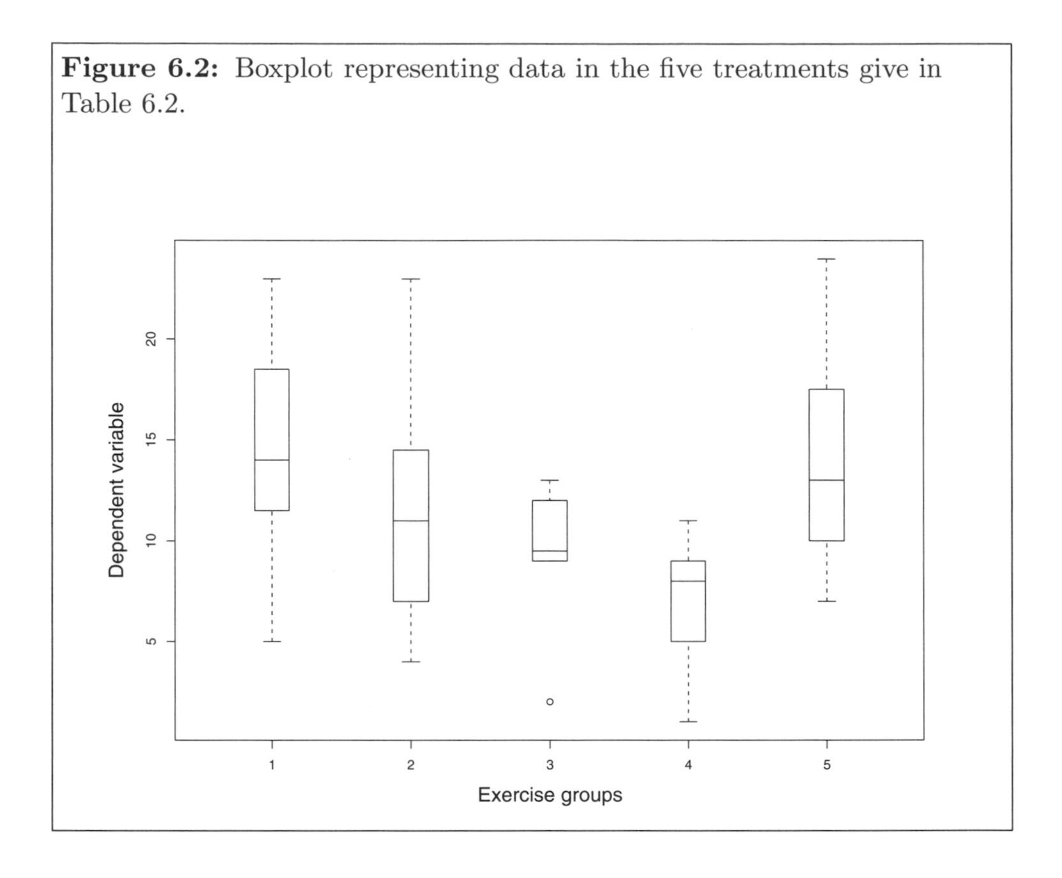

We need a statistical procedure to evaluate whether the likelihood that the difference in means across the five exercise programs could have been produced by chance under the null hypothesis that the five treatments have the same population mean. The analysis of variance is one such procedure that will help us answer this question. In order to perform the analysis of variance we need to know the variability within treatments and also the variability across treatments. We will illustrate the computation in detail so the reader can understand how the terms in an analysis of variance are defined.

The within-treatment sum of squares is a pooled sum of squares based on the n observations within each treatment. It is a measure of the variability within a treatment. As such, it is computed with respect to each separate treatment mean. For example, for each treatment group separately we have

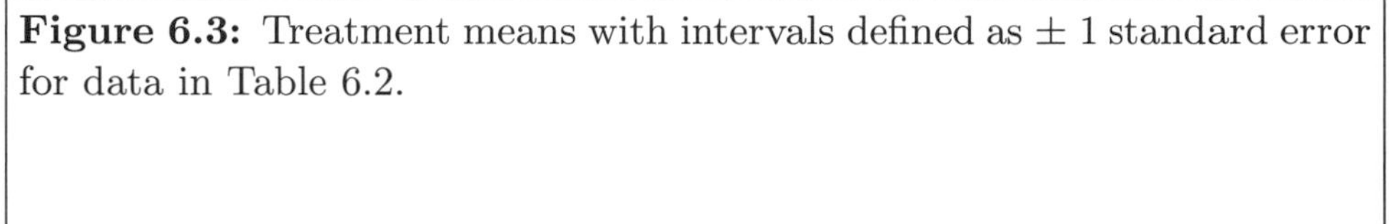

Figure 6.3: Treatment means with intervals defined as ± 1 standard error for data in Table 6.2.

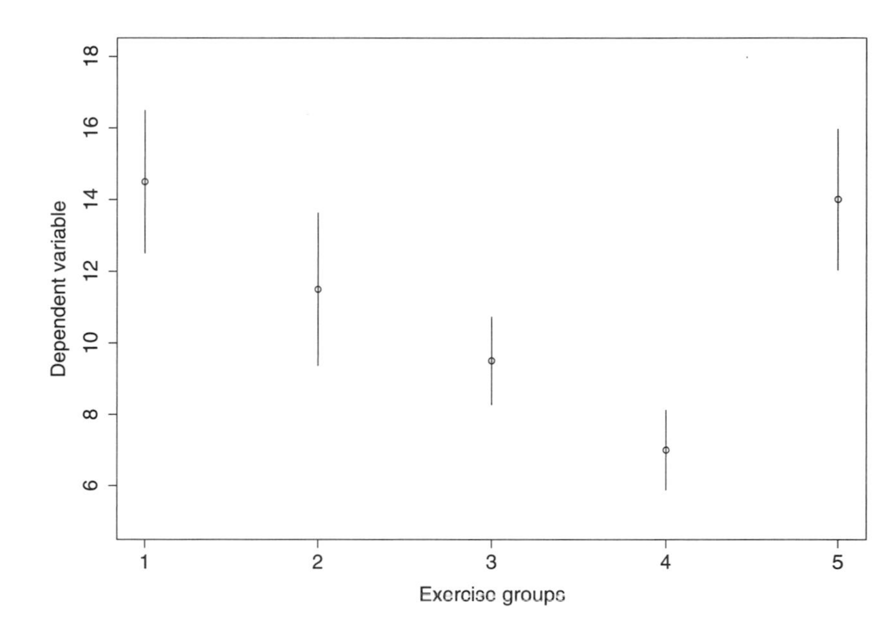

the following sum of squared discrepancies from the respective treatment mean:

$$\sum_{i=1}^{n} (X_{1i} - \bar{X}_{1.})^2 = (16 - 14.5)^2 + (18 - 14.5)^2 + \ldots + (19 - 14.5)^2 = 222$$
$$\sum_{i=1}^{n} (X_{2i} - \bar{X}_{2.})^2 = 254$$
$$\sum_{i=1}^{n} (X_{3i} - \bar{X}_{3.})^2 = 84$$
$$\sum_{i=1}^{n} (X_{4i} - \bar{X}_{4.})^2 = 70$$
$$\sum_{i=1}^{n} (X_{5i} - \bar{X}_{5.})^2 = 216$$

and the sum of these five sums of squares is equal to 846 (that is, $222 + 254 + 84 + 70 + 216$).

We now turn to the treatment sum of squares. Recall that the treatment sum of squares is a measure of the variability of the treatment means around the grand mean. For this example we have

$$n\sum_{j=1}^{k}(\bar{X}_{j.} - \bar{X}_{..})^2 = 8\left[(14.5 - 11.3)^2 + (11.5 - 11.3)^2 + \cdots + (14 - 11.3)^2\right]$$
$$= 314.4$$

We verify Equation 6.4 by showing that the within-treatment sums of squares plus the treatment sums of squares equals the total sums of squares. The total sums of squares is the sum of squared deviations of all observations from the grand mean $\bar{X}_{..}$, which in the example equals 11.3. The total sums of squares is $(16-11.3)^2+(18.5-11.3)^2+\cdots+(16-11.3)^2+(24+11.3)^2 = 1160.4$. The sum of the within-treatment sum of squares and the treatment sum of squares, $846 + 314.4$, is also equal to 1160.4. Thus, we verify that Equation 6.4 works as we said it would.

The results of all of these calculations are summarized in what is called a **source table** (other names are ANOVA table and ANOVA summary table). The source table for this example is displayed in Table 6.3. For the treatment sum of squares, we have $k - 1 = 5 - 1 = 4$ degrees of freedom, and dividing the treatment sum of squares by its corresponding degrees of freedom, we obtain a variance estimate called the treatment mean square, or MS_T. Similarly, if we divide the within-treatment sum of squares by its degrees of freedom, $k(n-1)$, we obtain another variance estimate called the within-treatment mean square, or MS_W. These two mean squares for the ongoing example are shown in Table 6.3.

Table 6.3: Analysis of variance source table for the one-way between subjects design—original data in Table 6.2.

Source of variation	Sum of squares	*df*	Mean square	F
Treatments	314.4	4	78.60	3.25
Within treatments	846.0	35	24.17	
Total	1160.4	39		

6.5 Test of Significance for a One-Way Between-Subjects Design

Here comes the major innovation of the analysis of variance (and the reason for calling this technique an analysis of variance rather than an analysis of means). Recall in Chapter 5 we presented a test for the equality of two variances. We will now use the F test not for purposes of checking the equality of variance assumption but for purposes of evaluating the equality of means across multiple treatment conditions. Look at the second-to-last column of the source table in Figure 6.3 (that is, a sum of squared deviations from the mean divided by the degrees of freedom). The two mean square (MS) terms are themselves variances. If the means across the five conditions are equal, then the two MS variance terms will tend to be similar values leading to a ratio of one. Thus, all the calculations so far in this chapter have been setting up a model where we can compare variances calculated in two different ways in order to assess the equality of means across five conditions.

For a one-way between-subjects design, we define F as the ratio

$$F = \frac{MS_T}{MS_W} \tag{6.5}$$

which is a ratio of two variance estimates. This F will have $k-1$ degrees of freedom for the numerator and $k(n-1)$ degrees of freedom for the denominator. The source table conveniently organizes the sum of squares, the degrees of freedom, and the MS terms so that we can keep track of the bookkeeping. For the source table in Table 6.3, we have

$$F = \frac{78.60}{24.17} = 3.25$$

with 4 and 35 degrees of freedom.

To evaluate the statistical significance of the obtained value of $F = 3.25$, we enter the column of the F table with degrees of freedom corresponding to the numerator of the F ratio and find the row entry corresponding to the degrees of freedom in the denominator. We do not have a row entry corresponding to 35 degrees of freedom in Table B.2.1 for $\alpha = 0.05$. In this setting we can move to a more conservative entry in the table (fewer degrees of freedom than we actually have). We see that the critical value for 4 and 34 degrees of freedom is $F = 2.65$. Thus, because our obtained value of $F = 3.25$ exceeds the critical value of 2.65, the F test is statistically significant at $\alpha = 0.05$.

The general null hypothesis tested by Equation 6.5 is that there are k independent random samples from the same normally distributed population. This assumption also implies that all k treatment population variances are identical. That is, the population variance of each treatment group is identical. This is analogous to the equality of variance assumption discussed in Chapter 5 for the case of two groups now generalized to any number of groups two or greater.

Under the null hypothesis, each of the k treatment means is assumed to have the same population mean:

$$E\left(\bar{X}_{k.}\right) = \mu \text{ for all } k \tag{6.6}$$

or, stated differently, the five population means are equal to each other:

$$\mu_1 = \mu_2 = \mu_3 = \mu_4 = \mu_5$$

If the equality of population variances across the k treatments holds, then MS_W is an unbiased estimate of the common population variance σ^2. If the k groups have the same population variance and Equation 6.6 (the k population means are equal) are both true, then it can be shown that MS_T is also an unbiased estimate of σ^2. However, if Equation 6.6 is not true (that is, the means of the five treatments differ), then the expected value of MS_T will exceed the expected value of MS_W. Thus, the ratio of MS_T to MS_W serves as a test statistic for the null hypothesis that the k population treatment means are identical.

The tabled values of $F = MS_T/MS_W$ are those that would be expected to occur, at various levels of significance, when both the equality of population variance and Equation 6.6 (equality of population means) are true (that is, under the null hypothesis of equal population mean and the assumption of equal population variances across the treatments). Consequently, if the obtained value of F is greater than the tabled value for some defined level of significance, so that we may regard it as an improbable value when the general null hypothesis is true, we reject the null hypothesis that the k population means are identical. In doing so, we conclude that the treatment means are not all estimates of the same common population mean μ, under the assumption that the treatment variances are homogeneous.

6.5.1 Structural Model

Stated still another way, Equation 6.6 in the context of the equality of the population variance assumption suggests a model that each observation is determined by a common population mean μ plus a random noise parameter ε. In symbols, the null hypothesis of equal population means across the treatments implies a simple additive model for each observation

$$X_{ij} = \mu + \epsilon_{ij} \tag{6.7}$$

for each participant j in treatment i. That is, other than random error the observations of all kn participants are modeled as a function of a constant, the grand mean of all kn observations. This is the equation that results if the treatment has no effect on the means.

Note that a separate random noise parameter is defined for each participant in each treatment. In this situation, MS_W and MS_T both estimate the population variance σ^2. However, the situation when Equation 6.6 is false but the equality of population variances holds implies that for each observation there is an extra term α_i that influences the scores in a given treatment

$$X_{ij} = \mu + \alpha_i + \epsilon_{ij} \tag{6.8}$$

Note that the α in this equation is not the same thing as the Type I error criterion α used elsewhere in this book. Here the α denotes the effect of treatment and models the differences in population means.

The difference between the two models (Equations 6.7 and 6.8) is the presence of the treatment term α_i, which indexes the additive effect of treatments. If all k treatments produce the same effect on the dependent variable, then the treatment αs are taken to be equal to zero; but when the treatments differ from one another on the dependent variable, then the treatment αs have different values. We can show that MS_W estimates σ^2 but MS_T estimates the sum of two terms—the common population variance σ^2 plus another term that is a function of the α_i terms; that is, the expected value of the MS_T term is

$$E(MS_T) = \sigma^2 + \frac{n\sum_{1}^{k}\alpha_i^2}{k-1} \tag{6.9}$$

because the k different treatment α values in Equation 6.8 contribute to the observed variability across the treatments. If the null hypothesis of equal means is true, then all the treatment effects α_i are 0, and the second term on the right-hand side of Equation 6.9 is 0. Though one must realize that in a specific study even when the null hypothesis is true (at the level of the population means) because of sampling the observed treatment means will not all be exactly equal to one another. So, even when the null hypothesis is true, in a specific study the MS_T will not be exactly the same as the MS_W term. The F test will serve as a metric for deciding whether MS_T is sufficiently greater than MS_W to warrant rejecting the null hypothesis of equal population treatment means.

We point out an interesting interpretation of the F test that results from Equation 6.9. In expectation terms the F test can be written as follows, merely substituting in the previous equations:

$$F = \frac{E(MS_T)}{E(MS_W)} = \frac{\sigma^2 + \frac{n\sum \alpha_i^2}{k-1}}{\sigma^2}$$

The general characteristic of this formulation of expected mean squares in words is a ratio of parts that we "care about" and parts that are less important so that we essentially "don't care" about them; that is,

$$F = \frac{\text{error} + \text{treatment}}{\text{error}} = \frac{\text{don't care} + \text{care}}{\text{don't care}}$$

Much of the analysis of variance, as we will see in subsequent chapters, is about identifying the parts that are relevant to defining the treatment and the parts that are relevant to defining the error term.

6.5.2 Equality of Variance Assumption

We emphasize that the F test defined in Equation 6.5 is relatively insensitive to minor violations of the equality of variance assumption, provided that we have the same number of observations for each treatment. Consequently, if we have the same number of observations for each treatment and if we obtain a significant value of F, we will ordinarily be safe in concluding that the population means for the treatments are not all equal to the same value μ, even though it may also be true that the population variances for the treatments are not all equal to the same value σ^2. Box (1953) suggested that as long as the sample sizes are equal one can generally be safe in using analysis of variance even when the equality of variance is not met. For a different opinion on this matter see Wilcox (1993).

We discussed how to check the equality of two population variances in Chapter 5; we can use the same techniques when we have two or more treatment groups. For example, we can use the F test to compare variances, we can use the Levene test, we can use boxplots, etc. We recommend that researchers examine the variability within each treatment through the boxplot to check for possible violations of the equality of variance assumption. This recommendation should be especially heeded when the sample sizes across treatments differ greatly (for example, the ratio of the smallest sample size to the largest sample size is 6 or more). This "factor of six" rule is offered as a heuristic.

6.5.3 Equality between the Two-Sample *t* Test and the *F* Test for Two Groups

There is an interesting relation between the two-sample t test we presented in Chapter 4 and the generalization to more than two groups that we present in this chapter. In the special case of two groups, the analysis of variance F test presented in this chapter is equal to the square of the two-sample t test. That is, when there are two treatments, $F = t^2$.

To verify this claim we will use data from Chapter 4 (Table 4.1). Recall that in that chapter we found that $t = 2.86$. Table 6.4 presents the source table for the analysis of variance for those data. We see that $F = 8.18$,

Table 6.4: Example illustrating SPSS syntax and output for the relation between F and t for the special case of two groups.

SPSS Syntax

```
UNIANOVA
  data  BY group
  /METHOD = SSTYPE(3)
  /INTERCEPT = INCLUDE
  /CRITERIA = ALPHA(.05)
  /DESIGN = group .
```

SPSS Output

Source of variation	Sum of squares	*df*	Mean Square	F
Treatments	90.0	1	90.0	8.18
Within treatments	4180	38	11.0	
Total	508.0	39		

which is equal to $t^2 = 2.86^2$. In addition, Table 6.4 presents SPSS syntax for computing a between-subjects analysis of variance.

One special detail about how to enter data for a between-subjects analysis of variance is that only two columns of numbers are needed to send to a statistics package such as SPSS. One column codes which treatment the subject is in (for example, treatments 1 through 5), and the second column includes the observed data that will be analyzed by the analysis of variance. So, if there are five treatments with eight participants in each treatment, there will be $5 * 8 = 40$ rows in the data table, with two columns such that one column codes which treatment that particular participant was in and the second column contains the observed data for that participant.

6.6 Weighted Means Analysis with Unequal *n*'s

In general, it is desirable to have an equal number of observations for each of the k treatments in a randomized group design. If participants drop out

Table 6.5: Outcome of an experiment with three treatments and with an unequal number of participants.

	Treatments		
	1	2	3
	3	5	2
	6	4	3
	3	6	4
		4	3
		6	
$\bar{X}$	4.0	5.0	3.0

Table 6.6: Weighted means analysis for the data in Table 6.5.

Source of variation	Sum of squares	*df*	Mean square	F
Treatments	8.92	2	4.46	3.35
Within treatment	12.00	9	1.33	
Total	20.92	11		

of the study and if the loss of participants is unrelated to the treatments, it is still possible to compute the analysis of variance to determine whether the treatment means differ significantly.

In Table 6.5, we give the results of an experiment with three treatments with three, five, and four observations in each treatment, respectively. It will be convenient here and in subsequent chapters to designate the total sum of squares by SS_{tot}, the treatment sum of squares by SS_T, and the within-treatment sum of squares by SS_W. The source table for these data appears in Table 6.6.

The definition for the within-treatment sum of squares is the same as before (Section 6.4). For a particular treatment, subtract the treatment mean from each observation, square the difference, sum over the participants in a treatment, and finally sum over all treatments. The convenient way to compute the degrees of freedom associated with the within-treatment sum of squares is to subtract the number of treatments k from the total number of participants in all the treatments combined (that is, $N = n_1 + n_2 +$

$\ldots + n_k$). The definition for the treatment sum of squares is a little tricky because it must take into account the different number of participants in each treatment. The general definition for the treatment sum of squares is

$$\sum_{j=1}^{k} n_j \left(\bar{X}_{j.} - \bar{X}_{..}\right)^2$$

The sample size n is now subscripted by the treatment number to account for the varying sample size in each treatment. When all k sample sizes are equal, this definition for the treatment sum of squares is identical to the one presented in Section 6.4. The degrees of freedom associated with the treatment sum of squares is still $k - 1$.

For the test of significance of the null hypothesis that the three population means are identical ($\mu_1 = \mu_2 = \mu_3$), we have the usual F ratio

$$F = \frac{MS_T}{MS_W} = \frac{4.46}{1.33} = 3.35$$

with 2 and 9 degrees of freedom, and this value is not statistically significant at $\alpha = 0.05$ because it does not exceed the critical F value of 4.27 from Table B.2.1.

The analysis of variance that we have illustrated for unequal sample sizes is called a **weighted means analysis**. The technique weights each squared deviation $(\bar{X}_{k.} - \bar{X}_{..})^2$ by the corresponding number of observations in that treatment. There are other methods for dealing with an unequal number of participants that become relevant in designs with more than one treatment factor. These methods will be presented in a later chapter.

6.7 Summary

This chapter presented the basic idea of analysis of variance: a decomposition of the variability (sum of squares) into a portion that is between treatments and a portion that is within treatments. The between-treatment variance is influenced by both error and treatment effects, whereas the within-treatment variance is assumed to be influenced only by error. Hence, if there is no treatment effect (as is assumed under the null hypothesis), then these two variances are independent estimates of the same term. The F test presented here compares these two estimates of variance. A significant F test suggests that there is a treatment effect over and above the estimate of noise, where the treatment effect is interpreted as the means differ across the k treatments.

In addition to the decomposition of the sum of squares term, we also presented a structural model that is a foundation for the analysis of variance. Each observed score can be modeled as a sum of the grand mean μ, a specific treatment effect α, and random noise ϵ. The null hypothesis implies that all the treatment effects α are equal to zero, so that when the null hypothesis is true the treatment means are equal to the same constant, the grand mean. The F test in the analysis of variance assesses the degree to which the treatment effects α differ from one another.

The analysis of variance is a generalization of the independent samples t test presented in Chapter 4. It permits an analysis of two or more treatment means. The F test by itself, however, is not completely informative in cases with three or more treatment groups because it merely "signals" a statistically significant difference between treatment means. To conclude that the k treatment means are not all estimates of the same population value μ is alone not very satisfying in most research settings. We would like to know something more specific about the nature of the differences among the means and which treatments differ from one another. In Chapters 7 and 8, we consider procedures that will be useful in testing specific hypotheses about the various population means.

6.8 Questions and Problems

1. Data for three treatment groups are given here:

Treatment group		
1	2	3
7	2	17
25	4	18
24	22	5
11	21	27
18	11	3

(a) Compute the treatment means.

(b) Write the relevant null hypothesis that will be tested in this ANOVA.

(c) Compute the source table and the value of $F = MS_T/MS_W$.

(d) Explain in words what this F ratio implies about the treatment means.

(e) Check the assumption of homogeneity of variance by using boxplots.

2. Participants were assigned at random to one of six treatments. The data for the experiment are presented in the following table:

Treatment group					
1	2	3	4	5	6
9	9	6	6	12	10
12	10	12	9	6	6
7	8	9	12	8	8
14	3	9	7	7	12
5	7	8	5	3	9
8	8	10	5	13	4
7	5	2	8	7	8
7	9	10	9	13	7
8	3	9	6	6	6
3	8	12	3	6	2

(a) Compute the treatment means.

(b) Write the relevant null hypothesis that will be tested in this ANOVA.

(c) Compute the source table and the value of $F = MS_T/MS_W$.

(d) Explain in words what this F ratio implies about the treatment means.

(e) Check the assumption of homogeneity of variance by using boxplots.

3. Participants were assigned at random to one of three treatments. The outcome of the experiment is given here:

Treatment group		
1	2	3
12	0	4
12	0	4
19	2	0
24	4	8
12	4	6
11	5	4
19	5	5
22	0	0
11	1	9
11	3	7

(a) Find the mean and variance for each treatment.

(b) Create a boxplot for each of the three treatments.

(c) Compute the value of $F = MS_T/MS_W$.

(d) Transform the data to the scale $\sqrt{X + 0.5}$. Compute the mean and variance for each treatment on the transformed scale. Has the variance been stabilized by the transformation (check with a new boxplot on the transformed scale)?

(e) Compute the value of $F = MS_T/MS_W$ for the transformed data. Compare this F ratio to the F ratio obtained with the untransformed data.

4. Explain intuitively why it is possible to regard the treatment mean square, when the null hypothesis is true, as an estimate of the common population variance.

5. We have two treatments with $n_1 = n_2 = 5$. Participants were randomly assigned to each treatment. The outcome of the experiment is given here:

Treatment group	
1	2
3	7
5	5
2	6
1	3
4	4

(a) Test the null hypothesis that $\mu_1 = \mu_2$ by using the t test.

(b) For the same data, compute $F = MS_T/MS_W$. Verify that $t^2 = F$.

6. We have six treatments with five participants randomly assigned to each treatment. Complete the analysis of variance, and report whether the observed F ratio is statistically significant at $\alpha = 0.05$. Explain in words what this F test tells us about the six treatment means.

Treatment group					
1	2	3	4	5	6
4	5	6	2	2	3
4	3	9	4	7	4
2	5	6	6	10	8
2	2	3	8	5	9
4	3	4	4	6	10

7

Pairwise Comparisons

7.1 Introduction

In the preceding chapter we presented the analysis of variance for determining whether treatment means differ statistically from each other. Recall that a statistically significant F test tells us that we can reject the null hypothesis that the treatment means are equal to each other. Such an F test is known as an **omnibus test**. A statistically significant omnibus test does not tell us which treatment means differ statistically from each other, which is usually the very question a researcher wants to answer. So, we need additional procedures to test specific research questions about differences, or patterns, among treatment means.

Consider the example comparing the five exercise programs presented in the preceding chapter. Suppose that a researcher analyzes those data by performing all possible comparisons between pairs of two treatments, using the pooled standard error. That is, the researcher performs a two-sample t test between exercise programs 1 and 2, another two-sample t test between exercise programs 1 and 3, a third two-sample t test between exercise programs 2 and 3, etc., until all possible comparisons between any two treatments are completed. Because the researcher assumes that the population variances for all five treatments are identical, he or she pools the standard errors across the five treatments. A simple way of estimating the pooled standard error is to take the square root of the mean square within treatment from the source table (that is, $\sqrt{MS_W}$). With five treatments, there are a total of 10 t tests that this researcher will perform. Does this procedure seem sensible to you? After all, this procedure will provide an answer to the question "Do the means across these five exercise programs differ from each other?" The researcher will know which means differ from each other by a series of all possible comparisons of two means.

This series of two-sample t tests is a sensible procedure as long as the assumption of equal population variance holds, except for one important detail: the chance of making at least one Type I error (that is, falsely rejecting the null hypothesis) is inflated. We develop this point below, but for now we want to return to the original reason for presenting this example. How should a statistically significant F test in the analysis of variance be interpreted? When the F test can be rejected, what reasonable statement can be made about which treatments produced different means?

The t tests for the 10 comparisons between any two of the five exercise programs (using a standard error pooled over the five treatments) can be computed using the formula

$$\frac{\bar{X}_i - \bar{X}_j}{\sqrt{MS_W(1/n_i + 1/n_j)}} \tag{7.1}$$

The subscripts i and j denote treatment. The 10 t values are presented in Table 7.1. Recall that in the exercise example we did not have reason to reject the homogeneity of variance assumption (that is, the five treatments appeared to have the same population variance σ^2) and that there were an equal number of participants in each of the five treatments.

It turns out that the omnibus F test is equal to the average of the square of all possible t tests between the treatments (that is, the average of the squared t's presented in Table 7.1). To verify, square the t's in Table 7.1

Table 7.1: Ten t tests between all possible pairs of $k = 5$ treatment means, using the pooled standard error.

Treatments	1	2	3	4
2	1.22			
3	2.03	0.81		
4	3.05	1.83	1.02	
5	0.20	-1.02	-1.83	-2.85

and take an average,

$$(1.22^2 + 2.03^2 + 3.05^2 + \ldots + (-1.83)^2 + (-2.85)^2)/10 = 3.25$$

which is the same value as the F observed in the preceding chapter.

This result gives a useful interpretation to the F test—it is an average of all possible comparisons between two treatments where the comparisons use the same pooled error term. This observation suggests that it is possible for a single pairwise comparison to be statistically significant, yet the omnibus F test may not reach statistical significance (the average is brought down by the nonsignificant t test). Thus, a significant omnibus F test tells us that the average pairwise comparison is statistically significant, but it does not inform us as to which mean or means differ from each other. Researchers are typically interested in making specific statements about treatments, and therefore the standard omnibus F test presented in Chapter 6 does not provide an answer to the researcher's question.

We will review methods for comparing several treatment means against each other using all possible pairwise tests. There is a wrinkle however that must be addressed. As the number of tests we perform increases, so does the chance of making a Type I error. Let's illustrate with an example. In an experiment involving k means (where k is 3 or more), if we test all pairwise differences between the means, we will make a total of $c = k(k-1)/2$ comparisons or tests of significance. For instance, in the exercise study $k = 5$ (five treatment means), so there are $c = 5(5-1)/2 = 10$ comparisons in all.

In this instance, if the general null hypothesis is true, we have the possibility of making anywhere from 0 to c Type I errors when testing the c comparisons.

Across the 10 comparisons, we let $P(E)$ represent the probability of at least one Type I error in the set of c comparisons. The emphasis here is on the phrase "at least one Type I error." Even though an individual test may have a small Type I error rate, the chances of making at least one Type I error increase in the aggregate. This problem is known as the **multiplicity problem**. In the special case of a study with two treatments (that is, $k = 2$), $c = 1$, so multiplicity is not an issue.

In this chapter, four methods are reviewed that could be used in testing the significance of all $c = k(k-1)/2$ pairwise comparisons for a set of k means. These methods differ in how they deal with the multiplicity problem. Some of these methods have more general applications; that is, they are not limited to making only pairwise comparisons. These more general applications will be discussed in subsequent chapters.

7.2 A One-Way Between-Subjects Experiment with 4 Treatments

Throughout this chapter we will use data from a single experiment to illustrate four techniques for performing pairwise comparisons. Consider a one-way between-subjects design with $k = 4$ treatments and with $n = 10$ participants randomly assigned to each treatment. The study returns to the problem of the farmer's ability to douse water. This experiment involves four groups: a standard water witching condition (real whalebone), a water witching condition with a plastic whalebone, a water witching condition with wooden sticks, and a water witching condition with no props (that is, no bones, no sticks). Participants were randomly assigned to conditions. The dependent variable is an accuracy score that measures for each participant whether they correctly identified those containers with water. Higher scores

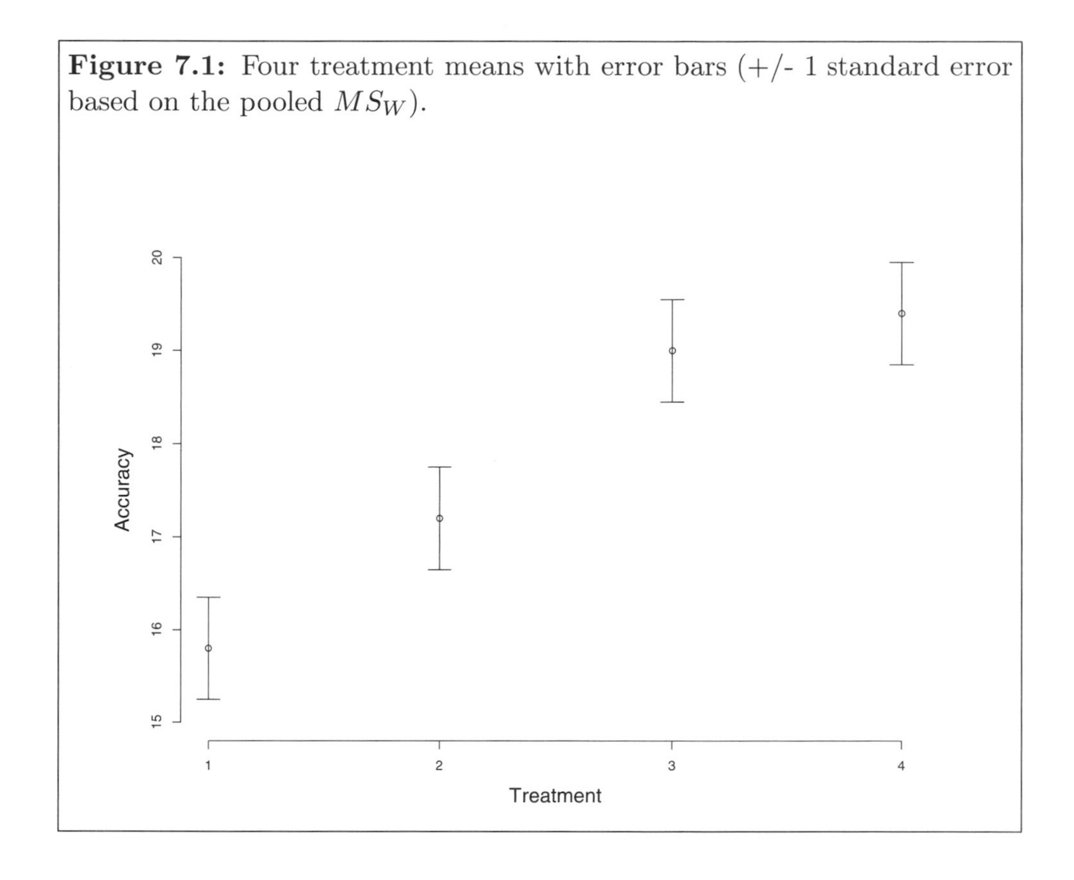

Figure 7.1: Four treatment means with error bars (+/- 1 standard error based on the pooled MS_W).

represent better accuracy. The statistical assumptions of equal variances and normality were checked and hold for these data.

The treatment means in this experiment are 15.8, 17.2, 19.0, and 19.4; see Figure 7.1. The summary of the analysis of variance for this experiment is given in Table 7.2. We have $F = MS_T/MS_W = 27.83/3.04 = 9.15$ with 3 and 36 degrees of freedom. For a change of pace we will use a strict Type I error criterion of $\alpha = 0.01$ in this example. The observed F of 9.15 exceeds the critical $F = 4.3957$ from Table B.2.2 in Appendix B (using the closest conservative degrees of freedom for the denominator of 35), so we reject the null hypothesis that the four population means are identical. Be sure you understand how the critical F is found in the F table.

Table 7.3 lists the values of the $k = 4$ treatment means in order of magnitude and also gives the differences between the pairs of means, $\bar{X}_{i.} - \bar{X}_{j.}$

Table 7.2: Summary of the analysis of variance for $k = 4$ treatments with $n = 10$ participants assigned at random to each treatment.

Source of variation	Sum of squares	*df*	Mean square	*F*
Treatments	83.50	3	27.83	9.15
Within treatments	109.44	36	3.04	
Total	192.94	39		

for $i > j$. This table provides another way of computing the treatment sum of squares. If we square the values of $\bar{X}_{i.} - \bar{X}_{j.}$, weight the squared differences by n_i and n_j (the number of observations on which each mean is based), sum those squared differences, and then divide by N (the total number of observations in the study), the result will also be equal to the treatment sum of squares. More generally, the formula can be written as

$$SS_T = \frac{\sum_i \sum_j \left[n_i n_j \left(\bar{X}_{i.} - \bar{X}_{j.} \right)^2 \right]}{N} \tag{7.2}$$

where one is careful to take the summation over $i > j$ to avoid double counting the treatment differences, N refers to the total sample size in the study, and n_i refers to the sample size in treatment i. For the differences in Table 7.3, we have

$$\begin{aligned} SS_T &= \frac{1}{N} \left[n_4 n_1 \left(\bar{X}_{4.} - \bar{X}_{1.} \right)^2 + n_4 n_2 \left(\bar{X}_{4.} - \bar{X}_{2.} \right)^2 + \cdots \right. \\ &\qquad \left. + n_2 n_1 \left(\bar{X}_{2.} - \bar{X}_{1.} \right)^2 \right] \\ &= \frac{1}{40} \left[(10)(10)(3.6)^2 + (10)(10)\,(2.2)^2 + (10)(10)\,(0.4)^2 + \cdots \right. \\ &\qquad \left. + (10)(10)\,(1.4)^2 \right] \\ &= \frac{100}{40} \left[(3.6)^2 + (2.2)^2 + (0.4)^2 + (3.2)^2 + (1.8)^2 + (1.4)^2 \right] \\ &= \frac{100}{40} (33.40) = 83.50 \end{aligned}$$

Table 7.3: Means and differences between the means for an experiment with $k = 4$ treatments.

	$\bar{X}_{1.}$	$\bar{X}_{2.}$	$\bar{X}_{3.}$	$\bar{X}_{4.}$
Means	15.8	17.2	19.0	19.4
$\bar{X}_{1.}$		1.4	3.2	3.6
$\bar{X}_{2.}$			1.8	2.2
$\bar{X}_{3.}$				0.4

which is equal to the treatment sum of squares presented in Table 7.2. Thus, the treatment sum of squares is equivalent to the weighted sum of squared differences among the treatment means.

7.3 Protection Levels and the Bonferroni Significant Difference (BSD) Test

The **Bonferroni test** is one of the simplest procedures for correcting the Type I error rate. It involves changing the α criterion that is used (all other computations remain the same). Suppose we have two means and test the difference between the two means with α as the level of significance for the test. Then, α is the probability of making a Type I error when the null hypothesis is true, and $1 - \alpha$ is the probability of not making a Type I error. We refer to the value of $1 - \alpha$ as the **protection level** of the test. When we have more than two means (so we are making $c > 1$ independent tests of significance, each with the same value of α), then the probability of making at least one Type I error across the c tests is greater than the α for a single test. The Bonferroni test uses a smaller α for each individual test so that the probability of making at least one Type I error is the desired level (usually 0.05).

More formally, the probability of making at least one Type I error is given by this equation:

$$P(E) = 1 - (1 - \alpha)^c$$

where c is the number of tests and α is the criterion used for an individual test. The last term in this equation, $(1 - \alpha)^c$, represents the probability of not making a Type I error on each of the c independent tests. We will find this term useful and call it the **protection level.** Note that, when there are only two means, $c = 1$ so the protection level is simply $1 - \alpha$, as described in the preceding paragraph.

The protection level $(1 - \alpha)^c$ and $P(E) = 1 - (1 - \alpha)^c$ are applicable when the c tests of significance are independent. However, a Bonferroni inequality can be used to set a **lower** bound for the protection level and consequently an **upper** bound for $P(E)$, regardless of whether or not the c tests are independent. If the c tests are each made with the same significance level α, then the Bonferroni inequality states that the protection level (the probability of not making any Type I errors) will be greater than or equal to $1 - c\alpha$. Consequently, $P(E)$, the probability of making at least one Type I error, will be less than or equal to $1 - (1 - c\alpha) = c\alpha$. This result provides a useful approximation for estimating the required α for the individual tests so that the aggregate has the desired Type I error rate.

This argument suggests that a simple procedure can be constructed by taking the desired Type I error rate and dividing by c to estimate the α level that should be used for individual tests. In our example with $k = 4$ treatments, there are $c = 6$ possible pairwise comparisons. If the investigator wants the probability of at least one Type I error to be 0.05 (under the null hypothesis), the six individual pairwise t tests can each be tested with $\alpha = 0.05/6 = 0.0083$. That is, the six pairwise comparisons, each evaluated at $\alpha = 0.0083$, yield an aggregate probability (a bound) of $\alpha = 0.05$ of at least one Type I error. The critical t value corresponding to a two-tailed $\alpha = 0.0083$ with 36 degrees of freedom is 2.79. Thus, any observed t test

Table 7.4: The six t values between all possible pairs of $k = 4$ treatment means, using the pooled standard error.

Treatments	1	2	3
2	1.795		
3	4.104	2.308	
4	4.617	2.821	0.513

between two means that exceeds the critical value of 2.79 will reject the null hypothesis under the Bonferroni criterion. Only three of the six tests in Table 7.4 reach statistical significance under this criterion.

An equivalent way to apply the Bonferroni correction is to evaluate the observed p value rather than the observed t value. That is, in the computer output one looks for observed p values less than the corrected α value to determine statistical significance. For example, in the preceding paragraph we determined that six individual pairwise tests each with 36 degrees of freedom can be evaluated against a two-tailed $\alpha = 0.0083$. One merely checks the column labeled "p value" in the computer output for any observed p values less than 0.0083. Those tests having observed p values less than the Bonferroni adjusted criterion are judged to be statistically significant.

The Bonferroni inequality provides a reasonable approximation for c independent tests, provided α is small and c is not large. For example, with $c =$ 10 independent tests, each made with $\alpha = 0.01$, we have $(1-0.01)^{10} = 0.9044$ as the protection level, and the Bonferroni inequality gives $1 - (10)(0.01) =$ 0.9000 as the lower bound for the protection level. However, for $c = 45$ independent tests, each with $\alpha = 0.01$, we have $(1 - 0.01)^{45} = 0.636$ as the protection level, whereas the Bonferroni inequality gives $1 - (45)(0.01) = 0.550$ as the lower bound for the protection level, a value that considerably underestimates the actual protection level for the 45 independent tests. As a result, the upper-bound Bonferroni inequality for $P(E) \leq 0.450$ is considerably greater than the correct value $P(E) = 0.364$.

The Bonferroni test is simple and easy to use. One merely adjusts the critical α based on the number of comparisons that will be made. In some situations the Bonferroni-based procedure may be the only option available for controlling the probability of making at least one Type I error. There are other procedures that provide more accurate corrections in the case of pairwise comparison than the approximation based on the Bonferroni inequality.

7.4 Fisher's Significant Difference (FSD) Test

A different attempt to provide a solution to the multiplicity problem was offered by Fisher (1942). **Fisher's significant difference** (FSD) test for pairwise comparisons requires that the F test in the analysis of variance be statistically significant before making any pairwise comparisons. If the omnibus $F = MS_T/MS_W$ is not significant, then no subsequent tests are made between any pairs of means. If the omnibus F is statistically significant, then proceed by testing every possible pair of treatment means. In our example, the observed F is 9.15, with 3 and 36 degrees of freedom, and is statistically significant with $\alpha = 0.05$. Consequently, under the FSD procedure we proceed to make pairwise comparisons between the means. The test provides some level of protection against an inflated Type I error rate because of its "stopping rule"—if the omnibus F test is not statistically significant, then no individual tests comparing pairs of means are allowed.

We provide a relatively simple way to use the FSD test. This simple procedure will also be applicable for other tests that we will present later. First, we need a formula for the standard error of a difference between two independent means. When sample sizes are equal, we have by Equation 7.1

$$s_{\bar{X}_i - \bar{X}_j} = \sqrt{\frac{2MS_W}{n}}$$

In this example with $MS_W = 3.04$ and treatment sample size $n = 10$, the estimate of the standard error of the difference between any two means i and j is

$$s_{\bar{X}_i - \bar{X}_j} = \sqrt{\frac{(2)(3.04)}{10}} = 0.78$$

There are 36 degrees of freedom for MS_W (denominator), and for a two-sided test of the null hypothesis $\mu_i = \mu_j$, with $\alpha = 0.05$, we will reject the null hypothesis if the obtained absolute value of t equals or exceeds

$$t = \frac{\bar{X}_{i.} - \bar{X}_{j.}}{0.78} = 2.03 \quad (7.3)$$

Consequently, if the absolute difference between any pairs of means equals or exceeds

$$\bar{X}_{i.} - \bar{X}_{j.} = (0.78)(2.03) = 1.58 \quad (7.4)$$

the null hypothesis will be rejected. This provides a useful way to perform the Fisher test. In this example, a difference between two means that exceeds 1.58 will be statistically significant by this criterion. This number serves as a metric against which one evaluates differences between means. It is analogous to "the chains" used in American football to indicate a first down—if the yards gained exceed the length of the 10-yard chain, then there is a first down. The FSD procedure sets the difference to beat in order to achieve statistical significance. Once one knows this difference, it isn't necessary to compute the individual t tests for each pairwise comparison. One can infer the same information of statistical significance under the FSD procedure either by examining the corrected t value (Equation 7.3) or the corrected difference between two means (Equation 7.4).

For the differences listed in Table 7.3, the FSD test results in the rejection of four of the six null hypotheses tested. That is, four of the absolute differences exceed the critical value of 1.58. We conclude, by statistical convention, that the following population means are not equal: $\mu_4 \neq \mu_1$, $\mu_3 \neq \mu_1$, $\mu_4 \neq \mu_2$, and $\mu_3 \neq \mu_2$. In this example the FSD test was more liberal than the Bonferroni test, which rejected three of the six pairwise comparisons.

By requiring that $F = MS_T/MS_W$ be significant prior to making pairwise comparisons, the FSD test attempts to control for the number of pairwise comparisons that are made. For example, if we have 100 experiments for which the null hypothesis is true, we would expect about five of the experiments to result in a significant value of $F = MS_T/MS_W$ with $\alpha = 0.05$. These five experiments should be the only ones in which we would then proceed to make pairwise comparisons. The FSD test, however, does not keep $P(E)$ equal to α for the family of $k(k-1)/2$ pairwise comparisons made within these five experiments when the general null hypothesis is true. Thus, the FSD test does not control the desired Type I error rate, and for that reason we do not recommend the employment of this test, which uses the omnibus F test as a "green light" to perform individual pairwise comparisons.

7.5 The Tukey Significant Difference (TSD) Test

John Tukey derived the sampling distribution for testing all possible pairwise means, and developed a statistical test that is now known as the Tukey range test, or the Tukey significant difference (TSD) test. (For a complete review of his contributions see Tukey, 1994.) The TSD test requires the use of a special table of critical values. Table B.6 in Appendix B gives the critical values of the studentized range q, for $\alpha = 0.05$ and a range of $k = 2$ to $k = 8$ means. The reason for a special table is that the table has built in the proper Type I error protection, given the error degrees of freedom and the number of means in the analysis of variance. For example, with $k = 4$ treatment means and with 36 degrees of freedom for the mean square within treatments, the value of q needed for statistical significance at $\alpha = 0.05$ is 3.8088.

We present the **Tukey test** in a manner parallel to the FSD test. When all the sample sizes are equal, the TSD test is given as

$$\frac{\bar{X}_i - \bar{X}_j}{\sqrt{MS_W/n}}$$

This absolute value is then compared to the critical value q from Table B.6. For the mean differences in the ongoing example (Figure 7.1), we conclude that $\mu_4 \neq \mu_1$, $\mu_3 \neq \mu_1$, and $\mu_4 \neq \mu_2$ because those are the only three comparisons that exceeded the critical value q. For example, the comparison between treatments 1 and 4 leads to an observed $q = \frac{19.4-15.8}{\sqrt{3.04/10}} = 6.529$, which exceeds the tabled q of 3.8088. Note that this is a different result than the one we observed for the FSD test, but this result coincides with the Bonferroni test.

The TSD test has the property that, if the difference between the largest mean and the smallest mean is not significant, then no other difference can be significant. Consequently, $P(E)$, the probability of at least one Type I error, will be equal to the probability of falsely declaring that the difference between the largest and smallest means in a set of k means is significant (hence the name "range test"). In the example, we have set this probability at $\alpha = 0.05$, and therefore $P(E) = 0.05$. We recommend the TSD test for comparing all possible pairwise means because it correctly controls the Type I error rate in a set of c pairwise comparisons.

7.6 Scheffé's Significant Difference (SSD) Test

For a set of $k > 2$ treatment means, there are many possible comparisons that can be made in addition to the set of all possible pairwise comparisons. In other words, the set of $c = k(k-1)/2$ pairwise comparisons is but a small subset of the total number of comparisons that could be made. An example of a comparison that is not pairwise would be the comparison between means 1 and 2 on one hand with means 3 and 4 on the other. This particular comparison involves four means rather than two. The test developed by Scheffé (1953) can handle comparisons over any number of means; such comparisons will be described in more detail in the next chapter. For now we will focus on the application of the Scheffé test to comparisons between two means in order to facilitate comparison with the other procedures presented in this chapter.

The **Scheffé significant difference** (SSD) test sets the protection level for the set of all possible comparisons, not just the subset of pairwise comparisons. Consequently, the SSD test requires a larger value for the difference between a pair of means, in order for the difference to be judged statistically significant, than the other methods we have described for testing pairwise comparisons. The Scheffé test tends to be too conservative when the investigator tests only the set of all possible pairwise comparisons. However, the Scheffé test is appropriate when the investigator is interested in pairwise comparisons as well as more general comparisons involving more complicated subsets of treatments (see Chapter 8).

If the omnibus $F = MS_T/MS_W$ in the analysis of variance is not statistically significant at the level of significance α, then no pairwise or nonpairwise comparison will be judged statistically significant by the SSD test. If the omnibus F is statistically significant, then at least one comparison on the k treatment means will also be statistically significant in terms of the SSD test, and there may, of course, be more than one comparison that is statistically significant. Unlike the FSD test that uses the statistical significance of the omnibus test as a decision rule, the Scheffé test is defined so it automatically will not yield statistically significant comparisons when the omnibus test is not itself statistically significant. The SSD test is not based on a decision rule to continue pursuing statistical tests based on the outcome of previous statistical tests (such as the FSD test), but the SSD test is based on a special sampling distribution that takes into account the maximum possible sum of squares for a comparison.

The SSD test proceeds as follows: if the omnibus $F = MS_T/MS_W$ is statistically significant, then we find a new critical value from the F tables in Appendix B (Table B.2),

$$t' = \sqrt{(k-1)\,F}$$

where F is the tabled F distribution for $(k-1)$ and the degrees of freedom associated with the within-treatment mean square MS_W. In our example, with $\alpha = 0.01$, and with 3 and 36 degrees of freedom, the tabled F value is 4.38. Then,

$$t' = \sqrt{(4-1)\,(4.38)} = 3.62$$

Thus, in this example an observed t of 3.62 or greater will be deemed statistically significant by the Scheffé criterion.

For a test of significance between any pair of means, we will reject the null hypothesis if the obtained absolute value of t equals or exceeds

$$t' = \frac{\bar{X}_{i.} - \bar{X}_{j.}}{0.78} = 3.62$$

or equivalently if the absolute value of $\bar{X}_{i.} - \bar{X}_{j.}$ equals or exceeds (0.78)(3.62) = 2.82. On the basis of the SSD test, we conclude that $\mu_4 \neq \mu_1$ and $\mu_3 \neq \mu_1$. Note that the Scheffé test is more conservative than Fisher's significant difference test and Tukey's significant difference test.

7.7 The Four Methods: General Considerations

In this chapter we reviewed four procedures for comparing all possible pairwise means for k treatments. The four procedures are quite simple. They use the same pooled error term MS_W, so all the tests presented in this chapter make the equality of variance assumption. The BSD test uses the t test with a corrected α criterion (that is, the desired probability of at least one Type I error is divided by the number of comparisons to be made). The BSD procedure makes the critical t value more conservative as compared to not performing the α correction. The FSD test uses the t test with the rule that the omnibus F test must be significant before the individual pairwise comparisons can be computed. The TSD test compares the observed t to a special table that deals with the sampling distribution of the range. The Scheffé test changes the critical values needed to reach significance in such a way that the researcher can perform an infinite number of tests.

The TSD, BSD, and SSD tests have the property that $P(E)$ for the set or family of $c = k(k-1)/2$ pairwise comparisons will be less than or equal to the desired value, regardless of the value of k. This statement is not true of the FSD test. In the example involving $k = 4$ means with $\alpha = 0.01$ for a single test, we have $P(E)$ approximately equal to 0.05 for the FSD test. However, what happens when we have $k = 10$ means? For the set of $c = 10(10-1)/2 = 45$ pairwise comparisons, with $\alpha = 0.01$ for a single test, $P(E)$ will be approximately equal to 0.23 for the FSD test, which is too high for an overall Type I error rate. Thus, we can reject the FSD test in most applications because it does not properly control $P(E)$. Levin, Serlin, and Seaman (1994) review some special situations where the FSD performs reasonably well.

All tests for pairwise comparisons that keep $P(E)$ less than or equal to 0.01 or 0.05 for the set of $c = k(k-1)/2$ pairwise comparisons are based on the assumption that the general null hypothesis is true. These tests are designed to protect the experimenter against Type I errors and not against Type II errors. If the general null hypothesis is false, so that real differences exist between the population means, then setting $P(E) \leq 0.01$ or $P(E) \leq 0.05$ for the complete family of $c = k(k-1)/2$ pairwise comparisons will serve to increase the probability of Type II errors. In other words, because the tests are conservative, some statistically significant differences between the means may be missed by these conservative statistical tests.

Which method an experimenter chooses for making all pairwise comparisons must inevitably depend on how the experimenter feels about the relative importance of Type I and Type II errors and what the experimenter believes to be true with respect to the values of the k population means. If the experimenter wishes to keep $P(E)$ less than or equal to 0.05 for the family of pairwise comparisons, then the obvious choice is the TSD test rather than the BSD or SSD test. Under most conditions when testing a complete set of pairwise comparisons, the TSD test will require smaller critical values for significance than those for either the BSD or the SSD test, and consequently the TSD will be more powerful. The SSD test will be useful when the experimenter wants to test more comparisons than those defined as pairwise comparisons, as shown in the next chapter. There are some complicated situations that we will encounter later in this book where a Tukey test analog is not available, and the practical alternative is the Bonferroni test.

The assumption of equal population variances is very important. One simulation study showed that even the TSD test does not perform well in terms of Type I and Type II error rates when there are unequal variances as well as unequal sample sizes across the treatment conditions (Keselman, Toothaker, & Shooter, 1975).

There are many different types of multiple comparison tests that are available. Some are variants of the tests presented here, such as the Newman–Keuls test, which is a weakening of the Tukey test to allow for different significance criteria to be used depending on which pairs of means are compared. We find that the small subset of tests we present cover the majority of applications in the social sciences and perform well to serve as "all-purpose tools." For a book-length treatment of multiple comparisons, different procedures and their pros and cons, the reader is referred to Hochberg and Tamhane (1987).

7.8 Questions and Problems

1. Consider a one-way between-subjects design with $n = 6$ participants assigned at random to each of eight treatments; $MS_W = 53.02$ with 40 degrees of freedom. The treatment means are given here:

$$\begin{array}{ll} \bar{X}_{1.} = 19.7 & \bar{X}_{5.} = 55.1 \\ \bar{X}_{2.} = 36.7 & \bar{X}_{6.} = 61.3 \\ \bar{X}_{3.} = 50.6 & \bar{X}_{7.} = 65.3 \\ \bar{X}_{4.} = 51.4 & \bar{X}_{8.} = 72.0 \end{array}$$

   Perform each of the four pairwise tests (BSD, FSD, Tukey and Scheffé) on these data.

2. What is meant by the protection level of a test?

3. Explain the concept of an omnibus test.

4. We have $k = 6$ treatments with $n = 10$ participants assigned at random to each treatment. The MS_W for the experiment is equal to 36.0. The treatment means for the experiment are given here:

$$\begin{array}{ll} \bar{X}_{1.} = 18.6 & \bar{X}_{4.} = 23.4 \\ \bar{X}_{2.} = 19.6 & \bar{X}_{5.} = 26.2 \\ \bar{X}_{3.} = 20.5 & \bar{X}_{6.} = 28.3 \end{array}$$

   Find the absolute difference between a pair of means $\bar{X}_{i.} - \bar{X}_{j.}$ that will be judged statistically significant with $\alpha = 0.05$ if pairwise comparisons are made by using the following:

   (a) The FSD test.

   (b) The BSD test.

(c) The SSD test.

(d) The TSD test.

5. The results of an analysis of variance involving $k = 8$ treatments with $n = 4$ observations for each treatment are given here:

Sources of variation	Sum of square	*df*	Mean square	F
Treatments	7803.16	7	1114.74	30.96
Within treatments	854.00	24	36.00	

The values of the eight treatment means for the experiment were as follows:

$$\begin{array}{ll} \bar{X}_{1.} = 24.7 & \bar{X}_{5.} = 60.1 \\ \bar{X}_{2.} = 41.7 & \bar{X}_{6.} = 66.3 \\ \bar{X}_{3.} = 55.6 & \bar{X}_{7.} = 70.3 \\ \bar{X}_{4.} = 56.4 & \bar{X}_{8.} = 77.0 \end{array}$$

Find the absolute difference between a pair of means $\bar{X}_{i.} - \bar{X}_{j.}$ that will be judged statistically significant with $\alpha = 0.05$ if pairwise comparisons are made by using the following:

(a) The FSD test.

(b) The BSD test.

(c) The SSD test.

(d) The TSD test.

6. Reanalyze Question 2 from Chapter 6 (page 187). This problem has six treatment groups. Test all possible pairwise comparisons with $\alpha = 0.05$, using each of the following tests:

(a) The FSD test.

(b) The BSD test.

(c) The SSD test.

(d) The TSD test.

8

Orthogonal, Planned and Unplanned Comparisons

8.1 Introduction

In this chapter we discuss in greater detail the nature of a comparison.[1] In the sections that follow, we will assume equal sample sizes to make the description simpler. That is, we have k means with an equal number of observations for each treatment. Later, we will generalize the discussion to cases where there are a different number of observations across conditions.

Table 8.1 repeats the analysis of variance summary table from the preceding chapter, where we had $k = 4$ treatments with $n = 10$ observations for each treatment. Although in that example the omnibus $F = MS_T/MS_W = 9.15$ was statistically significant, significance is not a necessary condition for testing orthogonal, planned comparisons. Comparisons can be tested directly without conducting the omnibus test. Indeed, a fledgling view among methodologists is that omnibus tests should be avoided because they do not provide information about specific patterns between treatment means. Com-

[1]Comparisons are also referred to as contrasts.

Table 8.1: Summary of the analysis of variance of a between-subjects design with $k = 4$ treatments and $n = 10$ participants randomly assigned to each treatment.

Source of variation	Sum of squares	*df*	Mean square	F
Treatments	83.50	3	27.83	9.15
Within treatments	109.44	36	3.04	
Total	192.94	39		

parisons provide one way to test specific research questions and hypotheses among a set of treatment means, so comparisons extend the omnibus test on treatment means presented in Chapter 6.

8.2 Comparisons on Treatment Means

A comparison involves quantifying a particular research question by taking a linear combination of treatment means. For instance, a researcher might be interested in comparing three treatments where clients received therapies to a fourth condition where clients were given a placebo therapy. This question can be worded as "Does the average of the three treatments that received therapy differ from the single group that did not receive therapy?" Table 8.2 shows three of the many comparisons that might be made on a set of $k = 4$ treatment means. The values of each comparison are called **coefficients** of the treatment means, and we will use a with appropriate subscripts, as shown on the right-hand side of the table, to represent coefficients. The first subscript refers to a particular treatment mean and the second subscript corresponds to a particular comparison, or research question. In this way we can tailor comparisons to specific research questions targeted to specific patterns of means.

The first comparison in Table 8.2 involves the difference between Treatment 1 and Treatment 2. This comparison corresponds to the research question "Is the mean for Treatment 1 statistically different from the mean for

Table 8.2: Three comparisons on $k = 4$ treatment means.

	Coefficients				Notation				Value
Comp.	$\bar{X}_{1.}$	$\bar{X}_{2.}$	$\bar{X}_{3.}$	$\bar{X}_{4.}$	$\bar{X}_{1.}$	$\bar{X}_{2.}$	$\bar{X}_{3.}$	$\bar{X}_{4.}$	$\sum_{j=1}^{k} a_{ji}^2$
d_1	1	-1	0	0	a_{11}	a_{21}	a_{31}	a_{41}	2
d_2	0	0	-1	1	a_{12}	a_{22}	a_{32}	a_{42}	2
d_3	1/2	1/2	$-1/2$	$-1/2$	a_{13}	a_{23}	a_{33}	a_{43}	1

Treatment 2?" Treatments 1 and 2 receive coefficients of 1 and -1, respectively, but Treatments 3 and 4 are assigned coefficients of 0 because those two means are irrelevant to this particular research question. Multiplying the treatment means by the coefficients in the first row, we obtain the comparison

$$\begin{aligned} d_1 &= (1)\bar{X}_{1.} + (-1)\bar{X}_{2.} + (0)\bar{X}_{3.} + (0)\bar{X}_{4.} \\ &= \bar{X}_{1.} - \bar{X}_{2.} \end{aligned}$$

The coefficients equal to zero eliminate the treatments that are not involved in that particular comparison. Note how the comparison yields a simple difference between the two means in question.

The second comparison in Table 8.2 involves the difference between the means of Treatment 3 and Treatment 4 because the coefficients are 0, 0, -1, 1. Multiplying the treatment means by the coefficients in the second row, we obtain the comparison

$$\begin{aligned} d_2 &= (0)\bar{X}_{1.} + (0)\bar{X}_{2.} + (-1)\bar{X}_{3.} + (1)\bar{X}_{4.} \\ &= \bar{X}_{4.} - \bar{X}_{3.} \end{aligned}$$

The third comparison in Table 8.2 involves a more complicated research question: Is the average of Treatments 1 and 2 statistically different from the average of Treatments 3 and 4? If we multiply the treatment means by the coefficients in the last row, we have the comparison

$$d_3 = \tfrac{1}{2}\left(\bar{X}_{1.} + \bar{X}_{2.}\right) - \tfrac{1}{2}\left(\bar{X}_{3.} + \bar{X}_{4.}\right)$$

or the difference between the average of the means for Treatments 1 and 2 and the average of the means for Treatments 3 and 4. The first two comparisons are pairwise comparisons, but the third is not. Thus, comparisons need not be limited to pairwise tests, and so the material in the present chapter generalizes the pairwise tests in Chapter 7. Many types of research questions can be converted into comparisons and tested in a very simple way.

Comparisons of the kind shown in Table 8.2 are linear functions of the treatment means. Any linear function of the treatment means such as

$$d_i = a_{1i}\bar{X}_{1.} + a_{2i}\bar{X}_{2.} + \cdots + a_{ki}\bar{X}_{k.}$$

is called a **comparison**, if at least two of the coefficients are not equal to zero and if the sum of the coefficients is equal to zero, that is, if

$$\sum_{j=1}^{k} a_{ji} = 0 \tag{8.1}$$

For Equation 8.1 to be true under the conditions stated, then it is obvious that for the sum of the coefficients to be 0 at least one of the coefficients must be negative and at least one must be positive.

Under the standard null hypothesis of the omnibus analysis of variance, all of the k treatment means have the same expected value μ. Then because the coefficients have the property that $\sum_{j=1}^{k} a_{ji} = 0$ for any comparison d_i, under the null hypothesis the expected value of d_i (the weighted sum of means where the values of the comparisons are the weights) will also be equal to zero. For example, if the comparison is (3, -1, -1, -1) so that on four treatment means we have

$$d_i = 3\bar{X}_{1i} - \left(\bar{X}_{2i} + \bar{X}_{3i} + \bar{X}_{4i}\right)$$

then under the null hypothesis the population value of d_i is 0.

This is a more general null hypothesis than the usual "population treatment means are equal" because the individual treatment means need not have identical population means in order for a weighted sum to be 0 under the null hypothesis. That is, the requirement under the null hypothesis for the comparison (3, −1, −1, −1) is

$$3\mu_1 - (\mu_2 + \mu_3 + \mu_4) = 0$$

There are many combinations of those four means that could result in a weighted sum of 0.

The data analyst can convert just about any research question about means into a comparison over those means. Thus, comparisons offer a direct way to test research questions. We now turn to describing how to perform statistical tests on comparisons and then return to the problem of how to convert a research idea into a comparison. We will also discuss constraints that are imposed on the number and types of comparisons one can make.

8.3 Standard Error of a Comparison

The estimated standard error of any comparison d_i, that is, the standard error of the corresponding weighted sum obtained by multiplying the means by the coefficients for the comparison, will be given by

$$s_{d_i} = \sqrt{MS_W \left(\frac{a_{1i}^2}{n_1} + \frac{a_{2i}^2}{n_2} + \cdots + \frac{a_{ki}^2}{n_k} \right)} \tag{8.2}$$

where MS_W is the mean square within treatments from the analysis of variance. Recall that MS_W is estimated as a pooled variance, so the homogeneity of variance assumption is applicable here as well. If the number of observations is the same for each mean, then Equation 8.2 may be written more succinctly as

$$s_{d_i} = \sqrt{\frac{MS_W}{n} \sum_{j=1}^{k} a_{ji}^2} \tag{8.3}$$

where n is the number of observations for a single mean. We note the special case that for any comparison between two means $\bar{X}_{l.}$ and $\bar{X}_{m.}$ the corresponding coefficients will be 1 and -1, and $\sum_{j=1}^{k} a_{ji}^2 = 2$. Then Equation 8.3 reduces to $\sqrt{\frac{2MS_W}{n}}$, which is similar to the usual standard error of the difference between two means when $n_1 = n_2 = n$. The difference however is in the computation of the MS_W term. When there are more than two treatment groups, all groups enter into the computation of the MS_W term, even though some treatment groups are weighted 0. This is justified under the equality of variance assumption. When the equality of variance assumption holds, then the pooled error term leads to a more powerful statistical test.

8.4 The *t* Test of Significance for a Comparison

Under the null hypothesis for the comparison, we have the population value of d equal to 0. Then, the statistical significance of the difference represented by any comparison d_i can be evaluated by finding the t value

$$t = \frac{d_i}{s_{d_i}} \tag{8.4}$$

The degrees of freedom for this t value is equal to the number of degrees of freedom associated with the mean square within treatments from the analysis of variance (that is, the denominator of the omnibus F test). This computed t value is compared to the critical value found in Table B.1 in Appendix B.

In Table 8.3, we give the means for the treatments of the analysis of variance reported in Table 8.1 and the coefficients for the three comparisons

Table 8.3: Application of the comparisons of Table 8.2. The means are those obtained in the experiment summarized in the source table shown in Table 8.1.

	$\bar{X}_{1.}$	$\bar{X}_{2.}$	$\bar{X}_{3.}$	$\bar{X}_{4.}$	
Comparison	17.2	19.4	15.8	19.0	Value of d_i
d_1	1	−1	0	0	−2.2
d_2	0	0	−1	1	3.2
d_3	1/2	1/2	−1/2	−1/2	0.9

of Table 8.2. Multiplying the means by the corresponding coefficients for each comparison, we obtain $d_1 = -2.2$, $d_2 = 3.2$, and $d_3 = 0.9$.

Summing the squares of the coefficients for each comparison, we have

$$\sum_{j=1}^{k} a_{j1}^2 = 2 \qquad \sum_{j=1}^{k} a_{j2}^2 = 2 \qquad \sum_{j=1}^{k} a_{j3}^2 = 1$$

From the analysis of variance in Table 8.1, the pooled error term is $MS_W = 3.04$. The standard errors given by Equation 8.3 for each of the three comparisons are as follows:

$$\begin{aligned} s_{d_1} &= \sqrt{\tfrac{3.04}{10}(2)} = 0.78 \\ s_{d_2} &= \sqrt{\tfrac{3.04}{10}(2)} = 0.78 \\ s_{d_3} &= \sqrt{\tfrac{3.04}{10}(1)} = 0.55 \end{aligned}$$

Dividing each d_i by its standard error, we obtain the corresponding t tests

$$\begin{aligned} t_1 &= \tfrac{-2.2}{0.78} = -2.82 \\ \mathrm{t}_2 &= \tfrac{3.2}{0.78} = 4.10 \\ \mathrm{t}_3 &= \tfrac{0.9}{0.55} = 1.64 \end{aligned}$$

Each of these t's has 36 degrees of freedom, the number of degrees of freedom associated with the MS_W from the analysis of variance. If we use a two-sided $\alpha = 0.05$ with a t critical of 2.028, the first two comparisons d_1 and d_2 are statistically significant, whereas the third d_3 is not.

Confidence limits for d_i may be established in the usual way as $d_i \pm ts_{d_i}$, where the t used in the confidence interval formula is the t from the Table for a two-tailed test at the desired interval. For a 95% confidence interval the two-tailed t-value corresponds to the tabled value for $\alpha = 0.05$ with degrees of freedom corresponding to the MS_W term. For each of the three comparisons in the example we have

$$
\begin{aligned}
-2.2 &\pm (2.028)(0.78) \\
3.2 &\pm (2.028)(0.78) \\
0.9 &\pm (2.028)(0.55)
\end{aligned}
$$

The lower and upper 95% confidence limits are, respectively, (−3.78, −0.62), (1.62, 4.78), and (−0.22, 2.02). Confidence intervals that do not include zero, the typical value of the null hypothesis, correspond to rejecting the null hypothesis in the context of a statistical test.

8.5 Orthogonal Comparisons

We now define the useful concept of orthogonality. If we make two comparisons d_i and d_j on the same set of k treatment means where all treatments have the same sample size, then d_i and d_j are said to be **orthogonal** if the sum of the products of the corresponding coefficients for the two comparisons is equal to zero. That is, two comparisons are orthogonal when the weights satisfy this equation

$$\sum_{t=1}^{k} a_{ti}a_{tj} = a_{1i}a_{1j} + a_{2i}a_{2j} + \cdots + a_{ki}a_{kj} = 0$$

The comparisons shown in Table 8.2 are **mutually orthogonal** because the sum of the products of the coefficients for all possible pairs of comparisons are equal to zero. For example, for comparisons d_1 and d_2, we have

$$(1)(0) + (-1)(0) + (0)(-1) + (0)(1) = 0$$

The sum of the products of the coefficients for comparisons d_1 and d_3, and, also for comparisons d_2 and d_3, totals zero. As we will see below, orthogonality permits an interesting connection between a set of comparisons and the sum of squares for treatment in the analysis of variance.

When sample sizes are unequal, then orthogonality between two comparisons should be defined as follows:

$$\sum \frac{a_{ti}a_{tj}}{n_t} = \frac{a_{1i}a_{1j}}{n_1} + \frac{a_{2i}a_{2j}}{n_2} + \cdots + \frac{a_{ki}a_{kj}}{n_k} = 0$$

Orthogonality can be given a geometric interpretation. Consider the two comparisons (1, 1) and (1, −1). If we plot these two points as vectors (an arrow from the origin to the point), we can see that the two comparisons are at 90 degrees to each other, as shown in Figure 8.1. These two comparisons are orthogonal because $(1)(1) + (1)(-1) = 0$. Indeed, orthogonality refers to comparisons that are at right angles when represented as vectors. When there are more than three treatments, the geometric picture is difficult to draw or see because we have trouble seeing in more than three spatial dimensions; but the concept of right angles extends to any number of treatments.

8.6 Choosing a Set of Orthogonal Comparisons

Orthogonality is a useful constraint on possible comparisons, but there are many sets of orthogonal comparisons that are possible. For example, with $k = 4$ means, the three sets of orthogonal comparisons given in Table 8.4 differ from one another and also from the set of orthogonal comparisons given in Table 8.2. Each of the three sets of comparisons given in Table 8.4

Figure 8.1: Geometric interpretation of orthogonality—the comparisons (1, 1) and (1, −1).

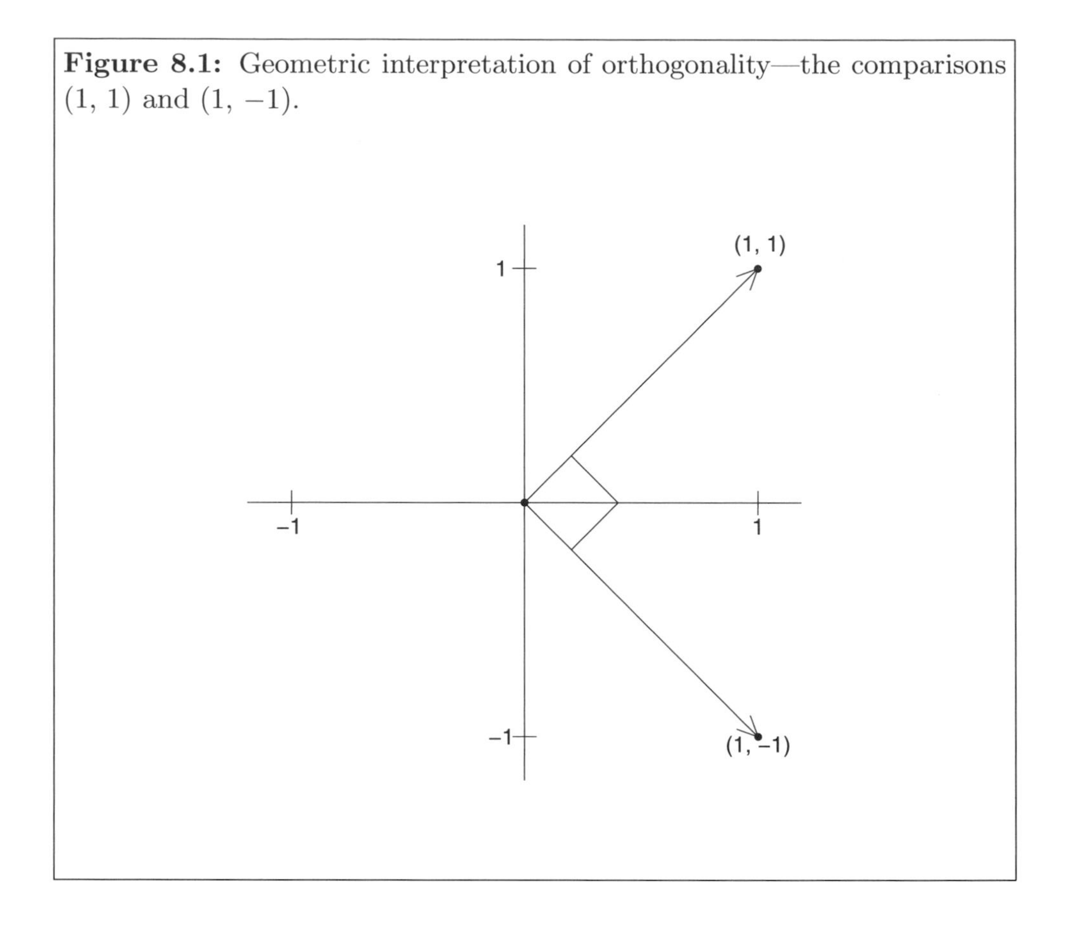

is orthogonal within the set, which can easily be verified by calculating the sum of the products of the corresponding coefficients for each possible pair of comparisons within each set. In each case, the sum of these products is equal to zero.

Because more than one set of orthogonal comparisons is possible for a given group of $k \geq 3$ means, the particular set of comparisons used in a study depends on the researcher's interests and should be planned at the same time the study is planned. Consider the particular three orthogonal comparisons shown in Table 8.2. Suppose the dependent variable of interest was a measure of maze performance and the four treatments were:

Group 1: a treatment tested after 12 hours of water deprivation

Table 8.4: Three different sets of orthogonal comparisons on $k = 4$ treatment means.

	$\bar{X}_{1.}$	$\bar{X}_{2.}$	$\bar{X}_{3.}$	$\bar{X}_{4.}$
Set 1	Coefficients			
d_1	−3	1	1	1
d_2	0	−2	1	1
d_3	0	0	1	−1
Set 2	Coefficients			
d_1	1	1	−1	−1
d_2	−1	1	−1	1
d_3	−1	1	1	−1
Set 3	Coefficients			
d_1	−3	−1	1	3
d_2	1	−1	−1	1
d_3	−1	3	−3	1

Group 2: a treatment tested after 24 hours of water deprivation

Group 3: a treatment tested after 12 hours of food deprivation

Group 4: a treatment tested after 24 hours of food deprivation

In this design the first comparison in Table 8.2 tests for the difference between the 12- and 24-hour water-deprived treatments; the second comparison tests for the difference between the 12- and 24-hour food-deprived treatments; and the third comparison tests for the difference between the average performance of the water-deprived and the food-deprived treatments. This third comparison tests whether the means of the water deprivation groups differ from the means of the food deprivation groups, ignoring the specific time interval of the deprivation (that is, 12 or 24 hours). We will return to examples such as these when we discuss factorial analysis of variance. These types of comparisons are given special names such as main effect comparison, or interaction comparison, or special main effect comparison, as we will see in later chapters.

We do not need to make all of the possible $k-1$ orthogonal comparisons in a given set. In some cases, the experimenter may only be interested in a few of the possible comparisons. Again, it is not necessary that the omnibus $F = MS_T/MS_W$ be statistically significant (or even tested, for that matter) prior to testing planned orthogonal comparisons. Researchers may test individual comparisons without testing the omnibus F test.

8.7 Protection Levels with Orthogonal Comparisons

If $k-1$ orthogonal comparisons are made on a set of k treatment means, the numerators of the t ratios will be independent due to orthogonality of the comparison. The t ratios themselves will not be independent because the tests of significance are all made by using a common denominator MS_W. However, if α is small, say 0.05 or 0.01 for a single test, and if k is not large, then the protection level, $(1-\alpha)^{k-1}$, based on the assumption that the k tests are independent, will be approximately equal to the lower-bound protection level, $1-(k-1)\alpha$, based on the Bonferroni inequality. For example, with $\alpha = 0.01$ for a single test and with $k = 16$, we have $(1-0.01)^{16-1} = 0.86$ as the protection level for a set of 15 independent tests. With the Bonferroni inequality, we have $1-(16-1)0.01 = 0.85$ as the lower-bound estimate of the protection level. For other values of α or k, the approximation may not work as well. Thus, for all practical purposes, the set of $k-1$ orthogonal comparisons can be regarded as independent in evaluating the protection level and $P(E)$.

8.8 Treatments as Values of an Ordered Variable

In some studies the treatments may consist of different values of an ordered variable. For example, we might test different treatments after 0, 6, 12, and 18 hours of food or water deprivation. In other cases, the treatments may consist of increasing intensities of shock, of increasing amounts of reward, of decreasing numbers of reinforcements, or of decreasing dosages of a drug.

If the treatments consist of different values of an ordered variable and the differences between the values are equal, then we may be interested in determining whether the treatment means are functionally related to the values of the treatment variable. We may, for example, be interested in testing whether the treatment means are linearly related to the values of the treatment variable or whether the treatment means deviate significantly from a linear relation (that is, deviate from a straight-line relation). If the deviations from linearity are statistically significant, then we may wish to determine whether there is a significant curvature in the trend of the means.

Assume, for example, that the treatments in an experiment consist of four equally increasing levels of reward, which we designate by 1, 2, 3, and 4. With $n = 10$ participants assigned to each treatment, assume that the analysis of variance for the experiment is as given in Table 8.1. The ordered treatment means, 15.8, 17.2, 19.0, and 19.4, represent the average performance on a game of skill at each of the four successive reinforcement levels. Figure 8.2 plots the treatment means against the levels of reinforcement. By visual inspection it appears that the trend of the means is approximately linear. We will next develop a statistical test to assess linearity and deviations from linearity.

8.9 Coefficients for Orthogonal Polynomials

To determine whether the linear component of the trend of the means is statistically significant and also whether the treatment means deviate significantly from linearity, we make use of a table of coefficients for orthogonal polynomials, Table B.5.[2] This table gives the coefficients to use for the linear, quadratic, and cubic components of the treatment sum of squares. The

[2]The coefficients for orthogonal polynomials given in Table B.5 are for the case of equal intervals in the values of the quantitative variable and for equal n's. If the intervals or n's

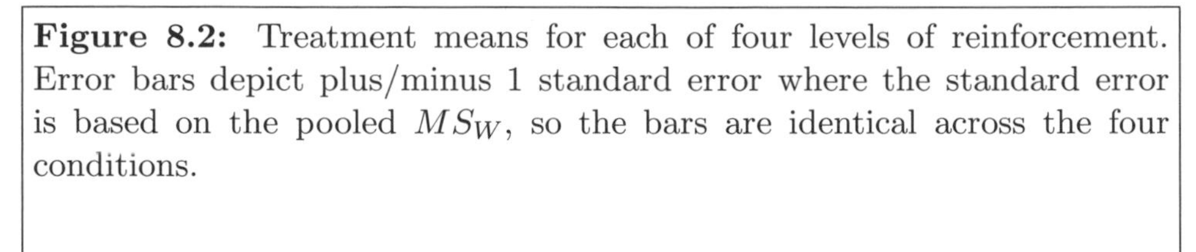

Figure 8.2: Treatment means for each of four levels of reinforcement. Error bars depict plus/minus 1 standard error where the standard error is based on the pooled MS_W, so the bars are identical across the four conditions.

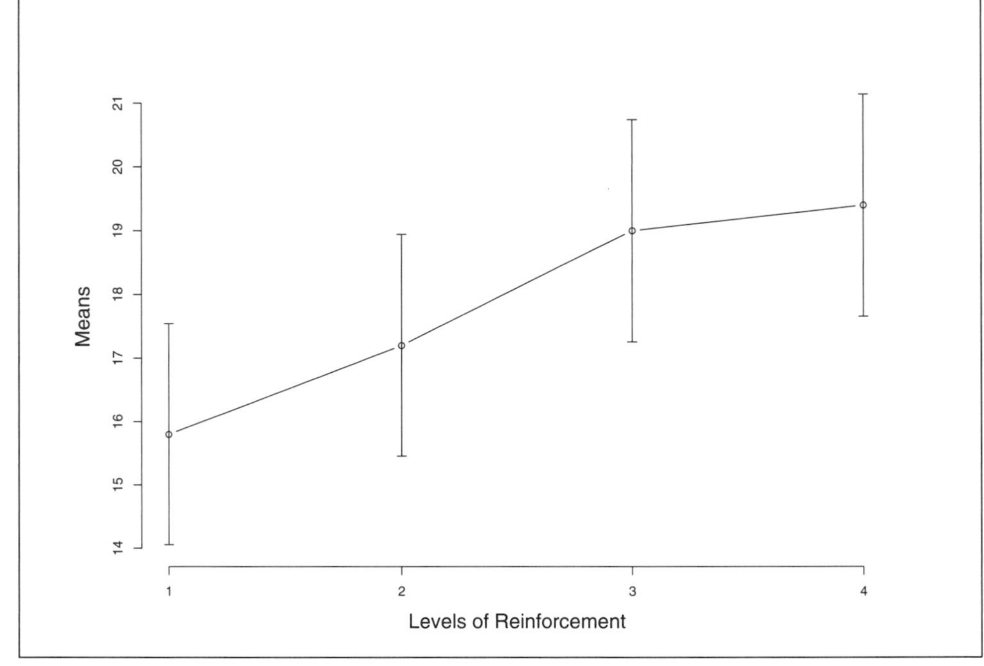

coefficients in each row of Table B.5 sum to zero, and for any fixed value of k the sum of the products of the coefficients for the linear and quadratic comparisons is also zero. This result is true also for the linear and cubic coefficients and for the quadratic and cubic coefficients. The linear, quadratic, and cubic comparisons, therefore, meet the requirements for mutual orthogonality discussed earlier. The coefficients for the linear, quadratic, and cubic components for $k = 4$ treatments are shown in Table 8.5.

As another example, if $k = 5$, the successive sets of coefficients would correspond to the linear, quadratic, cubic, and quartic components of the

---

are unequal, the coefficients given in Table B.5 should not be used. For procedures to be used with unequal intervals and/or unequal n's, see Grandage (1958) or Gaito (1965).

Table 8.5: Coefficients for the linear, quadratic, and cubic components for $k = 4$ treatments.

	Treatment means			
Comparison	15.8	17.2	19.0	19.4
Linear	−3	−1	1	3
Quadratic	1	−1	−1	1
Cubic	−1	3	−3	1

treatment sum of squares. Successive application of these coefficients would enable one to determine how well the trend of the treatment means is represented by a polynomial of the first, second, third, and fourth degree, respectively. Table B.5 in Appendix B gives only the coefficients for the linear, quadratic, and cubic components because seldom will the comparisons involving polynomials of degree greater than 3 be of interest. Coefficients for the higher-degree polynomials can be found in Fisher and Yates (1948) tables.

A graphical display of these polynomial comparisons may help illustrate the patterns they test. The coefficients for the linear component or comparison change signs only once, from minus to plus. These coefficients for $k = 4$ treatments are plotted in Figure 8.3(a), and the trend represented by the coefficients is a straight line. For the quadratic comparison, the coefficients change signs twice, from plus to minus to plus, and, as plotted in Figure 8.3(b), correspond to one reversal in the trend such as a U-shaped pattern. For the cubic coefficients, there are three sign changes in the coefficients, from minus to plus to minus to plus, and, as shown in Figure 8.3(c), these coefficients correspond to two reversals in the trend such as in an S-shaped, or Z-shaped, pattern. Thus, the number of sign changes in the coefficients indicates the degree of the polynomial.

Figure 8.3: Plots of linear (a), quadratic (b), and cubic (c) coefficients for orthogonal polynomials against equally spaced values of a quantitative variable. These plots show the characteristic pattern of each term in the polynomial: the linear checks for trends that do not have bends, the quadratic checks for trends with one bend, and the cubic checks for trends with two bends.

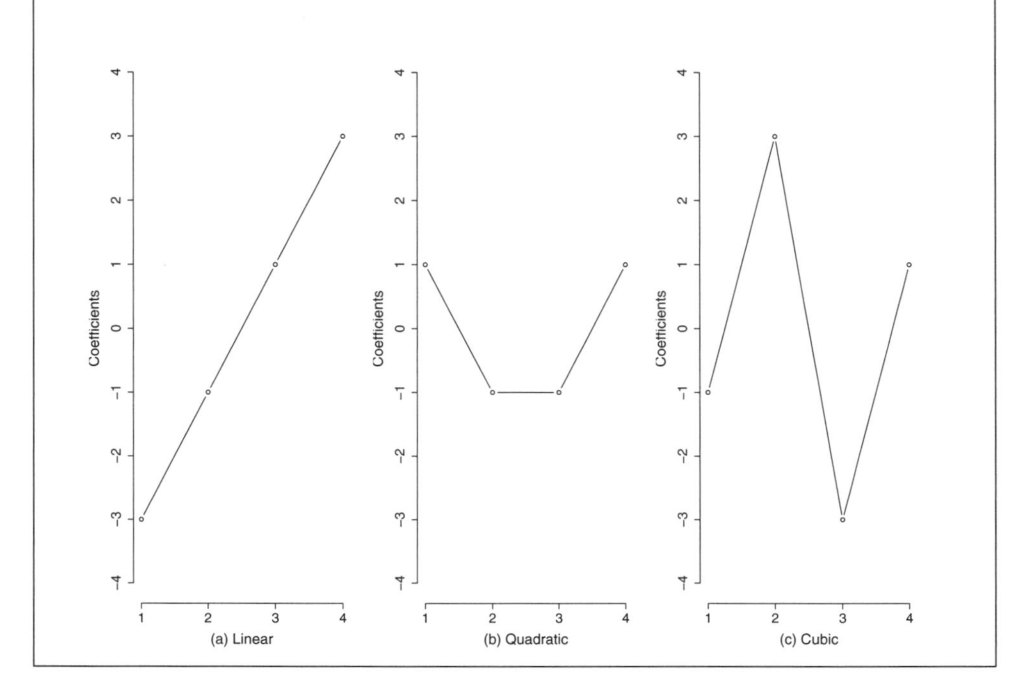

8.10 Tests of Significance for Trend Comparisons

The test of statistical significance uses the same equation for comparisons presented earlier in this chapter. Multiplying the treatment means by the coefficients for the linear comparison, as given in Table 8.5, we have for this comparison

$$L = (-3)(15.8) + (-1)(17.2) + (1)(19.0) + (3)(19.4) = 12.6$$

Then, with $\sum_{j=1}^{k} a_{jL}^2 = 20$, $n = 10$ observations for each treatment and $MS_W = 3.04$, we use Equation 8.4 to find the t value

$$t_L = \frac{12.6}{\sqrt{20\frac{3.04}{10}}} = 5.11$$

This observed t is compared to the critical t value from Table B.1 in Appendix B, based on 36 degrees of freedom (that is, the degrees of freedom associated with MS_W). The test confirms the visual inspection of Figure 8.2 that the four treatment means have a linear trend. The test rejects the null hypothesis that the slope of the line is zero (that is, rejects a horizontal line) because the observed t exceeds the $t_{critical} = 2.028$.

Similarly, the t test for the quadratic term has a numerator of

$$Q = (1)(15.8) + (-1)(17.2) + (-1)(19.0) + (1)(19.4) = -1$$

and a denominator of

$$\sqrt{4\frac{3.04}{10}} = 1.1027$$

because there are 10 participants per treatment, $MS_W = 3.04$, and $\sum_{j=1}^{k} a_{ji}^2$ $= 4$. The resulting t ratio is

$$\frac{-1}{1.1027} = -0.91$$

which in absolute value terms does not exceed $t_{critical}$. This failure to reject the null hypothesis for the quadratic comparison suggests there is little evidence in these data for a quadratic trend, at least up to the statistical power afforded by the present sample size.

Finally, the comparison corresponding to the cubic trend on the treatment means is

$$C = (-1)(15.8) + (3)(17.2) + (-3)(19.0) + (1)(19.4) = -1.8$$

with a resulting test statistic of

$$\frac{-1.8}{\sqrt{20\frac{3.04}{10}}} = -.73$$

The cubic trend is not statistically significant because the absolute value of the observed t of $-.73$ is not more extreme than the $t_{critical} = 2.028$. Thus, in this experiment the linear trend comparison is statistically significant with $\alpha = 0.05$, but the quadratic and cubic comparisons are not.

8.11 The Relation between a Set of Orthogonal Comparisons and the Treatment Sum of Squares

A set of orthogonal comparisons decomposes the sum of squares for treatments into smaller parts, each part representing the portion of sum of squares treatment attributable to that comparison. We illustrate this idea with the three orthogonal trend comparisons performed in the preceding section, but this idea will hold for any complete set of $k - 1$ orthogonal comparisons.

Recall that a complete set of orthogonal comparisons must involve $k-1$ comparisons, where k is the number of treatments and each pair of comparisons are orthogonal. The sum of squares for a single comparison i is defined as

$$SS_i = \frac{d_i^2}{\sum \frac{a^2}{n_i}}$$

Thus, for the linear, quadratic, and cubic comparisons in the preceding section we have

$$SS_L = \tfrac{12.6^2}{20}10 = 79.38$$
$$SS_Q = \tfrac{(-1)^2}{4}10 = 2.5$$
$$SS_C = \tfrac{(-1.8)^2}{20}10 = 1.62$$

Figure 8.4: Pie chart depicting the decomposition of sums of squares into treatments and within treatments, as well as the further decomposition of the sum of square treatments into orthogonal comparisons. The shaded regions together correspond to the entire treatment sum of squares (83.50), which is decomposed into separate portions by the particular orthogonal set of comparisons (in this example, the decomposition is based on the polynomial comparisons and their respective sum of squares SS_{linear}, $SS_{quadratic}$, and SS_{cubic}).

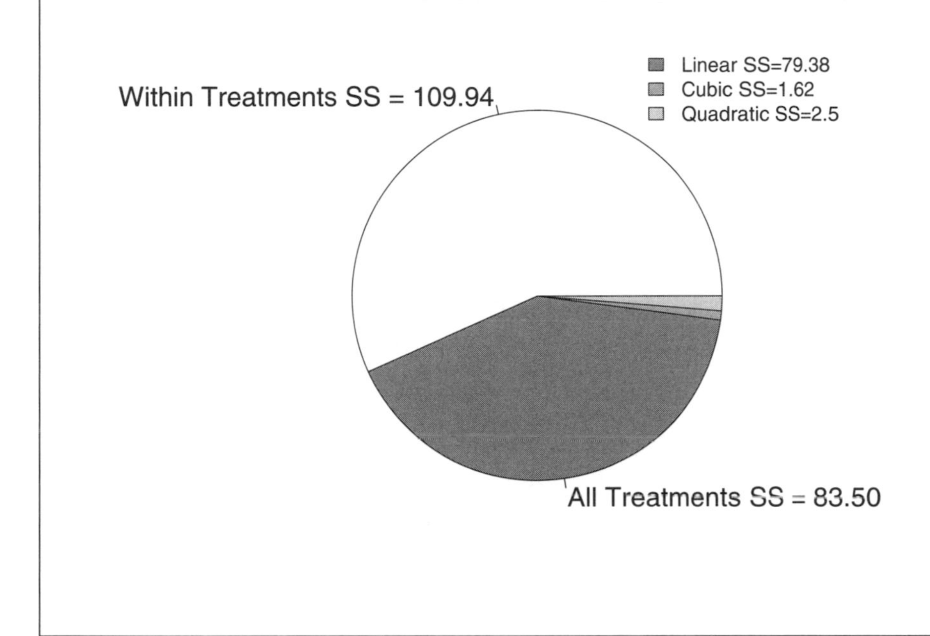

We now illustrate the decomposition of treatment sum of squares: the sum of the three sum of squares for each comparison (79.38 + 2.5 + 1.62) equals the treatment sum of squares from the analysis of variance, or 83.50, shown in Table 8.1. This is depicted graphically in Figure 8.4. Thus, a set of orthogonal comparisons decomposes the omnibus question into smaller chunks that test specific patterns in the treatment means.

8.12 Tests of Significance for Planned Comparisons

Planned comparisons provide information that is relevant to the interpretation of the outcome of a well-designed study. They are usually limited in number and planned prior to the examination of the data obtained in the experiment. In almost all cases, the comparisons will be based on theoretical or practical considerations of importance. In the drug experiment, for example, determining whether the combination of drugs A and B is or is not more effective than either drug A or drug B alone would be of practical, if not theoretical, importance. The comparison of the difference between the mean of drug A and the mean of drug B would also be of interest.

In testing planned orthogonal comparisons, one sets the protection level by the number of degrees of freedom associated with the MS_T (that is, $k - 1$). From an experimental point of view, it is difficult to perceive any great difference between testing $k - 1$ planned orthogonal comparisons on the one hand and $k - 1$ planned comparisons, not all of which are necessarily orthogonal, on the other hand. If the number of planned comparisons to be tested exceeds $k - 1$, then some experimenters may also consider it reasonable to perform these tests in the same manner in which they would perform planned orthogonal tests. Other experimenters may be more concerned about Type I errors and decide to use a more conservative test, such as the Bonferroni test, for planned but not necessarily orthogonal comparisons when the number of such comparisons is greater than $k - 1$. Recall that the Bonferroni procedure involves changing the criteria of the individual tests so that the overall Type I error remains at the desired level, say $\alpha = 0.05$. If one performs c comparisons, then each comparison can be tested using a criterion of $0.05/c$. Keep in mind that if the comparisons are not orthogonal, the Bonferroni procedure provides an upper-bound approximation.

Our suggestion is that if the experimenter is concerned about the Type I error rate, then he or she should replicate the study. Replication is a better way of dealing with concerns over Type I error rates than tinkering with the

α criterion level. But when replication is costly or not feasible (for example, when conducting a 30-year longitudinal study), then the Bonferroni correction provides one way to alleviate concerns about Type I errors that emerge when performing multiple tests. For a discussion of issues surrounding replication see Greenwald, Gonzalez, Harris, and Guthrie (1996).

8.13 Effect Size for Comparisons

In this section we present a measure of effect size for a specific comparison. The definitional formula is given by

$$r = \sqrt{\frac{SS_C}{SS_C + SS_W}} \tag{8.5}$$

where SS_C refers to the sum of squares for comparison (as given in Section 8.11) and SS_W refers to the sum of squares within treatments. For example, in Section 8.11 we presented the linear trend comparison, which had a sum of squares equal to 79.38. Recall that the sum of squares within treatments for this example was equal to 109.44. Thus, the effect size r for the linear comparison is equal to

$$0.648 = \sqrt{\frac{79.38}{79.38 + 109.44}}$$

The definition of effect size presented here compares the sum of squares of the specific comparison to the sum of squares within treatments. A more convenient and equivalent version of Equation 8.5 is

$$r = \sqrt{\frac{t^2}{t^2 + df}} \tag{8.6}$$

where the t corresponds to the observed t of the comparison and df corresponds to the degrees of freedom associated with the within-treatment MS_W

term. Both Equations 8.5 and 8.6 yield identical results. Equation 8.6 is more versatile because it can be applied readily to computer output and to published papers. Applying Equation 8.6 to the linear comparison example, we verify that it yields the same answer as Equation 8.5; that is, $t = 5.11$ and $df = 36$, thus

$$0.648 = \sqrt{\frac{5.11^2}{5.11^2 + 36}}$$

8.14 The Equality of Variance Assumption

In this chapter we reviewed the topic of comparisons under the assumption of equal variances. The homogeneity of variance assumption should be checked whenever comparisons are tested. Fortunately, there is a generalization of the Welch t test presented in an earlier chapter that permits testing of comparisons even when the pooling assumption may not be justified (Brown & Forsythe, 1974). The logic is the same as for the Welch t test—the degrees of freedom are adjusted to take into account the discrepancy in the treatment variances. This more general test of a comparison is now implemented in many statistical packages, usually under the label "separate variance" test to indicate that the variances are not pooled.

8.15 Unequal Sample Size

Unequal sample sizes are not a problem for the comparison test presented in this chapter (though an ANOVA purist would be careful to define orthogonality to take into account differences in sample size). The computation of the standard error uses Equation 8.2 to take into account the different sample sizes. Unequal sample sizes will become an issue when we discuss factorial designs, and we will return to this issue in a later chapter.

8.16 Unplanned Comparisons

We describe a procedure developed by Scheffé (1953) that can be used to test the significance of any and all comparisons on a set of k means, including comparisons that may be suggested after observing the treatment means themselves. In other words, we do not need to plan the comparisons in advance, nor do the comparisons need to be limited in number, nor do the comparisons need to be orthogonal. The Scheffé test provides such flexibility because of a very clever idea that Scheffé formalized. Scheffé's test can, of course, be used for testing planned or orthogonal comparisons as well as pairwise comparisons of the kind described in the preceding chapter. The test will be more conservative than the procedures described for testing planned or orthogonal comparisons; that is, larger observed differences from the null hypothesis will be required for statistical significance by the Scheffé criterion, so one may not want to use the Scheffé test in all settings. For example, if the investigator only wants to test all possible pairwise comparisons, then the Tukey test is sufficient.

We emphasize, however, that if the omnibus $F = MS_T/MS_W$ is not statistically significant at a given α criterion, then no comparison will be judged significant by the Scheffé test at the same α criterion. It is useless, in other words, to apply the Scheffé test to comparisons when the omnibus test is not significant. The rationale for this assertion will be explained in more detail later in the chapter. On the other hand, if $F = MS_T/MS_W$ is significant with, say, $\alpha = 0.05$, then there will be at least one comparison on the treatment means that will also be statistically significant by the Scheffé criterion. There may, of course, be more than one statistically significant comparison.

8.16.1 Some Examples of the Scheffé Test

Table 8.6 shows some of the many possible comparisons that could be made on a set of $k = 4$ treatment means. The first 6 comparisons are pairwise comparisons; the next 12 comparisons are between one mean and the average of two other means; the next 3 comparisons are between the average of two means and the average of two other means; and the last 4 comparisons are between one mean and the average of the other three means. Since there are at most $k - 1 = 3$ orthogonal comparisons possible, Table 8.6 has a high degree of redundancy across the 25 comparisons that are shown.

The comparisons shown in Table 8.6 do not exhaust all the possibilities. For example, the table does not show the linear, quadratic, and cubic comparisons that could be performed on $k = 4$ treatment means. Nor does the table show comparisons of the kind

$$2 \quad 1 \quad -3 \quad 0$$

or

$$2 \quad 7 \quad -6 \quad -3$$

of which there are an unlimited number.

We do not recommend that experimenters perform many comparisons simply because it is possible. Rather, we are pointing out that the Scheffé test gives the researcher the opportunity to test as many comparisons as he or she wants. The price paid, however, is a very conservative criterion for statistical significance. Further, there may be problems in interpretation when there are many comparison tests. Some of those comparisons will be redundant (that is, nonorthogonal); thus two comparisons may turn out to be significant because they overlap in how they weight the treatment means. For instance, these two comparisons on four treatment means overlap on the first two treatments: (1, −1, 0, 0) and (1, −1, 1, −1). If the four means are (6, 2, 3, 3), then both comparisons will be significant because they pick up the 6 versus 2 difference in the first two treatment means. The Scheffé test merely controls the Type I error rate; it does not identify which contrasts represent meaningful research questions—that task is left to the researcher.

Table 8.6: Some possible comparisons on $k = 4$ treatment means.

Comparison	$\bar{X}_{1.}$ 17.2	$\bar{X}_{2.}$ 19.4	$\bar{X}_{3.}$ 15.8	$\bar{X}_{4.}$ 19.0	$\sum a^2$	d
1 vs. 2	1	−1	0	0	2	−2.2
1 vs. 3	1	0	−1	0	2	1.4
1 vs. 4	1	0	0	−1	2	−1.8
2 vs. 3	0	1	−1	0	2	3.6
2 vs. 4	0	1	0	−1	2	0.4
3 vs. 4	0	0	1	−1	2	−3.2
1 vs. 2 + 3	2	−1	−1	0	6	−0.8
1 vs. 2 + 4	2	−1	0	−1	6	−4.0
1 vs. 3 + 4	2	0	−1	−1	6	−0.4
2 vs. 1 + 3	−1	2	−1	0	6	5.8
2 vs. 1 + 4	−1	2	0	−1	6	2.6
2 vs. 3 + 4	0	2	−1	−1	6	4.0
3 vs. 1 + 2	−1	−1	2	0	6	−5.0
3 vs. 1 + 4	−1	0	2	−1	6	−4.6
3 vs. 2 + 4	0	−1	2	−1	6	−6.8
4 vs. 1 + 2	−1	−1	0	2	6	1.4
4 vs. 1 + 3	−1	0	−1	2	6	5.0
4 vs. 2 + 3	0	−1	−1	2	6	2.8
1 + 2 vs. 3 + 4	1	1	−1	−1	4	1.8
1 + 3 vs. 2 + 4	1	−1	1	−1	4	−5.4
1 + 4 vs. 2 + 3	1	−1	−1	1	4	1.0
1 vs. 2 + 3 + 4	3	−1	−1	−1	12	−2.6
2 vs. 1 + 3 + 4	−1	3	−1	−1	12	6.2
3 vs. 1 + 2 + 4	−1	−1	3	−1	12	−8.2
4 vs. 1 + 2 + 3	−1	−1	−1	3	12	4.6
D_1	5	−4	0	0	−30	
D_2	0	0	8	−3	120	
D_3	11	11	−9	−9	−300	

Note: Columns 2–5 list the coefficients for each of four treatments, column 6 lists the sum of the squared coefficients, and column 7 lists the value of d, the sum of the products of the comparison coefficient with the respective treatment mean.

8.16.2 The Scheffé Test for Comparisons

If comparisons are made on the treatment means, then, as we saw before, the standard error of the comparison will be given by

$$s_{d_i} = \sqrt{MS_W \sum \frac{a_i^2}{n_i}} \tag{8.7}$$

The test of significance for the comparison is then made by finding

$$t = \frac{d_i}{s_{d_i}} \tag{8.8}$$

The numerator of Equation 8.8 (that is, d_i) is computed by multiplying the comparison coefficient with the respective treatment mean and summing the products.

The Scheffé test uses a special criterion to compare the observed t ratio of the comparison. The t defined by Equation 8.8 is evaluated for significance by comparing it with the Scheffé criterion

$$t' = \sqrt{(k - 1)\,F} \tag{8.9}$$

where k is the number of treatments and F is the critical value from Table B.2 in Appendix B for $(k - 1)$ numerator degrees of freedom and the degrees of freedom for the denominator corresponding to MS_W.

Confidence limits for the comparison value d_i can also be constructed under the Scheffé framework by the formula

$$d_i \pm t' s_{d_i} \tag{8.10}$$

Defining confidence intervals in this manner for a set of comparison values d_i yields what is known as a **simultaneous confidence interval**. The Type I error rate for the set of confidence intervals is controlled by the Scheffé criterion.

Because there is an MS_W term in Equation 8.7, the equality of variance assumption is invoked as usual (that is, the homogeneity of variance assumption is what justifies the pooling of the treatment variances into MS_W). The Scheffé test has been generalized in a manner that relaxes the equality of variance assumption (Brown & Forsythe, 1974). However, this generalized Scheffé test has not been widely implemented in standard statistical packages.

8.16.3 Properties of the Scheffé Test

The Scheffé test is a statistical test that permits the investigator to examine the data and to make an unlimited number of comparisons. Regardless of the number of comparisons tested, the protection level remains greater than or equal to $1 - \alpha$, and $P(E)$ remains less than or equal to α. That is, if comparisons are tested with the Scheffé procedure, the probability of making a Type I error will be less than or equal to α—regardless of how many comparisons one chooses to make.

As we pointed out earlier, the Scheffé test has the property that if the omnibus $F = MS_T/MS_W$ is not statistically significant at α, then no comparison that can be made on the k treatment means will be statistically significant. In the example, if the F test for the treatment mean square had not been equal to or greater than $F = 2.86$ (which is the critical value for 3 and 36 degrees of freedom with $\alpha = 0.05$), then there would not be any comparison on the $k = 4$ means that would result in $t \geq t' = 2.93$ (that is, no comparison would reach statistical significance). If the omnibus $F = MS_T/MS_W$ is statistically significant with $\alpha = 0.05$, then there will be at least one comparison that can be made on the treatment means that will also be significant, and, of course, there may be more than one comparison that will be statistically significant. It does not follow, however, that comparisons found to be statistically significant will necessarily be those that are of interest to the experimenter or even correspond to meaningful research questions.

The Scheffé test is based on Scheffé's observation that it is possible to find a single comparison that yields the same sum of squares as the overall treatment sum of squares. Recall that orthogonal comparisons decompose the treatment sum of squares. Scheffé showed that it is always possible to find a comparison that completely exhausts the sum of squares treatment SS_T. That is, there exists a comparison with sum of squares SS_C equal to the entire sum of squares for treatments SS_T. This comparison is called the "maximum comparison" because a comparison cannot be greater than this single comparison (otherwise it would be greater than the treatment sum of squares, which it cannot be). Scheffé derived the sampling distribution for this maximum comparison and thus proved the test that we now refer to as the Scheffé test. Because this test is based on sampling distribution of the maximum comparison, it has the property that it can be used for any number of comparisons, and it will automatically provide a correction for the Type I error problem.

8.17 Summary

The comparison of treatment means described in this chapter is an important tool in research. Frequently, a researcher conducts a study to test a particular hypothesis about a pattern of treatment means. The omnibus test we discussed in Chapter 6 is useless to the researcher who is not interested in the global question of whether the means differ but instead is interested in testing specific hypotheses about the treatment means. As long as the research question can be operationalized in terms of a weighted sum of treatment means, then a comparison to test that predicted pattern directly is most useful.

The Scheffé test is the ideal test for the experimenter who does not have planned comparisons and who wishes to explore thoroughly the outcome of an experiment, making any and all comparisons suggested by the data. The experimenter can do so knowing that, regardless of the number of comparisons made, $P(E)$ will be less than or equal to the chosen value of α, say $\alpha = 0.05$. Of course, this flexibility in statistical tests comes at the price of a very conservative criterion for statistical significance. This may lead to lower statistical power.

In the case where an experimenter will only test all possible pairwise comparisons, then we recommend the Tukey test (Chapter 7) over the Scheffé test. The Tukey test directly takes into account the sampling distribution relevant to pairwise comparisons and will not be as conservative as the Scheffé test, which takes into account the sampling distribution for a much larger number of possible comparisons.

8.18 Questions and Problems

1. A study includes a control group and 5 treatment groups, with 10 observations for each group. We have $MS_W = 36.00$ for the within-treatment mean square. The means for the six groups are given here:

Control	A	B	C	D	E
18.6	20.5	23.4	19.6	28.3	26.2

   Perform a comparison that tests whether the control group differs from the average of the five treatments. Explain why this comparison is not identical to a two-sample t test that compares the control group to all other participants (that is, calling participants in the other five treatments a single group).

2. We have a between-subjects experimental design in which the treatments consist of three equally spaced intervals of testing. One group is tested for retention of learned material after 12 hours, another group after 24 hours, and the third group after 36 hours. The means for the groups are 11.0, 9.0, and 5.0, respectively. We have $n = 10$ participants in each group, and MS_W is equal to 20.0 with 27 degrees of freedom.

(a) Do the treatment means differ significantly from one another? Explain how you interpret this question, and justify the particular statistical test you perform.

(b) Test the linear trend component of the means for significance. What is the effect size of the linear component?

3. In an experiment involving $k = 4$ treatments, $n = 10$ participants were assigned at random to each treatment. We have $MS_W = 16.0$. Find the standard error for each of the following comparisons:

(a)	1	1	−1	−1
(b)	1	−1	0	0
(c)	3	−1	−1	−1

4. What is meant by a comparison on a set of k means?

5. What is the condition for two comparisons to be called orthogonal?

6. We have $k = 5$ treatments with $n = 8$ participants assigned at random to each treatment. If the omnibus $F = MS_T/MS_W$ is not significant with $\alpha = 0.05$, explain why there will not be a comparison on the treatment means that will be significant according to the Scheffé test.

7. We have $n = 10$ participants assigned at random to each of $k = 8$ treatments. We wish to make $c = 5$ unplanned comparisons and to have $P(E) \leq 0.05$. What value of t (that is, $t_{critical}$) will be required for significance for these comparisons, using Scheffé's test?

8. Suppose that in the experiment described in Problem 7 the $c = 5$ comparisons are planned comparisons, decided upon prior to conducting the experiment. We are concerned, however, about Type I errors and want to have $P(E) \leq 0.05$ for the set of $c = 5$ comparisons. If the Bonferroni t statistic is used to control $P(E)$, what value of t will be required for significance in testing each of the five comparisons? Which test, Scheffé or Bonferroni, is more conservative in this example?

9. Use data presented as part of Question 2 in Chapter 6 (page 187), which presented raw data for six groups. Imagine that these six conditions are six levels of increasing dose of a drug, with Treatment 1 receiving the lowest dose and Treatment 6 receiving the highest dose.

   (a) Construct a set of polynomial comparisons for these six treatment groups.

   (b) Compute the t test for each of these comparisons.

   (c) Show that the sum of squares across the set of orthogonal comparisons equals the sum of squares from the one-way ANOVA.

9

The 2^k Between-Subjects Factorial Experiment

9.1 Introduction

Experimenters are sometimes concerned with the influence of two or more independent variables, usually called **factors**, on the means of the dependent variable. The number of ways a particular factor is varied is referred to as the number of levels of the factor. A factor that is varied in two ways has two levels; a factor that is varied in three ways has three levels; and so on. When the treatments consist of all possible combinations of levels from different factors, the experiment is described as a **complete factorial experiment**.

In this chapter, we will introduce the analysis of variance for a 2^k between-subjects factorial experiment. A 2^k between-subjects factorial experiment is one in which there are k factors where each factor has two levels. Although the example will be confined to a 2^3 factorial experiment, the technique can be generalized to any 2^k factorial experiment where k is greater than or equal to 2. The method of analysis will be similar to the one-way between-subjects design presented in Chapter 6. The two-sample t test from

Chapter 4 is a special case of the design presented here because it is a 2^1 design (that is, $k = 1$ because there is only one factor). The special case of 2^k makes it relatively easy to explain the relevant concepts. In later chapters we will generalize to factors that have an arbitrary number of levels (not just factors with two levels).

9.2 An Example of a 2^3 Factorial Experiment

As an illustration suppose that the dependent variable is a measure of memory, the number of correctly recalled items. The researcher wants to test the effect of three factors. One factor of interest is the number of times the material is studied. This factor is varied in two ways (that is, has two levels): presenting the material once for study and presenting the material twice for study. We designate this factor as A. The two levels of Factor A are represented by A_1, corresponding to one presentation, and A_2, corresponding to two presentations. A second factor of interest is the mode of presentation, and this factor is also varied in two ways. In one condition, the participants themselves read the passage, and we refer to this method as the visual mode of presentation. In the second condition the passage is read to participants. We call this method the auditory mode of presentation. We designate the mode of presentation as the B factor and the two levels as B_1, corresponding to the visual mode, and B_2, corresponding to the auditory mode. A third factor of interest is the time of testing, and this factor is also varied in two ways. We designate this factor as C, and we let C_1 correspond to an immediate test and C_2 to a delayed test 24 hours after the study session.

When working with factorial designs, it is convenient to specify levels of a factor with subscripts of different numbers such as B_1. It gets cumbersome to specify level names such as B_{auditory} or C_{delayed}, even though such labeling may be clearer. It is best to invest time to learn the notation, and then general designs will become easier to write and think about.

A particular treatment will be obtained by selecting one level from each of the three factors. For example, one treatment will be $A_1B_1C_1$ and represents a treatment consisting of a single presentation using the visual mode with an immediate test. Be sure you understand how the subscripts for A, B, and C denote single presentation, visual mode, and immediate test, respectively. The total number of different treatments for this design will be $2 \times 2 \times 2 = 8$, and they are all listed in Table 9.1. In this example there are $n = 10$ observations (participants) for each treatment. Thus there are a total of $8 \times 10 = 80$ participants in this study, and because each treatment has a different group of 10 participants it is called a between-subjects factorial design. The data for this hypothetical experiment are given in Table 9.2.

The SPSS file is organized differently from how the data are presented in Table 9.2. Rather than organizing the data by eight columns, it is necessary to specify each factor by its code so that the SPSS file would have only four columns of numbers: one column denotes the two levels of factor A (a column containing 1 or 2), one column of codes denotes the two levels of factor B, one column denotes the two levels of factor C, and a fourth column for the observed data. So the SPSS file would consist of four columns and 80 rows. The first row would be 1, 1, 1, 8, the second row would be 1, 1, 1, 7, the third row would be 1, 1, 1, 4, ..., the final row would be 2, 2, 2, 6 (because the 80th observation belongs to level two of factor A, level two of factor B, level two of factor C, and is an observation of 6).

The boxplot is presented in Figure 9.1. The sample size is relatively small; the outliers and variance differences are not extreme enough to cause concern. The SPSS syntax for the boxplot is

```
EXAMINE
 VARIABLES=data BY A by B by C
 /PLOT BOXPLOT
 /COMPARE GROUP
 /NOTOTAL.
```

Figure 9.1: SPSS boxplots for data in Table 9.2.

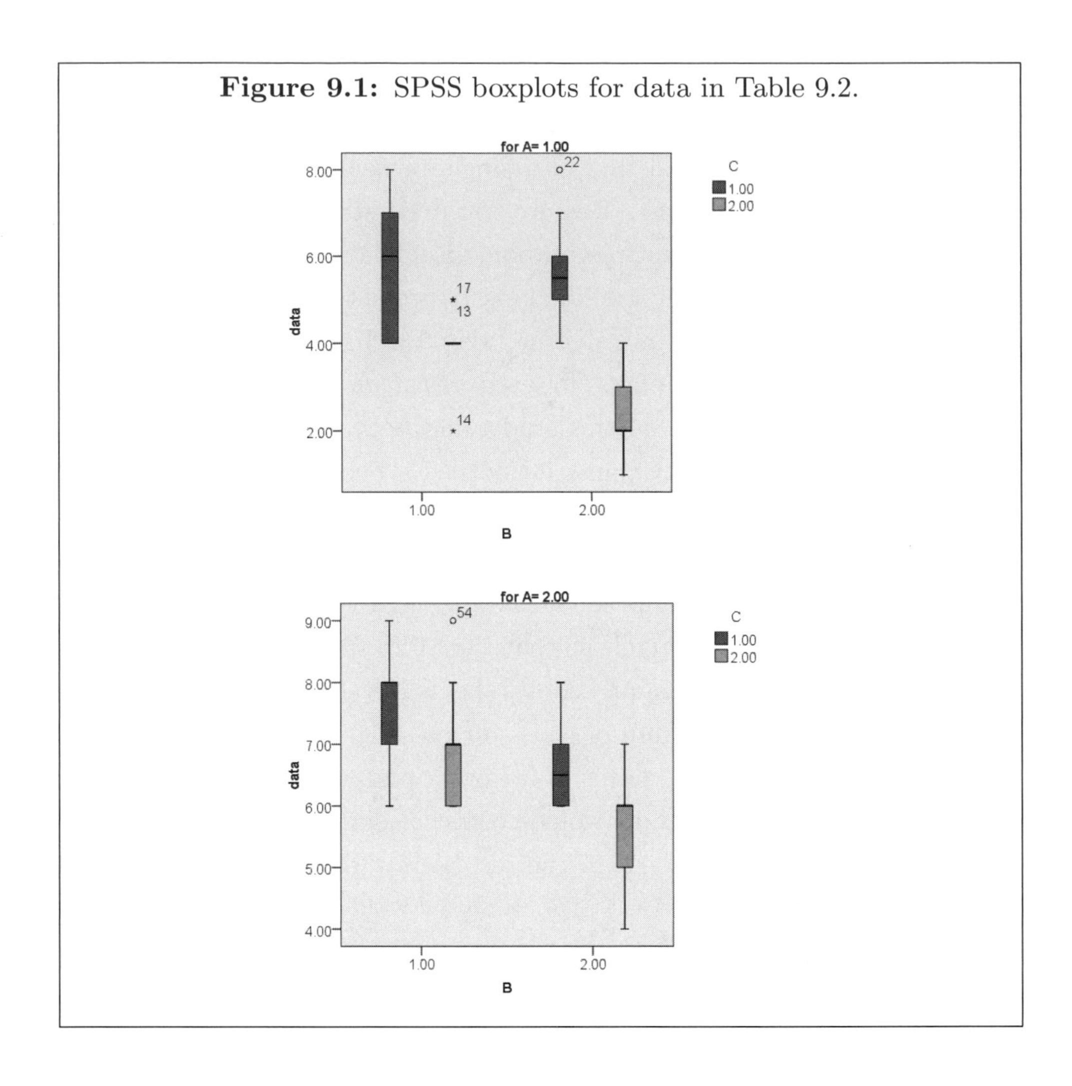

In this experiment, we regard the levels of the factors as having been selected for investigation because they were of experimental interest. The levels of the factors are regarded as fixed and not as representing a random sampling from a larger population of levels. We are not concerned, for example, with being able to generalize beyond the particular number of presentations, the particular modes, or the particular times actually investigated. We will be interested however in generalizing the result to a population of participants. Thus, we conceptualize participants as representing a sample from a larger population. The appropriate error term for

Table 9.1: The eight treatment combinations of the 2^3 factorial experiment.

Treatment	Number	Mode	Time	Number of participants
$A_1B_1C_1$	One	Visual	Immediate	10
$A_1B_1C_2$	One	Visual	Delayed	10
$A_1B_2C_1$	One	Auditory	Immediate	10
$A_1B_2C_2$	One	Auditory	Delayed	10
$A_2B_1C_1$	Two	Visual	Immediate	10
$A_2B_1C_2$	Two	Visual	Delayed	10
$A_2B_2C_1$	Two	Auditory	Immediate	10
$A_2B_2C_2$	Two	Auditory	Delayed	10

all tests of significance will be the within-treatment mean square. We now turn to the computation of the within-treatment mean square term for a 2^3 factorial design and check the equality of variance assumption.

9.3 Assumption of Homogeneity of Variance

We can compute an error term within each of the eight treatments to assess the variability of the scores with respect to each one's treatment mean. In the present example with eight treatments we have (using the formula for SS_W presented in Chapter 6):

$$\begin{aligned}
SS_{W_1} &= \sum(X_i - 6)^2 &= 22.00\\
SS_{W_2} &= \sum(X_i - 4)^2 &= 6.00\\
SS_{W_3} &= \sum(X_i - 5.7)^2 &= 12.10\\
SS_{W_4} &= \sum(X_i - 2.3)^2 &= 6.10\\
SS_{W_5} &= \sum(X_i - 7.7)^2 &= 10.10\\
SS_{W_6} &= \sum(X_i - 6.9)^2 &= 8.90\\
SS_{W_7} &= \sum(X_i - 6.7)^2 &= 6.10\\
SS_{W_8} &= \sum(X_i - 5.7)^2 &= 8.10
\end{aligned}$$

Table 9.2: Outcome of a $2 \times 2 \times 2$ factorial experiment with $n = 10$ participants assigned at random to each treatment combination.

	A_1				A_2			
	B_1		B_2		B_1		B_2	
	C_1	C_2	C_1	C_2	C_1	C_2	C_1	C_2
	8	4	4	4	9	7	7	7
	7	4	8	2	9	7	6	6
	4	5	7	2	8	6	7	5
	6	2	5	2	8	9	6	5
	6	4	6	I	8	7	6	5
	4	4	6	3	8	7	6	4
	4	5	5	2	7	6	6	6
	6	4	6	2	8	6	7	6
	8	4	5	2	6	8	8	7
	7	4	5	3	6	6	8	6
$\bar{X}$	6.0	4.0	5.7	2.3	7.7	6.9	6.7	5.7

Adding the above sums of squares, we have an estimate of the within-treatment sum of squares:

$$22.00 + 6.00 + 12.10 + 6.10 + 10.10 + 8.90 + 6.10 + 8.10 = 79.40$$

Under the assumption that the population variance is the same for all treatment groups, the separate variance estimates will all be estimates of the same parameter σ^2. Dividing each of the sums of squares by $n - 1 = 9$ (that is, the respective degrees of freedom for each treatment), we obtain, as the separate σ^2 estimates, 2.44, 0.67, 1.34, 0.68, 1.12, 0.99, 0.68, and 0.90. An examination of the equality of variance assumption (as described in Chapter 5 with a boxplot) shows that there is not sufficient evidence to reject the null hypothesis of homogeneous variances. The data, in other words, do not provide evidence against the hypothesis that the sample variances are all estimates of a common population variance. The pooled within-treatment mean squares term is simply the sum of all eight sums of squares (79.40) divided by the appropriate degrees of freedom, that is, $k(n-1) = 8(10-1) =$

72, and for this example equals

$$\begin{aligned} MS_W &= SS_W/df \\ &= 79.40/72 \\ &= 1.10 \end{aligned}$$

9.4 Factorial Data as a One-Way Between-Subjects Design

Before illustrating the 2^3 factorial design for this example, it will be instructive to analyze these data using the technique presented in Chapter 6. We will act as though there is a single factor with eight treatments and, for now, ignore the specific factorial combination that is present in this example. Thus, we model the observations as follows:

$$X_{ij} = \mu + \alpha_i + \epsilon_{ij} \tag{9.1}$$

where μ is the grand mean, the α terms represent the eight treatment effects, and the ϵ terms (all 80 of them in this example represent the random component of the model). Of course, this model doesn't consider the three factors each with two levels, but we want to present this model to provide a comparison for the factorial design.

The resulting source table is presented in Table 9.3. The MS_W term involves pooling the eight within-treatment variances, and, as we saw in the preceding section, is equal to 1.10. Recall that the treatment sum of squares is a measure of variability of each treatment mean around the grand mean (that is, the mean of all observations regardless of which treatment they came from). The treatment sum of squares is given by

$$n \sum (\bar{X}_i - \bar{X}_{..})^2$$

Table 9.3: Analysis of variance showing the treatment sum of squares and the within-treatment sum of squares for the data in Table 9.2.

Source of variation	Sum of squares	*df*	Mean square	*F*
Treatments	209.35	7	29.91	27.19
Within treatments	79.40	72	1.10	
Total	288.75	79		

when all treatments have the same sample size. In this example, the grand mean (the mean of the observations across all treatments) is 5.625, and the sum of squares treatment is

$$10[(6-5.625)^2 + (4-5.625)^2 + \cdots + (5.7-5.625)]^2 = 209.35$$

The treatment mean square term MS_T is simply the sum of squares treatment divided by the degrees of freedom ($k-1=7$) and is equal to 29.91.

Until this point our analysis has resulted in a partitioning of the total sum of squares and degrees of freedom into two parts. (See Figure 9.2.) One part, SS_T, is associated with differences among the eight treatment means and is based on $k-1=7$ degrees of freedom. The other part, SS_W, is associated with the variation within each of the treatment groups and has $k(n-1)=72$ degrees of freedom. Testing the treatment mean square for significance, we have $F = 29.91/1.10 = 27.19$ with 7 and 72 degrees of freedom. By comparison with the critical value of F from the F table (Table B.2.1 in Appendix B), we find that for 7 and 72 degrees of freedom the observed $F = 27.19$ is significant with $\alpha = 0.05$, and we conclude that the treatment means differ significantly from each other. However, this conclusion is not satisfying on at least two accounts: (1) it does not tell us which treatment means differ from each other; and (2) it does not take into account the special factorial structure in this design. The first concern is the issue we saw before (Chapter 6) that the omnibus F test is not informative about which treatment means differ from which. This can be handled by either *post hoc* pairwise comparisons of means (in Chapter 7) or contrasts (Chapter 8). We

Figure 9.2: Pie chart showing the decomposition of the sum of squares for the omnibus one-way design.

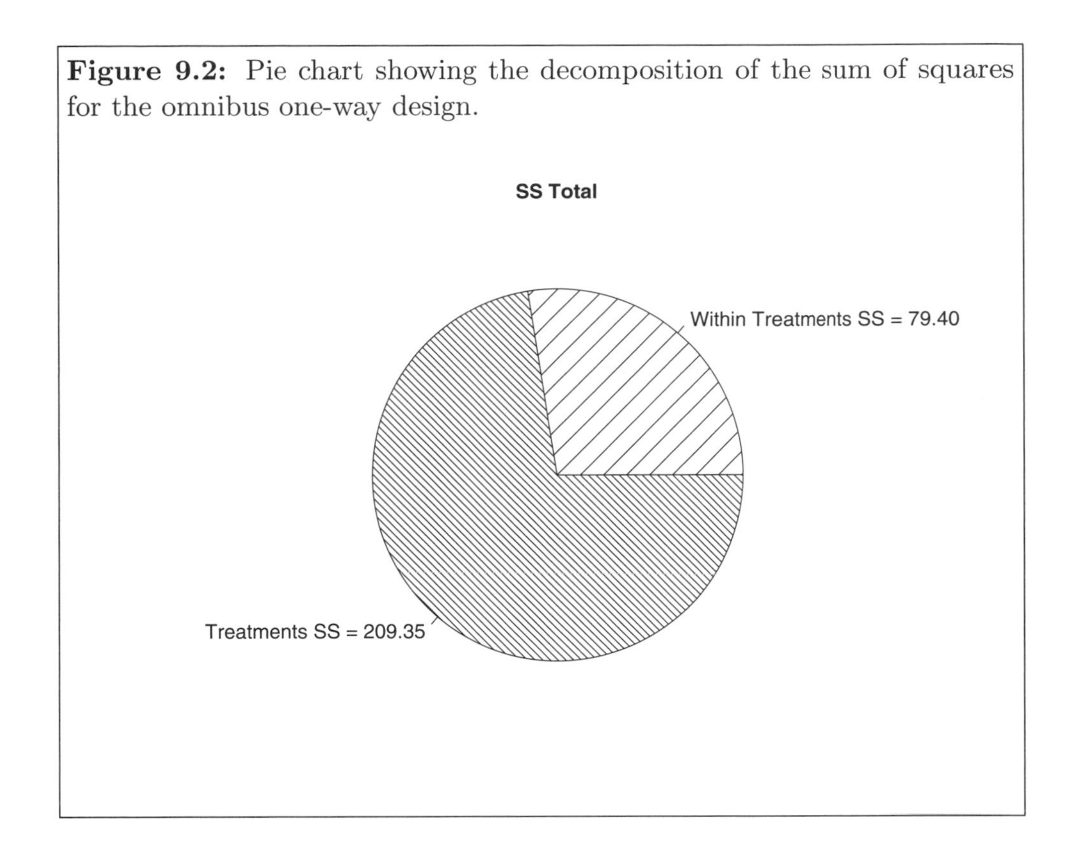

now turn to the second concern: using the factorial structure in the design to construct particular comparisons that provide information about the relative contributions of each factor individually (and in combination) to the treatment sum of squares.

9.5 Partitioning the Treatment Sum of Squares

The source table presented in Table 9.3 does not take advantage of the special factorial structure in the example. In fact, as the chapters on pairwise and planned comparisons suggested, the source table in Table 9.3 merely presents an omnibus test, which may not be useful to a researcher asking specific questions about patterns among the treatment means. We introduced the technique of planned comparisons as a way to ask specific questions on the means of a dependent variable.

The factorial design in this example implies a specific set of comparisons. For instance, a researcher may be interested in testing whether the mean for the participants who studied the material once (A_1) differs from the mean for the participants who studied the material twice (A_2), ignoring all other factors. Examination of Table 9.1 shows that there were four treatment conditions that had A_1 and four treatment conditions that had A_2. This research question about A_1 versus A_2 can be tested using the planned comparison 1, 1, 1, 1, -1, -1, -1, -1 (where the ordering of the eight treatments left to right corresponds to the ordering of the eight treatments in Table 9.1 from top to bottom). We are simply applying the technique of planned comparisons presented in Chapter 8.

Recall from Chapter 8 that the sum of squares for a comparison is given by

$$SS_i = \frac{d_i^2}{\sum \frac{a^2}{n_i}}$$

where d_i refers to the value that results from multiplying the treatment means by their respective coefficients and summing the products. Thus, for the comparison between the means of A_1 and A_2, we have

$$\begin{aligned} d &= (1)6 + (1)4 + (1)5.7 + (1)2.3 + (-1)7.7 + (-1)6.9 \\ &\quad +(-1)6.7 + (-1)5.7 \\ &= -9 \end{aligned}$$

which leads to a sum of squares for the comparison equal to $\frac{(-9)^2}{\sum \frac{a^2}{10}} = 101.25$. The test of significance for this comparison requires the computation of MS_C (the subscript denoting "comparison"), which is equal to 101.25 because comparisons have one degree of freedom. The observed F value is $MS_C/MS_W = 101.25/1.10 = 92.05$, a value that is statistically significant at $\alpha = 0.05$ because it exceeds the critical F value of 3.974.

Table 9.4: Complete source table for the factorial experiment of Table 9.2.

Source of variation	Sum of squares	*df*	Mean square	F
A: number	101.25	1	101.25	92.05
B: mode	22.05	1	22.05	20.05
C: time	64.80	1	64.80	58.91
$A \times B$: number $\times$ mode	0.05	1	0.05	0.05
$A \times C$: number $\times$ time	16.20	1	16.20	14.73
$B \times C$: mode $\times$ time	3.20	1	3.20	2.91
$A \times B \times C$:	1.80	1	1.80	1.64
Error: within treatments	79.40	72	1.10	
Total	288.75	79		

We just addressed one question a researcher may be interested in, but in a factorial design there is a family of questions that can be asked. This family of questions corresponds to a particular set of orthogonal comparisons. Two other questions of interest are: Does the mean for the participants who received a visual presentation (B_1) differ from the mean for the participants who received an auditory presentation (B_2)? and Does the mean for the participants who received an immediate test (C_1) differ from the mean for the participants who received a delayed test (C_2)? These two research questions correspond to the comparisons 1, 1, −1, −1, 1, 1, −1, −1 and 1, −1, 1, −1, 1, −1, 1, −1, respectively, using the ordering of treatments presented in Table 9.1.

The sum of squares and F value for these two tests appear in Table 9.4. The three comparisons considered so far are called **main effects** because they consider only one factor at a time (that is, they collapse information of the other factors).

We have identified three comparisons that correspond to the main effects of factors A, B, and C. The reader can verify that these three comparisons are orthogonal. With eight treatment groups we know that there can be up to $k - 1$ orthogonal comparisons. Given that three comparisons have already been earmarked as main effects, we can find four more mutually orthogonal comparisons. A convenient way to find these additional four comparisons in a way that all seven will be orthogonal is to create them by multiplying the comparisons we already have. For example, we can multiply in a special way the comparison for the main effect of A (1, 1, 1, 1, −1, −1, −1, −1) and the comparison for the main effect for B (1, 1, −1, −1, 1, 1, −1, −1). A new comparison can be formed by multiplying the two comparisons in the following way:

$$(1)(1),\ (1)(1),\ (1)(-1),\ (1)(-1),\ (-1)(1),\ (-1)(1),\ (-1)(-1),\ (-1)(-1)$$

which yields the comparison 1, 1, −1, −1, −1, −1, 1, 1. What is the meaning of this new comparison? One way to think about this comparison is to realize that it was created by combining information from factors A (one or two presentations) and B (auditory or visual presentation). By looking at the first four coefficients (and Table 9.1 to check on which conditions are being coded), it is clear that they are comparing the difference between B_1 and B_2 only for level A_1. The last four coefficients create a comparison between B_1 and B_2 but this time for level A_2. Because the first four coefficients have a different sign than the last four coefficients, the entire comparison tells whether the difference between B_1 and B_2 observed in A_1 differs from the difference between B_1 and B_2 observed in A_2.

This type of comparison is called an **interaction**—more specifically, a two-way interaction because it involves a combination of two factors. Interactions have this pattern of testing a difference between two (or more) differences, and consequently interactions are useful to test hypotheses about moderation: Does a difference between two or more levels of one factor vary as a function of a second factor?

The sum of squares for the interaction comparison between factor A and factor B is given by the same formula (after all, it is a comparison like any other comparison):

$$SS_i = \frac{d_i^2}{\sum \frac{a^2}{n_i}}$$

and for this example the sum of squares is equal to 0.05, as is the mean square term because comparisons have one degree of freedom in the numerator. An

F ratio is constructed in the usual way by dividing the mean square for the comparison by the MS_W term. In this example, $MS_W = 1.10$, and the mean square for the comparison MS_C is equal to 0.05. The resulting F ratio is 0.04545, which is not statistically significant at $\alpha = 0.05$.

It is possible to create two more interaction comparisons by performing analogous calculations. One interaction comparison can be made by multiplying the main effect for Factors A and C, and another interaction comparison by multiplying the main effect for Factors B and C.

So far we have identified a total of six comparisons—three main effects and three two-way interactions. But we know there can be up to 7 comparisons $(k - 1)$ that constitute a complete orthogonal set in this example. The seventh comparison can be found by multiplying the coefficients of all three main effect comparisons for Factors A, B, and C in the following manner:

$$(1)(1)(1),\ (1)(1)(-1),\ (1)(-1)(1),\ (1)(-1)(-1),$$
$$(-1)(1)(1),\ (-1)(1)(-1),\ (-1)(-1)(1),\ (-1)(-1)(-1)$$

which yields the new comparison 1, −1, −1, 1, −1, 1, 1, −1. This new comparison is a three-way interaction because it uses information from all three factors. A simple way to state this three-way interaction is that it compares the two-way interaction between factors B and C at level A_1 with the two-way interaction between factors B and C at level A_2. There are other ways of verbally expressing this three-way contrast, as we will show below graphically.

9.6 Summary of the Analysis of Variance

The seven comparisons described in the preceding section are summarized in Table 9.5. The values of F that have been entered in the source table (Table 9.4) were obtained by dividing the mean squares for each comparison

Table 9.5: A set of orthogonal comparisons for the $2 \times 2 \times 2$ factorial.

	Mean	A	B	C	$A \times B$	$A \times C$	$B \times C$	$A \times B \times C$
$A_1B_1C_1$	6	1	1	1	1	1	1	1
$A_1B_1C_2$	4	1	1	-1	1	-1	-1	-1
$A_1B_2C_1$	5.7	1	-1	1	-1	1	-1	-1
$A_1B_2C_2$	2.3	1	-1	-1	-1	-1	1	1
$A_2B_1C_1$	7.7	-1	1	1	-1	-1	1	-1
$A_2B_1C_2$	6.9	-1	1	-1	-1	1	-1	1
$A_2B_2C_1$	6.7	-1	-1	1	1	-1	-1	1
$A_2B_2C_2$	5.7	-1	-1	-1	1	1	1	-1
d		-9.0	4.2	7.2	-0.2	3.6	-1.6	-1.2
$\sum a^2$		8	8	8	8	8	8	8
SS_C		101.25	22.05	64.80	0.05	16.20	3.20	1.80

by the error mean square (that is, MS_W). So each test in Table 9.4 is based on the same error term and has 1 and 72 degrees of freedom.

An important observation about the sum of squares comparisons SS_C listed in Table 9.4 is that their sum is equivalent to the sum of squares for treatment SS_T presented in Table 9.3. That is, $101.25 + 22.05 + 64.80 + 0.05 + 16.20 + 3.20 + 1.80 = 209.35$. Thus, the set of seven orthogonal comparisons provides a perfect decomposition of the omnibus sum of squares. This is depicted in the pie chart in Figure 9.3. Any set of seven orthogonal comparisons would have also achieved a perfect decomposition, but this particular set of orthogonal comparisons is special because it corresponds to what most investigators are interested in comparing when they conduct a factorial design. That is, the comparisons provide information about main effects (the factors in isolation) and interactions (simultaneous effects of two or more factors). So, a 2^k between-subjects factorial design can be thought of as a one-way between-subjects design tested with a special set of orthogonal comparisons.

Figure 9.3: Pie chart showing how orthogonal contrasts decompose the sum of squares treatment. The sum of squares for one of the effects, the A $\times$ B interaction, is so small that it does not show up well in the figure.

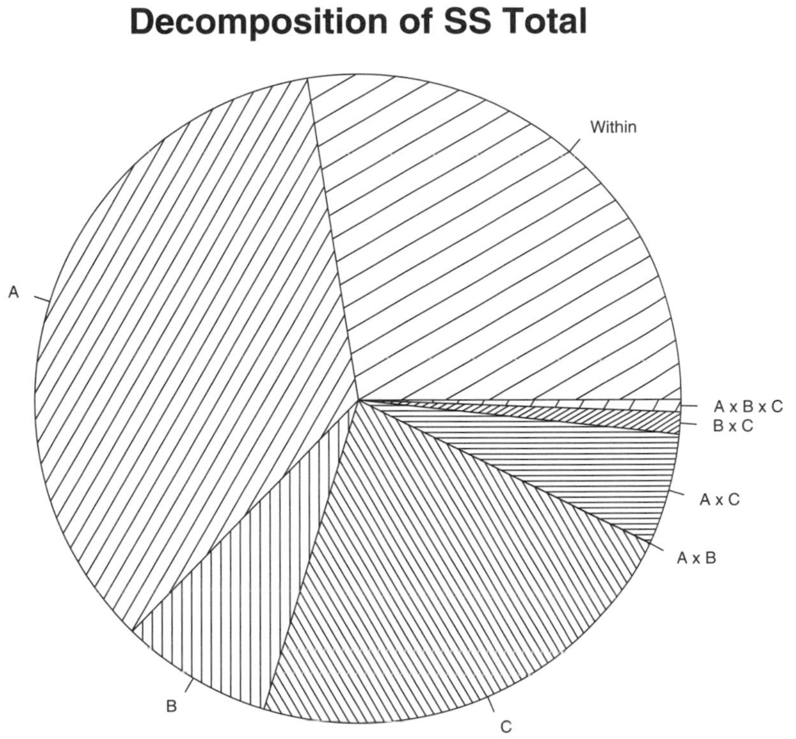

9.7 Graphs That Depict the Interactions

One way of examining the nature of a two-factor interaction is to present the interaction graphically. We place one of the factors, say B, on the X axis (horizontal) and graph the means for each level of A. For the example under discussion, the graphs for A_1 and A_2 are given in Figure 9.4. Each line in the figure corresponds to a different level of A. If the lines for A_1 and A_2 were exactly parallel, then the SS_C for the $A \times B$ interaction would be zero. The fact that the lines are very nearly parallel, within the limits of random

Figure 9.4: Means for levels of A at each level of B.

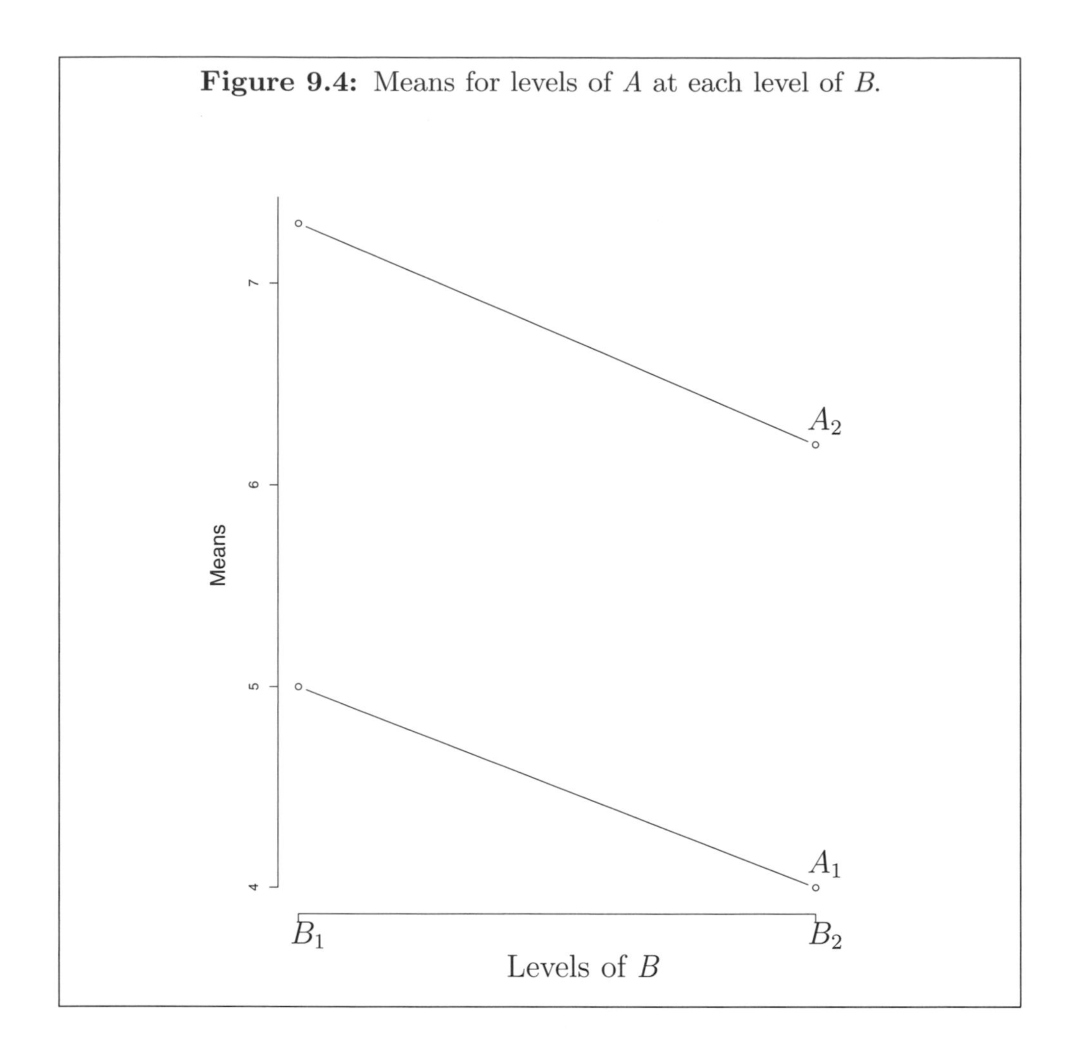

variation, corresponds to the observation that the $A \times B$ interaction is not statistically significant. Compare, however, the corresponding graph for the $A \times C$ interaction in Figure 9.5. Here, we have taken C for the X axis and plotted the means for A_1 and A_2. Note that the lines for A_1 and A_2 are not parallel. The fact that the $A \times C$ interaction is statistically significant is equivalent to the observation on the graph that the lines A_1 and A_2 are not parallel within the limits of random variation.

Figure 9.6 gives the graph for the $B \times C$ interaction, where we have chosen B for the X axis. The $B \times C$ interaction mean square is not signif-

Figure 9.5: Means for levels of A at each level of C.

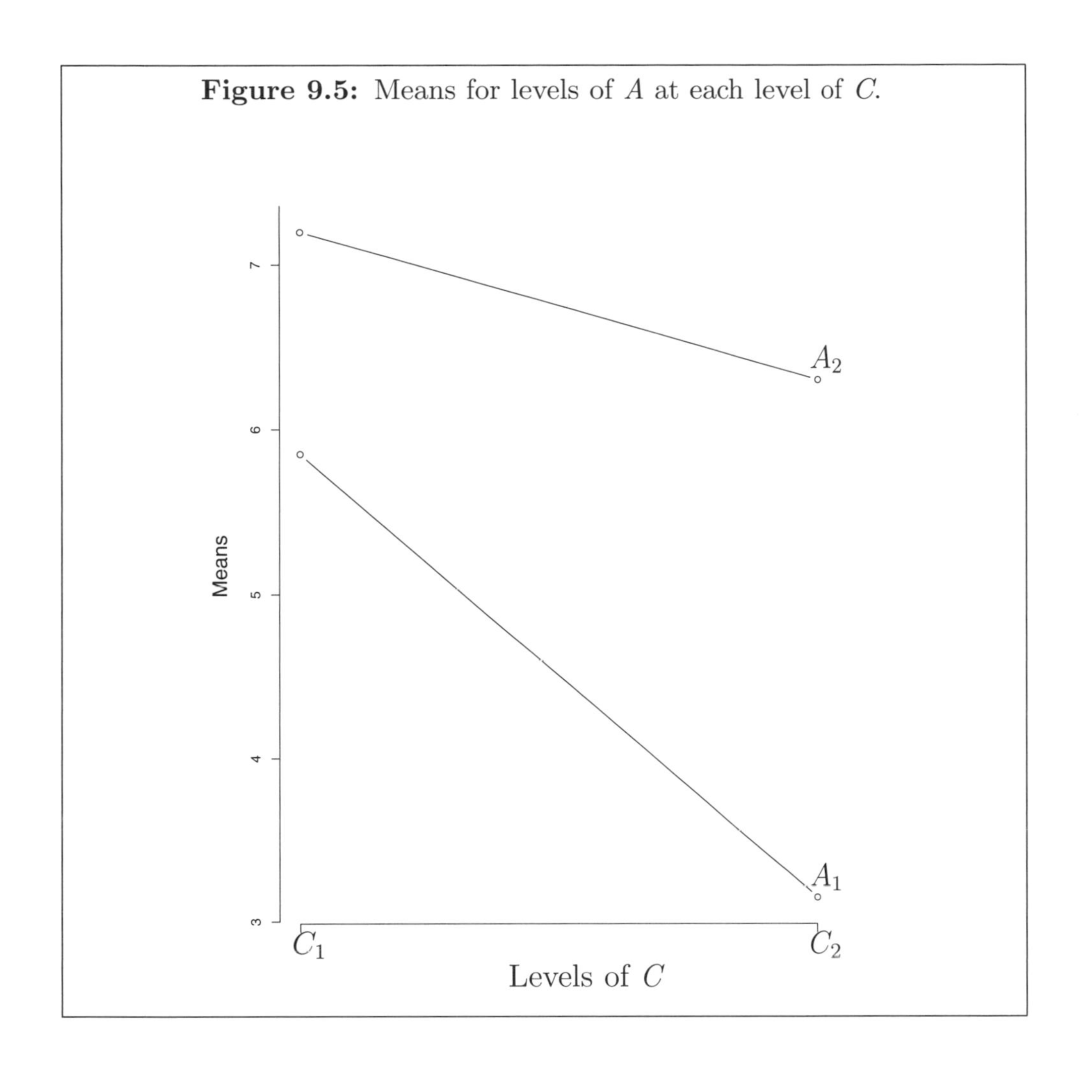

icant, and the lines for C_1 and C_2 are parallel within the limits of random variation. Which factor is chosen for the X axis and which factor is chosen to identify the different lines is arbitrary, though different choices may highlight patterns differently.

The graph of the three-way interaction involves plotting the two-way interaction separately at each level of the third factor. For example, Figure 9.7 shows the two-way interaction between Factors B and C for level A_1, and the two-way interaction between Factors B and C for level A_2. The graph suggests that the two-way interaction pattern between Factors B and

Figure 9.6: Means for levels of C at each level of B.

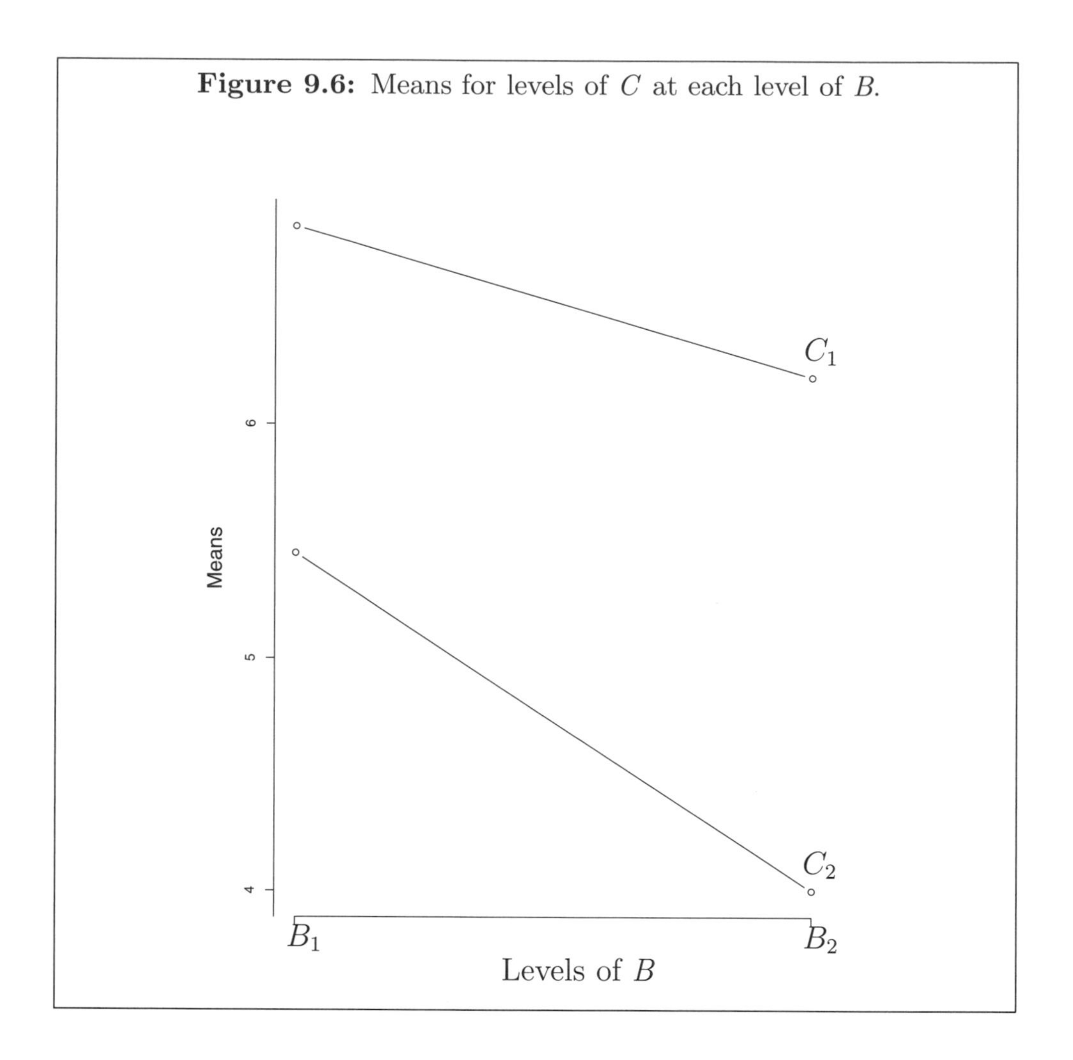

C at A_1 differs from the two-way interaction pattern between Factors B and C at A_2, but the three-way interaction test was not statistically significant, suggesting that the nonparallel lines are within the limits of random variation.

A purist will object to our use of line graphs because the treatments represent discrete levels of factors, such as auditory and visual mode. Some researchers prefer to display such graphs as bar plots. As an illustration we show the $A \times C$ interaction plot (Figure 9.5) as a bar plot in Figure 9.8. It

Figure 9.7: Means for the three-way interaction.

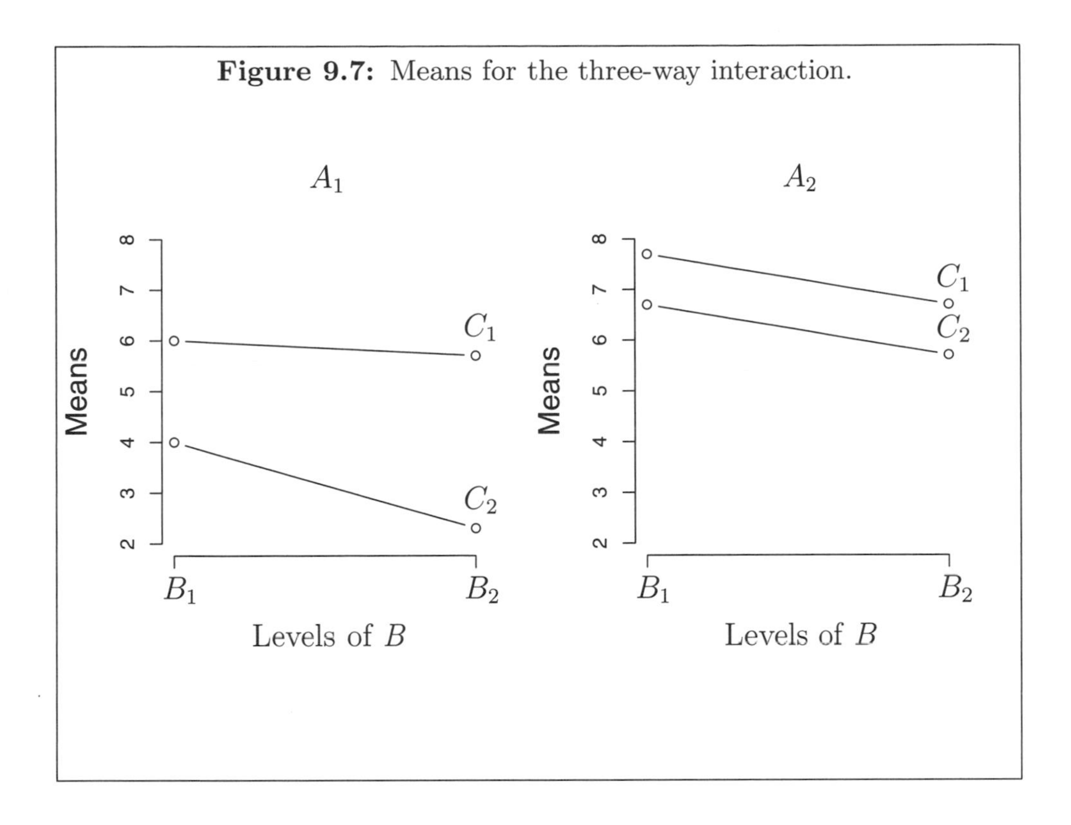

is also useful to include confidence bars (or plus/minus one standard error) around either the points in a line graph or the bars in a bar graph.

9.8 Other 2^k Factorial Experiments

The technique presented here that a 2^k design may be decomposed into individual single degree of freedom comparisons generalizes to any 2^k factorial design where $k \geq 2$. For instance, if $k = 4$, there will be a total of $2 \times 2 \times 2 \times 2 = 16$ treatments. With 16 treatments there are a total of 15 orthogonal comparisons. These 15 comparisons involve four main effects (one for each factor), six two-way interactions (the six ways to multiply all pairs of the four main effect comparisons), four three-way interactions, and

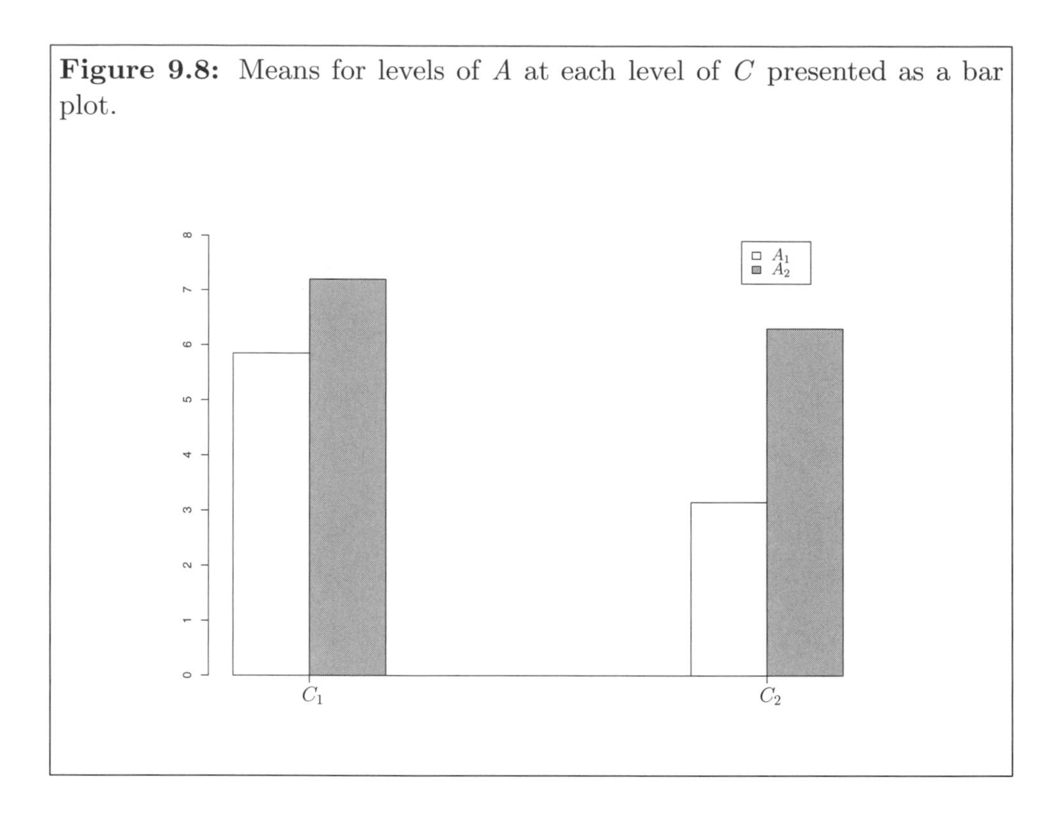

Figure 9.8: Means for levels of A at each level of C presented as a bar plot.

one four-way interaction. The logic is analogous to the previous 2^3 example, but there are more terms to keep track of when coding the comparison. The four-way interaction is identical to the product of the four main effect comparisons. Note that as the number of factors increases, so does the required number of participants to maintain the equivalent sample size in each cell. That is, if a researcher wanted $n = 10$ participants per treatment, then a 2^3 factorial design would require 80 participants, a 2^4 factorial design would require 160 participants, and a 2^5 factorial design would require 320 participants.

9.9 Notation and Sums of Squares for a Factorial Experiment

We consider now a general notation for the factorial experiment. This notation system will be useful in later chapters. We let a = the number of levels of Factor A, b = the number of levels of Factor B, c = the number of levels of Factor C, and n = the number of observations for each treatment combination. The number of treatment combinations will be $k = abc$, and when there are an equal number of observations n in each treatment, the total number of observations will be kn. A specific observation will be denoted X_{abcn}, with the understanding that when a, b, c, and n are used as subscripts they index a particular level of a factor or a particular participant. Thus with A at two levels, B at two levels, C at two levels, and $n = 10$, a as a subscript can take values of 1 or 2, b can take values of 1 or 2, c can take values of 1 or 2, and n can take values of 1 to 10. Thus, X_{1126} would correspond to the sixth observation of the first level of A, the first level of B, and the second level of C.

Then, using the dot notation, we can write the following identity:

$$\begin{aligned}
X_{abcn} - \bar{X}_{....} = {} & (X_{abcn} - \bar{X}_{abc.}) \\
& + (\bar{X}_{a...} - \bar{X}_{....}) \\
& + (\bar{X}_{.b..} - \bar{X}_{....}) \\
& + (\bar{X}_{..c.} - \bar{X}_{....}) \\
& + (\bar{X}_{ab..} - \bar{X}_{a...} - \bar{X}_{.b..} + \bar{X}_{....}) \\
& + (\bar{X}_{a.c.} - \bar{X}_{a...} - \bar{X}_{..c.} + \bar{X}_{....}) \\
& + (\bar{X}_{.bc.} - \bar{X}_{.b..} - \bar{X}_{..c.} + \bar{X}_{....}) \\
& + (\bar{X}_{abc.} + \bar{X}_{a...} + \bar{X}_{.b..} + \bar{X}_{..c.} \\
& - \bar{X}_{ab..} - \bar{X}_{a.c.} - \bar{X}_{.bc.} - \bar{X}_{....})
\end{aligned}$$

which states that the deviation of an observation from the overall mean can be expressed as the sum of the eight terms on the right.

If we square both sides of the above expression and sum over all observations, we find that the products of all terms on the right sum to zero. Thus with $k = abc$ (that is, the product of the number of levels of each factor), we have

$$\begin{aligned}
\sum_{1}^{kn}(X_{abcn} - \bar{X}_{....})^2 = & \sum_{1}^{kn}(X_{abcn} - \bar{X}_{abc.})^2 \\
& + bcn\sum_{1}^{a}(\bar{X}_{a...} - \bar{X}_{....})^2 \\
& + acn\sum_{1}^{b}(\bar{X}_{.b..} - \bar{X}_{....})^2 \\
& + abn\sum_{1}^{c}(\bar{X}_{..c.} - \bar{X}_{....})^2 \\
& + cn\sum_{1}^{ab}(\bar{X}_{ab..} - \bar{X}_{a...} - \bar{X}_{.b..} + \bar{X}_{....})^2 \\
& + bn\sum_{1}^{ac}(\bar{X}_{a.c.} - \bar{X}_{a...} - \bar{X}_{..c.} + \bar{X}_{....})^2 \\
& + an\sum_{1}^{bc}(\bar{X}_{.bc.} - \bar{X}_{.b..} - \bar{X}_{..c.} + \bar{X}_{....})^2 \\
& + n\sum_{1}^{k}(\bar{X}_{abc.} - \bar{X}_{a...} - \bar{X}_{.b..} + \bar{X}_{..c.} \\
& - \bar{X}_{ab..} - \bar{X}_{a.c.} - \bar{X}_{.bc.} - \bar{X}_{....})^2
\end{aligned}$$

The term on the left is the total sum of squares SS_{total}. The successive terms on the right give the sum of squares within treatments (SS_W), the sum of squares for A, the sum of squares for B, the sum of squares for C, the $A \times B$ sum of squares, the $A \times C$ sum of squares, the $B \times C$ sum of squares, and the $A \times B \times C$ sum of squares. Thus, the logic presented in Chapter 6 to partition the total sum of squares into two parts (treatment and within treatment) can be generalized to a partition of the total sum of squares involving main effects, interactions, and within treatment terms.

Stated another way, each observation X_{abcn} is assumed to be a function of all the main effects, all the two-way interactions, all the three-way interactions, etc., until all the terms in the 2^k factorial design are exhausted. Each of these terms corresponds to a particular portion of the treatment sum of squares SS_T, and in a 2^k factorial design each term also corresponds to a particular comparison.

The structural model for the factorial analysis of variance can be written as a decomposition of terms, as we saw in Chapter 6. In a factorial design with three factors each observation is decomposed into the following terms:

$$X_{ijk} = \mu + \alpha_i + \beta_j + \gamma_k + \alpha\beta_{ij} + \alpha\gamma_{ik} + \beta\gamma_{jk} + \alpha\beta\gamma_{ijk} + \epsilon_{ijk}$$

Aside from the grand mean μ and the error term ϵ there are seven terms. The three single terms (terms with one Greek letter) refer to the three main effects, the three double terms (two Greek letters) refer to the two-way interactions, and the single triple term (three Greek letters) refers to the three-way interaction. As in Chapter 6 the model is additive. The subscripts refer to the levels of each factor, so when factor A has two levels subscript i is either 1 or 2, when factor B has two levels subscript j is either 1 or 2, etc. The interaction terms have as many subscripts as factors in the interaction (for example, the $\alpha\beta\gamma$ term has three subscripts to denote the composition of all three factors).

SPSS offers several commands for factorial ANOVA. One is the UNIANOVA command. The syntax for this command, with data arbitrarily labeled "data" and factors arbitrarily labeled A, B, and C is

```
UNIANOVA data  BY A B C
  /DESIGN = A B C A*B A*C B*C A*B*C .
```

9.10 Summary

In this chapter we presented an example of a 2^k between-subjects factorial design and showed how to decompose the sum of squares for treatment SS_T into separate components involving specific comparisons. The factorial structure of this design permits the testing of factors in isolation (main effects) and in combination with other factors (interactions). The assumption of homogeneity of variance was invoked to permit the pooling of the within-treatment sum of squares. We showed how the factorial structure of the design can be addressed through individual comparisons, and thus the ideas presented here are a natural extension of those presented in Chapter 8 for comparisons in a one-way between-subjects design.

9.11 Questions and Problems

1. Using the data in Table 9.2, verify with a statistics computer program that the seven orthogonal comparisons presented in the text for the three main effects, three two-way interactions, and the three-way interaction perfectly decompose the sum of squares from treating the design as a one-way ANOVA.

2. We have two factors, A and B, each with two levels. For each treatment combination, we have $n = 8$ participants assigned at random. The data are given in the table below. Complete the analysis of variance source table.

A_1		A_2	
B_1	B_2	B_1	B_2
8	5	10	5
6	8	9	7
9	10	4	3
9	7	8	5
8	10	8	3
7	7	4	5
6	8	3	5
3	5	6	8

3. In a $2 \times 2 \times 2$ factorial experiment, we have the following measures:

A_1				A_2			
B_1		B_2		B_1		B_2	
C_1	C_2	C_1	C_2	C_1	C_2	C_1	C_2
8	5	10	5	7	6	5	2
6	8	9	7	10	8	7	7
9	10	4	3	6	7	4	5
9	7	8	5	7	6	7	7
8	10	8	3	5	8	6	5
7	7	4	5	7	9	8	9
6	8	3	5	6	8	10	6
3	5	6	8	10	9	6	6

Complete the analysis of variance. Test all main effects and interactions, using $\alpha = 0.05$.

4. Describe an experiment in which a statistically significant two-factor interaction might be expected. Describe the nature of the interaction you expect, and explain why you think this interaction might occur.

5. Consult a recent issue of an empirical journal. Find a study in which a statistically significant two-factor interaction is reported. Describe the interaction. Can you offer some explanation as to why it occurred?

6. We have a $2 \times 2 \times 2$ factorial experiment with $n = 10$ participants assigned at random to each treatment combination. The data are presented below. Complete the analysis of variance. Test all main effects and interactions, using $\alpha = 0.05$.

A_1				A_2			
B_1		B_2		B_1		B_2	
C_1	C_2	C_1	C_2	C_1	C_2	C_1	C_2
7	6	10	3	10	1	1	9
7	8	1	2	2	4	10	6
4	1	3	2	8	3	10	5
1	7	9	5	10	4	6	6
5	7	9	2	3	6	7	3
9	7	4	4	1	9	3	9
6	8	1	7	4	3	8	5
8	7	3	1	3	4	8	3
9	9	10	8	4	4	3	7
8	3	3	9	1	10	5	2

7. With a 2^3 factorial experiment, it is possible for the sums of squares for all main effects and two-factor interactions to be equal to zero and for the three-factor interaction sum of squares to be greater than zero. Make up a simple numerical example where this result is the case. Draw the plots to verify that only the three-factor interaction displays a pattern but no other interaction or main effect is present.

10

Between-Subjects Factorial Experiments: Factors with More Than Two Levels

10.1 Introduction

A factorial experiment is not limited to factors having only two levels, as described in the preceding chapter. Factorial experiments may involve factors each with several levels. For example, participants in a memory study may be given 5 minutes, 10 minutes, or 15 minutes to memorize a transcript. In this case, the total time to study is a factor with three levels. If a factor has three or more levels, then the sum of squares for the main effect of this factor will have more than 1 degree of freedom. As we saw in Chapter 8, when a factor has three or more levels, then it will require more than one comparison to define the relevant sum of squares. It follows that the interactions involving such a factor will also have more than 1 degree of freedom, and those interactions will require more than one interaction comparison. In this chapter we generalize the technique for 2^k factorial experiments to

Table 10.1: Outcome of a $4 \times 3 \times 2$ factorial experiment. Each cell entry is the mean of $n = 5$ observations.

		A_1	A_2	A_3	A_4
	B_1	4.2	6.6	7.0	8.8
C_1	B_2	4.8	6.4	7.2	8.2
	B_3	3.6	6.0	7.4	9.8
	B_1	3.2	5.6	7.0	7.2
C_2	B_2	2.8	5.2	6.6	6.8
	B_3	2.6	4.6	5.4	6.6

permit factors having more than two levels. We will show that an analogous procedure of finding individual comparisons to code the main effects and interactions can be used but that when there are more than two levels for a factor there will be more than one comparison needed to code that factor.

10.2 An Example of a $4 \times 3 \times 2$ Factorial Experiment

Consider an experiment with three factors. We provide the sketch of this design, merely describing the number of factors, the number of levels for each factor, and the sample size in each of the treatments. Suppose Factor A has four levels, Factor B has three levels, and Factor C has two levels. This yields a total of $(4)(3)(2) = 24$ treatment combinations. For this example, we assume that a between-subjects design is used with $n = 5$ observations for each treatment, yielding a total of 120 observations. Table 10.1 displays the 24 treatment means.

For the moment we will ignore the factorial structure of this experimental design. This design can be treated as a one-way between-subjects design with 24 treatments, using the technique presented in Chapter 6. In such a case the total sum of squares can be partitioned into two pieces: sum of squares treatments and sum of squares within treatments. The definitional formulas are the same as those presented in Chapter 6, but in this case the

Table 10.2: Source table for $4 \times 3 \times 2$ factorial experiment treated as a one-way ANOVA with 24 levels.

Source	SS	*df*	MS	*F*
Treatment	395.17	23	17.18	4.52
Within	364.80	96	3.80	

computations will be time-consuming because there are 24 treatments. The structural model underlying this 24-group analysis follows Chapter 6 for this design:

$$X_{ij} = \mu + \alpha_i + \epsilon_{ij} \tag{10.1}$$

with the grand mean denoted by μ, the 24 treatment effects denoted by α (subscript i has values 1 to 24), and error term ϵ_{ij} (an epsilon corresponding to each observation ij).

Recall that the within-treatment sum of squares is equal to the sum of squared differences between each score and its respective treatment mean; the treatment sum of squares is equal to the sum of squared differences between the treatment mean and the grand mean, the mean of all 120 participants. Applying these definitions to the 24 means in Table 10.1 leads to a sum of squares treatment equal to 395.17, with $k - 1 = 24 - 1 = 23$ degrees of freedom (see Table 10.2). The within-treatment sum of squares was also computed using the formula presented in Chapter 6 and is equal to 364.80 with $k(n - 1) = 24(5 - 1) = 96$ degrees of freedom. The corresponding F value is $MS_T/MS_W = 17.18/3.80 = 4.52$, which is not statistically significant by conventional standards ($\alpha = 0.05$).

Obviously, an omnibus test having 23 degrees of freedom will not lead to an informative statistical test. When statistical significance is observed, the investigator will only know that a difference exists for at least one linear combination of the 24 treatment means. The investigator will likely want to know specifically which means differ from which means (or which comparisons are statistically significant). This information is not given by the omnibus F test just computed.

The solution to finding which treatment means differ from which means comes from applying the concepts we have developed in the preceding chapters. With 24 treatment means we know that an orthogonal set of comparisons must contain a total of 23 comparisons (that is, number of treatment means minus 1). The investigator can test any set of 23 orthogonal comparisons. However, the particular structure of the present factorial design suggests one natural set of 23 orthogonal comparisons. As we saw in the preceding chapter, a factorial design can be partitioned into main effects and interactions. In the example with three factors, there will be three main effects (one for each factor A, B, and C), three two-way interactions (all possible combinations of two factors, $A \times B$, $A \times C$, and $B \times C$), and one three-way interaction ($A \times B \times C$). Each of these main effects and interactions can be understood from its component orthogonal comparisons. Thus, by making use of the factorial structure of the design, we can choose a specific set of 23 comparisons that provides useful information about the factors and their combinations.

10.3 Partitioning the Sum of Squares into Main Effects and Interactions

The treatment sum of squares presented above with 23 degrees of freedom can be partitioned into main effects and interactions as follows, with the number in the right-most column indicating the degrees of freedom associated with each term and lowercase letters referring to the number of levels (for example, the variable a refers to the number of levels in Factor A, which in this example is four):

	Factor	*df*	Definition of *df*
Main effects	A	3	$a-1$
	B	2	$b-1$
	C	1	$c-1$
Two-factor interactions	$A \times B$	6	$(a-1)(b-1)$
	$A \times C$	3	$(a-1)(c-1)$
	$B \times C$	2	$(b-1)(c-1)$
Three-factor interaction	$A \times B \times C$	6	$(a-1)(b-1)(c-1)$
Within treatment	Error	96	

The degrees of freedom for the error term in this example is 96. Recall that the degrees of freedom for error are given by the number of participants minus the number of treatments, or $N - T = 120 - 24 = 96$.

The degrees of freedom for the separate categories such as main effects indicate the number of comparisons that will be needed to specify those effects. So, the main effect for Factor A will require three orthogonal comparisons. The sum of all the degrees of freedom associated with treatments $(3 + 2 + 1 + 6 + 3 + 2 + 6)$ equals 23, the number of orthogonal comparisons needed for 24 treatments. The concept here is the same as in previous chapters—comparisons decompose the sum of squares treatment.

In the special case where a, b, and c are all equal to 2, then this design becomes a 2^3 factorial design and is identical to the design reviewed in the preceding chapter. The only difference is that now we allow for the possibility that a factor (or factors) can have more than two levels. This will in turn influence the number of comparisons that are needed to construct each main effect and corresponding interaction terms involving those comparisons. Table 10.3 gives the complete source table for this design. In the following sections, we discuss the meaning of each line in Table 10.3. A pie chart is presented in Figure 10.1 to illustrate the sum of squares decomposition in Table 10.3.

Table 10.3: Analysis of variance of the $4 \times 3 \times 2$ factorial experiment.

Source of variation	SS	*df*	MS	F
A	310.57	3	103.52	27.24
B	4.07	2	2.04	
C	56.04	1	56.04	14.75
$A \times B$	6.73	6	1.12	
$A \times C$	5.76	3	1.92	
$B \times C$	5.06	2	2.53	
$A \times B \times C$	6.94	6	1.16	
Within treatments	364.80	96	3.80	
Total	759.97			

10.4 Orthogonal Partitioning for Main Effects

One way to think about the main effects in a factorial design is that each factor is treated as though it is its own one-way between-subjects design. For example, Factor A has four levels. We know that any factor with four levels requires a set of three orthogonal comparisons in order to create a complete partition of the sum of squares. Factor B has three levels and consequently will require two orthogonal comparisons. Factor C has two levels, and, as illustrated in the preceding chapter, can be understood with just one comparison. Thus, the degrees of freedom associated with the main effect of each factor (number of levels minus 1) give the number of orthogonal comparisons.

We first show how three orthogonal comparisons on the treatment means are needed for Factor A. Because Factor A has four levels, the 24 means in Table 10.1 can be collapsed to reflect the four levels of Factor A; that is, the 24 means can be condensed to four means, one mean for each level of Factor A, with each of these four means representing 30 participants (5 participants in each of six treatments). Figure 10.2 presents a graph of the four marginal means associated with Factor A. One possible orthogonal set of comparisons on the main effect for Factor A is made up of these three comparisons: (1, −1, 0, 0), (1, 1, −2, 0), and (1, 1, 1, −3). The interpretation of these three comparisons is: (1) the difference between the mean

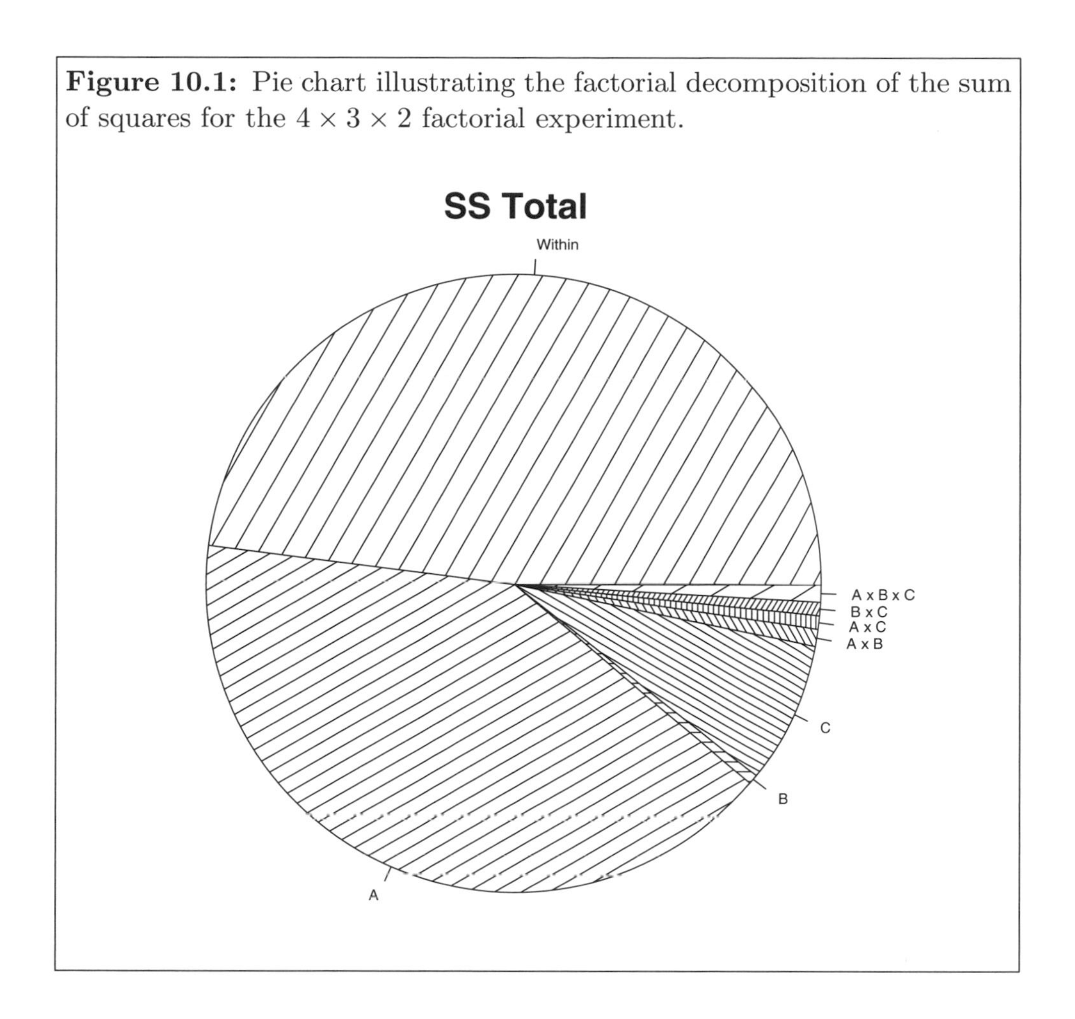

Figure 10.1: Pie chart illustrating the factorial decomposition of the sum of squares for the $4 \times 3 \times 2$ factorial experiment.

for A_1 and the mean for A_2, (2) the difference between the sum of the two means for A_1 and A_2 and twice the mean for A_3, and (3) the difference between the sum of the three means for A_1, A_2, and A_3 and three times the mean for A_4, respectively. To make this concrete, suppose A represents four levels of a treatment such that A_1 represents receiving alcohol, A_2 represents receiving aspirin, A_3 represents receiving both alcohol and aspirin, and A_4 represents not receiving alcohol or aspirin. The dependent variable could be performance on a task involving hand-eye coordination. The first comparison applied to this factor tests whether the effects of alcohol alone differ from the effects of aspirin alone, the second comparison tests whether there is a

Figure 10.2: Trend of the marginal means for Factor A.

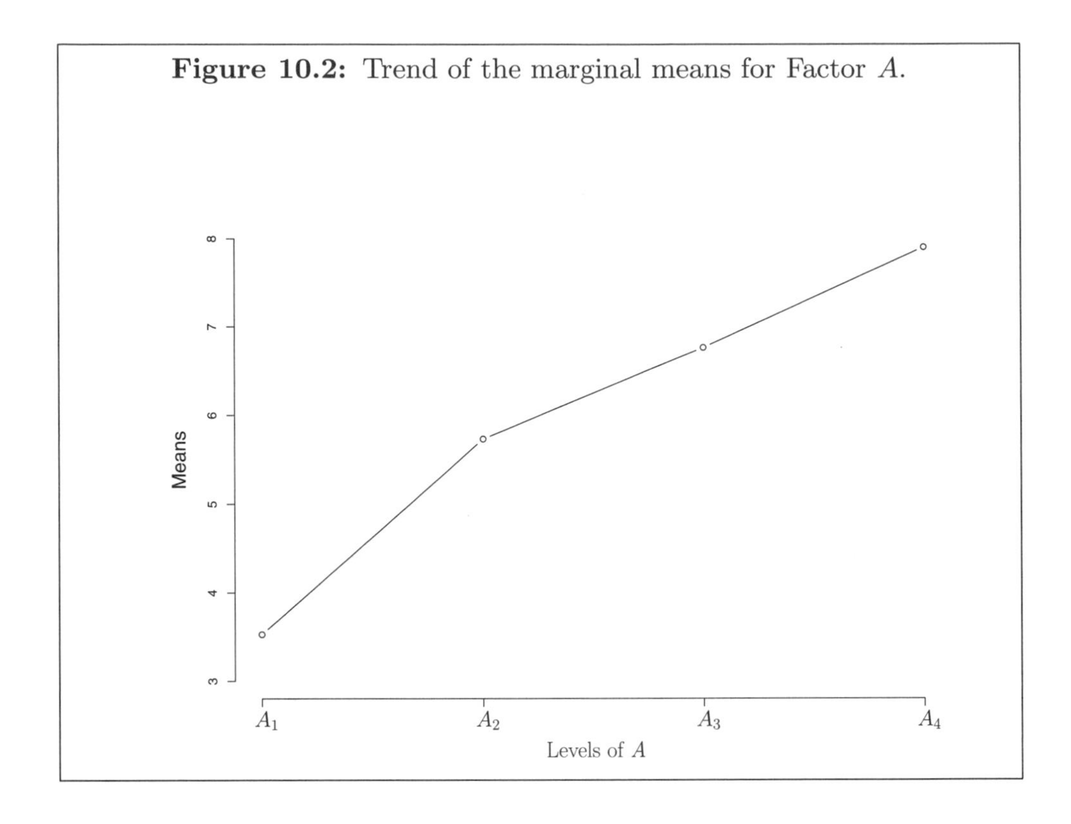

synergistic effect of the two drugs (that is, does receiving both alcohol and aspirin concurrently, A_3, differ from the combined effects of receiving alcohol and aspirin separately, A_1 and A_2), and the third comparison tests whether the three treatments that received a drug differ from the single treatment, the control condition A_4, that did not receive a drug. Of course, any set of three orthogonal comparisons will accomplish the goal of defining the main effect; here we gave an example of one such orthogonal set. As always, the specific choice of orthogonal comparisons depends on the research questions you are asking because comparisons are associated with research questions.

We compute the marginal means for Factor A by averaging the six means in each level of Factor A, which we can justify in this example because all treatments have equal sample sizes. For example, the average of the six means in A_1 is $(4.2 + 4.8 + 3.6 + 3.2 + 2.8 + 2.6)/6 = 3.53$. The marginal means for A_2, A_3 and A_4 are 5.73, 6.77, and 7.9, respectively. Multiplying the marginal means for each level of Factor A by the corresponding coefficients for each comparison, we obtain

$$D_1 = (1)\,(3.53) \;+\; (-1)\,(5.73) \;+\; (0)\,(6.77) \;+\; (0)\,(7.9) = -2.2$$
$$D_2 = (1)\,(3.53) \;+\; (1)\,(5.73) \;+\; (-2)\,(6.77) \;+\; (0)\,(7.9) = -4.28$$
$$D_3 = (1)\,(3.53) \;+\; (1)\,(5.73) \;+\; (1)\,(6.77) \;+\; (-3)\,(7.9) = -7.67$$

Each of the A means is based on $bcn = (3)(2)(5) = 30$ observations. For the first comparison, we have $\sum a_{j1}^2 = 2$, for the second comparison we have $\sum a_{j2}^2 = 6$, and for the third comparison we have $\sum a_{j3}^2 = 12$. For these three comparisons, we have the following mean square terms (taking a few more decimal places in the values of the comparison to minimize roundoff error):

$$MS_{D_1} = \frac{D_1^2(bcn)}{\sum a_{j1}^2} = \frac{(-2.2)^2(30)}{(2)} = 72.60$$
$$MS_{D_2} = \frac{D_2^2(bcn)}{\sum a_{j2}^2} = \frac{(4.2667)^2(30)}{(6)} = 91.02$$

and

$$MS_{D_3} = \frac{D_3^2(bcn)}{\sum a_{j3}^2} = \frac{(7.6667)^2(30)}{(12)} = 146.95$$

with each comparison having 1 degree of freedom. Consequently, each of these mean square terms is equal to the respective sum of squares term, and we also note that

$$72.60 + 91.02 + 146.95 = 310.57$$

which is equal to the total sum of squares for Factor A as given in Table 10.3 and shown in a pie chart in Figure 10.3. The equivalence between the sum of the comparison sum of squares and the overall sum of squares for Factor A occurs because the three comparisons are orthogonal.

We collapsed the 24 treatment means into four means that represent the four levels of Factor A. That is, we collapsed treatments to compute the mean for all participants in treatment A_1, the mean for all participants in

Figure 10.3: Pie chart illustrating a particular choice of three orthogonal comparisons applied to Factor A.

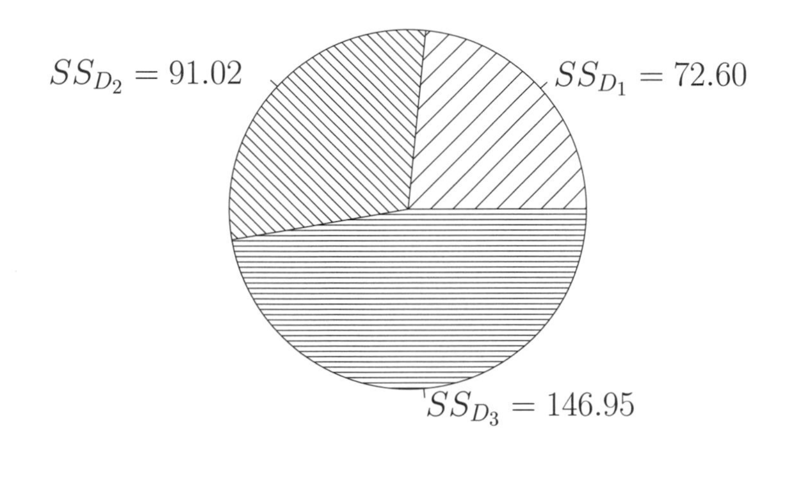

treatment A_2, the mean for all participants in treatment A_3, and the mean for all participants in treatment A_4. The comparison was then defined on these four collapsed means. Because the sample sizes are equal, the same result would have been obtained had the coefficients been applied to all 24 treatment means separately. For example, the first comparison could be written over the 24 groups in Table 10.1 (reading row by row, left to right) as (1, −1, 0, 0, 1, −1, 0, 0, 1, −1, 0, 0, 1, −1, 0, 0, 1, −1, 0, 0, 1, −1, 0, 0). This comparison with 24 coefficients is identical to the simpler comparison (1, −1, 0, 0) applied to the four treatment means of Factor A. Applying the comparison to the 24 treatment means yields $(1)(4.2) + (-1)(6.6) + \ldots + (0)(6.6) = 13.2$. In this case the sum of squares is based on 24 means, each with 5 observations, so the sum of squares becomes

$$\frac{D^2(n)}{\sum a^2} = \frac{13.2^2(5)}{12} = 72.60$$

which is the identical sum of squares as with the comparison based on the four collapsed means. The reader should check this equivalence for him or herself on the remaining comparisons.

The main effects for Factors B and C are found in a similar manner. Factor B requires two orthogonal comparisons; Factor C requires only one comparison. An example of an orthogonal set of two comparisons for Factor B could be $(1, -1, 0)$ and $(1, 1, -2)$. We collapse the two other factors (Factors A and C) and reduce the means to three, representing the six treatments that received B_1, the six treatments that received B_2, and the six treatments that received B_3. A comparison for Factor C could be $(1, -1)$, where the comparison is against the two levels of Factor C. To reiterate, because the cell sample sizes are equal, the identical sum of squares would be obtained if the comparisons were applied to all 24 treatments. So, for example, in the case of Factor C, the main effect could be equivalently represented by the comparison $(1, 1, 1, 1, 1, 1, 1, 1, 1, 1, 1, 1, -1, -1, -1, -1, -1, -1, -1, -1, -1, -1, -1, -1)$ on the 24 treatment means in Table 10.1.

The main effects for Factors A and B have more than one degree of freedom and, thus, are omnibus tests. Fortunately, we have the individual comparisons that lead to each main effect. Being able to have an automatic decomposition of omnibus main effects is the primary advantage of thinking about factorial designs in terms of component comparisons. Each individual comparison can be tested by dividing the sum of squares for the comparison by the within-treatment mean square. For example, to test the first comparison for Factor A (the difference between the treatment means A_1 and A_2), simply divide the sum of squares comparison 72.60 by MS_W, which in this experiment is 3.80. This yields an observed F value of $72.60/3.80 = 19.11$ with 1 degree of freedom for the numerator and 96 degrees of freedom for the denominator. This test is statistically significant at $\alpha = 0.05$, so we reject the null hypothesis that the population mean for treatment A_1 equals the population mean for treatment A_2.

Most statistics programs will print the omnibus source table by default. The analyses would be more informative if the investigator selected the specific set of orthogonal comparisons that are appropriate to the particular design and partitioned the omnibus test into single comparisons that are meaningful for the particular research questions under investigation.

The investigator should not perform three separate one-way between-subjects ANOVAs (one for each of the three factors) in order to test the main effects. The reason three separate one-way ANOVAs is inappropriate is because the error terms for each of these three designs (that is, MS_W) will be incorrect. The correct MS_W error term, when the equality of population variance is justified, should be based on the 24 treatments. The MS_W term for an individual one-factor ANOVA would be inflated because it would include sum of squares attributable to the other two factors in the experimental design.

10.5 Orthogonal Partitioning for Interactions

As in the preceding chapter on 2^k designs, the interaction terms are created by multiplication of the corresponding main effect comparisons. So, the interaction comparisons are completely determined by the initial choice of the main effect comparisons. First, let's make sure we understand why there will be a total of 17 orthogonal comparisons that code the different types of interactions. Recall that there are three comparisons that make up the main effect for Factor A, two that make up the main effect for Factor B, and one that makes up the main effect for Factor C. That implies that the $A \times B$ interaction will require six comparisons (three pairwise combinations from Factor A, crossed with two comparisons from Factor B), the $A \times C$ interaction will require three comparisons (the three comparisons for Factor A, each paired with the sole comparison for Factor C), the $B \times C$ comparison will require two comparisons, and the three-way interaction $A \times B \times C$ will require six (the three comparisons for Factor A, crossed with the two comparisons for Factor B, crossed with the sole comparison for Factor C).

Figure 10.4: Trend of the Factor A means for each of the two levels of Factor C.

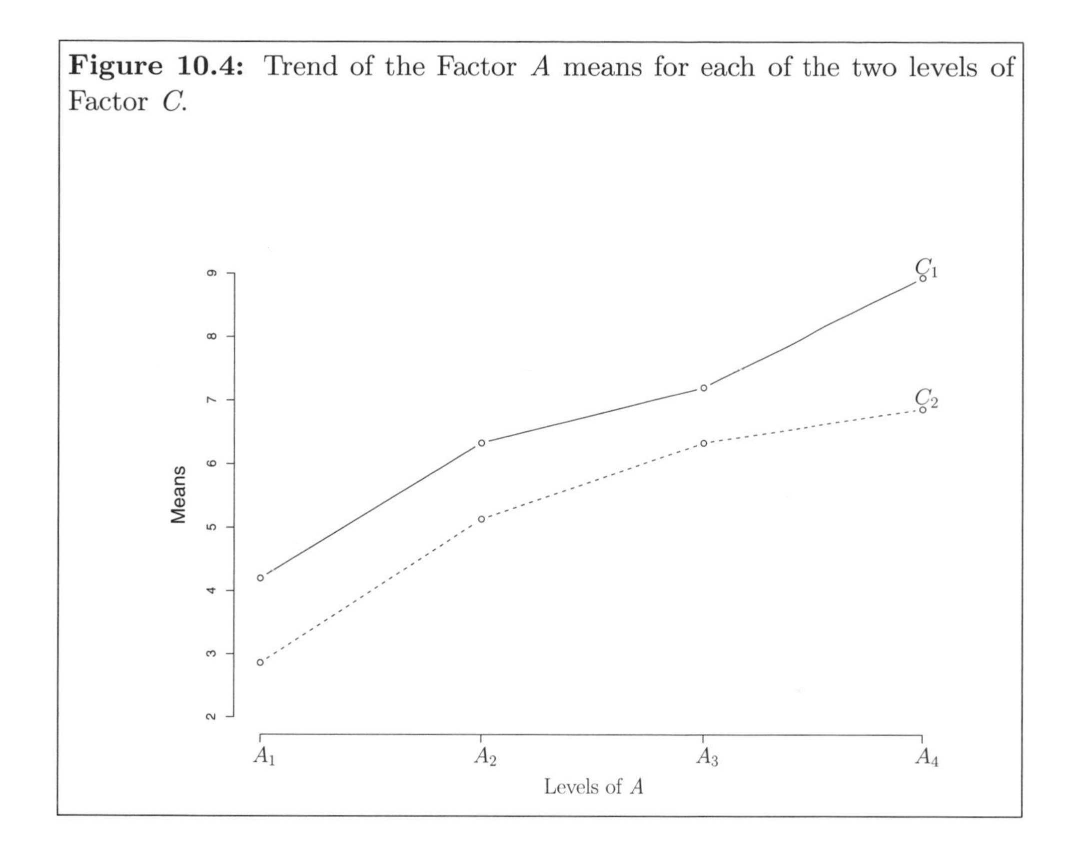

Let us examine the two-way interaction between Factors A and C. The graph of the eight collapsed means (the four means for Factor A corresponding to C_1 and the four means for Factor A corresponding to C_2) is presented in Figure 10.4. Examination of the graph suggests that an interaction is unlikely because the two curves are nearly parallel. Each point is based on 15 participants (collapsing over the three levels of Factor B).

Recall that we chose the comparison $(1, -1, 0, 0)$ as one of the comparisons that made up the orthogonal set for Factor A and the comparison $(1, -1)$ as the comparison for the two levels of Factor C. One way to think about the interaction between these two comparisons is that we need to apply the comparison for Factor A separately to each level of Factor C in such a manner that we can determine whether the comparison for Factor A applied to one level of Factor C differs from the same comparison applied to the other level of Factor C. So, by applying the comparison (1, -1, 0, 0) to the four means of Factor A for only those participants who also had treatment C_1 and the comparison (-1, 1, 0, 0)—note the change in sign—to the four means of Factor A for those participants who also had treatment C_2, we are testing whether the difference between the means for treatments A_1 and A_2 for those participants in C_1 differs from the difference between the means for treatments A_1 and A_2 for those participants in C_2. This is the reason an interaction is frequently referred to as "a difference between two differences." One way to conceptualize this two-way interaction comparison is that one multiplies the comparison coefficient for C_1 (the number 1) by the comparison for Factor A (that is, 1, -1, 0, 0) and multiplies the comparison coefficient for C_2 (the number -1) by the comparison for Factor A.

The complete comparison on the eight means in Figure 10.4 (reading first the A means corresponding to C_1 and then the A means corresponding to C_2) is (1, -1, 0, 0, -1, 1, 0, 0). Again, in words this comparison is testing whether there is a difference between two differences, or whether the difference observed between the means A_1 and A_2 for participants in treatment C_1 differs from the difference observed between the means A_1 and A_2 for participants in treatment C_2.

The sum of squares for the comparison is given by the formula

$$\frac{D^2(bn)}{\sum a^2} = \frac{0.1333^2(15)}{4} = 0.06667$$

The test of significance is found by dividing the sum of squares for the comparison by the MS_W term from the source table, yielding $F = 0.06667/3.8 = 0.0175$ with 1 and 56 degrees of freedom. This test is not statistically significant at the $\alpha = 0.05$ level. We should not be surprised at the lack of statistical significance because the two curves in Figure 10.4 are nearly parallel and the cell sample size is relatively small.

One would apply the same logic to the remaining interaction comparisons. The remaining two comparisons for the two-way interaction between

A and C are (1, 1, −2, 0, −1, −1, 2, 0) and (1, 1, 1, −3, −1, −1, −1, 3). Each of these comparisons is applied to the eight collapsed means relevant to this two-way interaction. The sum of squares for these two remaining comparisons that make up the $A \times C$ interactions, given the particular choices made earlier for the main effect comparison, are

$$\frac{D^2(bn)}{\sum a^2} = \frac{0.8^2(15)}{12} = 0.80$$

and

$$\frac{D^2(bn)}{\sum a^2} = \frac{2.80^2(15)}{24} = 4.90$$

The sum of squares for these three comparisons equals the omnibus sum of squares for the $A \times C$ interaction presented in Table 10.3 (that is, $0.06667 + 0.80 + 4.90 = 5.77$, up to roundoff error).

The remaining two-way interactions—$A \times B$ and $B \times C$—are also partitioned in a similar manner through their corresponding main effect comparisons. The reader should check his or her understanding and verify the partition of the sum of squares for the remaining two-way interactions and the three-way interaction.

The design considered here can be conceptualized as having 24 treatments. Therefore, it can be partitioned into a set of 23 (that is, 24 − 1) comparisons, as we saw in Chapter 8. The factorial design presents one such set of orthogonal comparisons, and this particular set makes sense if one has designed a study with the factorial structure in mind. Thus, the factorial structure of the design implies a particular set of comparisons.

10.6 Effect Size for Comparisons in a Factorial Design

As we mentioned in Chapter 8, a simple effect size measure for a comparison is given by

$$r = \sqrt{\frac{SS_C}{SS_C + SS_W}}$$

where SS_C is the sum of squares for a particular comparison and SS_W is the sum of squares within treatment. This formula can be used for any comparison (main effect or interaction comparison). Similarly, the identical effect size measure can be obtained directly from the t test through this formula:

$$r = \sqrt{\frac{t^2}{t^2 + df}}$$

where t is the observed value of the t test for the comparison (recall that for comparisons $F = t^2$) and df is the degrees of freedom associated with the within-treatment term (that is, number of subjects − number of treatment combinations).

10.7 Performing Multiple Tests

We find it useful to think about factorial designs in terms of their elementary comparisons, with orthogonal comparisons acting as the building blocks. Remember that as long as each comparison is planned (which is usually the case in a factorial design), then there is no need to worry about the increase in Type I error rate when performing multiple tests, especially when the important results are replicated. Of course, if an investigator is concerned about the large number of tests that are performed in factorial designs, then he or she can perform the Bonferroni correction we discussed in Chapter 7. In the example presented in this chapter, there were a total of 23 comparisons, so the Bonferroni correction would be based on a total of 23 tests. This leads to a conservative test. A different strategy (perhaps a more sensible strategy in the context of a factorial design) is to apply the Bonferroni separately to each "family" of tests (where a family is defined as a main effect for a given factor or an interaction on a particular set of factors). For instance, in the example presented in this chapter the main effect for Factor A had four levels, so it contained three orthogonal comparisons. The main effect on Factor A consists of a family of three orthogonal comparisons, and the investigator could apply a Bonferroni on each family (for example, the three comparisons on Factor A could be tested with $\alpha/3$ rather than α). Similarly, the two-way interaction between Factor A and Factor B consists of six comparisons, so applying the Bonferroni criterion to these six tests involves using $\alpha/6$. The example contains a total of seven "families" (that is, main effect on A, main effect on B, main effect on C, the two-way interaction between A and B, the two-way interaction between A and C, the two-way interaction between B and C, and the three-way interaction), and a separate Bonferroni would be applied to each family consisting of 2 or more comparisons. There is also the possibility of performing the Scheffé test in the context of a factorial design. The Scheffé could be applied to all comparisons (in the example of this chapter, the 23 comparisons), or the Scheffé could be applied separately to each family consisting of two or more comparisons. We discuss these post hoc tests in the next chapter. Our own preference however is to perform statistical tests of planned comparisons without correction. We admit that we are in the minority among methodologists on this particular point.

10.8 The Structural Model and Nomenclature

It turns out that the structural model for the $4 \times 3 \times 2$ is written exactly the same way as the 2^3 factorial design in the preceding chapter. That is, the structural model is

$$X_{ijk} = \mu + \alpha_i + \beta_j + \gamma_k + \alpha\beta_{ij} + \alpha\gamma_{ik} + \beta\gamma_{jk} + \alpha\beta\gamma_{ijk} + \epsilon_{ijk}$$

Besides the grand mean and the error term, there are seven key elements: three main effects, three two-way interactions, and one three way-interaction. The differing number of levels for each factor is absorbed into the subscripts.

The literature has given names to different types of comparisons, or contrasts. For example there are polynomial contrasts, Helmert contrasts, difference contrasts, dummy contrasts and effects contrasts. For us, the particular names are not important and it isn't clear that memorization of such terminology is useful. A comparison is made informative by examining its coefficients, and seeing which treatment conditions are compared to which. Some contrasts have special names such as simple main effects. A **simple main effect** is a contrast that applies to only one level of an interaction contrast. For example, in a 2×2 factorial design the typical interaction contrast would be (1, −1, −1, 1), but the simple main effect contrast could be defined separately at each level of the other factor as in the comparison (1, −1, 0, 0) and the comparison (0, 0, −1, 1). Again, just a name; the informative value of the comparison is seen directly through the coefficients it uses to weight the treatment means.

A prescription that we follow is that, if there are particular comparisons we want to make, we generate an orthogonal set of comparisons that includes those comparisons of interest. If we, in addition, want to test other comparisons (for example, break up an interaction comparison into smaller subparts using simple main effects), then we worry about the increase in the Type I error rate. We could apply a Bonferroni correction, or if appropriate use the Tukey test (for example, when the comparisons involve pairs of cell or marginal means) or a Scheffé test.

Sometimes our research questions do not lend themselves to conversion to a set of orthogonal comparisons. In some cases this could be a clue that we should clarify our research hypotheses in order to make them orthogonal and nonoverlapping. But sometimes research questions are just nonorthogonal. For example, if the researcher wants to test all possible pairwise comparisons between means in an experimental design, then necessarily those pairwise comparisons are nonorthogonal. We don't see anything fundamentally wrong with testing nonorthogonal comparisons (just as we don't see anything wrong with testing all possible pairwise comparisons of cell means). The main issues are that (1) the elegance of the decomposition of the experimental design into nonoverlapping sums of squares is lost when testing nonorthogonal comparisons; (2) two separate hypotheses might be more related than the researcher thinks, so the researcher should be careful when providing a list of statistically significant results because the list may contain partial redundancies; and (3) concern about the multiplicity of statistical tests arises—hence the need for procedures like the Tukey test in the case of pairwise comparisons.

10.9 Summary

This chapter did not present much new information. We simply combined ideas from Chapters 6, 8, and 9. From Chapter 6 we took the notion of partitioning the sum of squares of treatments, from Chapter 8 we took the notion of orthogonal contrasts, and from Chapter 9 we took the notion of a factorial design. The combination of these ideas permits an analysis of variance on a factorial design where each factor has two or more levels. We hope the reader can see how the basic building blocks of experimental design can be used to make sense of complicated designs.

10.10 Questions and Problems

1. A factorial experiment involves two factors, A and B, with A varied in four ways and B varied in three ways. The treatment combinations are replicated with $n = 5$ observations for each. The raw data are given here:

A_1			A_2			A_3			A_4		
B_1	B_2	B_3	B_1	B_2	B_3	B_1	B_2	B_3	B_1	B_2	B_3
38	54	65	24	21	35	36	35	35	45	45	34
45	34	86	43	67	45	81	36	65	55	98	65
22	54	62	56	98	76	22	54	67	34	65	65
23	23	26	75	46	89	23	65	76	34	34	43
45	32	42	43	55	98	45	78	55	45	54	36

Graph the means and complete the analysis of variance source table for all main effects and interactions.

2. We have a factorial experiment in which A is varied in two ways and B is varied in three ways. The raw data are given here:

A_1			A_2		
B_1	B_2	B_3	B_1	B_2	B_3
19	43	53	30	49	64
10	42	51	24	43	61
21	41	57	25	49	68
15	44	57	30	53	56
20	42	68	28	44	60
24	49	60	34	46	55
16	46	48	31	46	54
22	39	47	32	56	68
18	48	60	27	54	59
18	39	60	32	53	57

Graph the means and complete the analysis of variance source table for all main effects and interactions. Select a set of orthogonal comparisons for the main effect of B. Show the decomposition of the sum of squares by drawing a pie chart.

3. Describe an experiment in which one might expect to find a significant three-factor interaction. Explain why you would expect to find a three-way interaction.

4. We have an experiment where A is varied in four ways, B is varied in three ways, and C is varied in two ways. To each treatment combination, $n = 5$ participants were assigned at random. The treatment means for each of the treatment combinations are given here:

		A_1	A_2	A_3	A_4	$\bar{X}$
	B_1	12.0	18.0	18.8	17.2	16.5
C_1	B_2	10.8	18.4	14.8	19.6	17.0
	B_3	14.0	15.2	16.0	12.0	14.3
	B_1	11.6	14.2	15.6	16.8	14.6
C_2	B_2	15.2	16.4	14.8	12.8	14.8
	B_3	13.2	11.2	14.2	15.6	13.6
$\bar{X}$		12.8	15.6	16.5	15.6	15.1

Graph the means in a manner that displays whether there is a three-way interaction. In this experiment, MS_W is equal to 11.48 with 96 degrees of freedom. Complete the analysis of variance source table for the omnibus tests (for example, the three main effects, the three two-way interactions, and the one three-way interaction).

5. For the experiment described in Problem 4, compute the three comparisons shown below for the main effect of A:

Comparison	A_1	A_2	A_3	A_4
D_1	1	1	−1	1
D_2	1	−1	0	0
D_3	0	1	1	−1

Calculate the significance test for each of the three comparisons. Also, test whether the comparison D_1 interacts with the main effect for C.

11

Between-Subjects Factorial Experiments: Further Considerations

11.1 The Scheffé Test for Comparisons

If we want to explore the outcome of a factorial experiment thoroughly, making comparisons on the treatment means, including those that are planned and those that are suggested by the data, then the recommended procedure to correct for multiple tests is the Scheffé test. In this chapter we extend the test presented in Chapter 8 for the one-way between-subjects design. Recall that when Factor A has more than two levels, then we may evaluate the comparisons on the marginal means of Factor A in terms of the Scheffé critical value

$$F' = (a - 1)\,F$$

where a is the number of levels for Factor A and F is the tabled value at the α significance level with degrees of freedom $a - 1$ for the numerator and degrees of freedom associated with MS_W for the denominator. To illustrate, consider the (1, −1, 0, 0) comparison on Factor A in the example presented in the preceding chapter. The observed F for that comparison was 19.10, which is statistically significant using the Scheffé test because the observed F exceeds the Scheffé critical $F' = (4 - 1) \times 3.84 = 11.52$.

We may make as many comparisons as we wish on the marginal means of Factor A, and if the F ratios for these comparisons are evaluated in terms of F', the probability that all statements regarding statistical significance are correct will be greater than or equal to $1 - \alpha$. It is useless, however, to apply the Scheffé test when the main effect is not significant, because in this case no comparison on that main effect will be found that will result in a statistically significant F'. Recall that the Scheffé test has the property that if the main effect is not statistically significant, then none of the contrasts on that main effect will be significant either.

Similar considerations apply to comparisons on the means for other factors that have more than two levels. For example, if the main effect for Factor B is statistically significant at some significance level α, then comparisons on the marginal means for Factor B can be tested against the Scheffé critical value

$$F' = (b - 1)\,F$$

where b is the number of levels for Factor B, and F is the tabled value at the significance level α with numerator degrees of freedom $b-1$ and denominator degrees of freedom associated with MS_W. Again, the probability that all statements regarding statistical significance are correct for the family of comparisons made on the means of Factor B will be greater than or equal to $1 - \alpha$.

If a two-factor interaction, say MS_{AB}, is statistically significant at the α significance level, then comparisons over the $A \times B$ interaction means can be evaluated in terms of the Scheffé critical value

$$F' = (a - 1)(b - 1)\,F$$

where F is the tabled value at the α significance level with $(a - 1)(b - 1)$ numerator degrees of freedom and denominator degrees of freedom associated with MS_W. The observed F is compared to the Scheffé critical value F' rather than the tabled value of F. If the $A \times B$ interaction is not statistically significant, then no comparison involving the $A \times B$ interaction will be judged statistically significant in terms of F'.

Similarly, if a three-factor interaction, say MS_{ABC}, is statistically significant at the α significance level, then the three-way interaction comparisons involving the $A \times B \times C$ interaction means can be evaluated by the Scheffé critical value

$$F' = (a - 1)(b - 1)(c - 1)F$$

where F is the tabled value at the α significance level with $(a - 1)(b - 1)(c - 1)$ and the degrees of freedom for MS_W. If MS_{ABC} is not statistically significant, then no comparisons involving the $A \times B \times C$ interaction will be judged significant in terms of the critical value F'.

Note the general pattern in the definition of F': the usual critical value F is multiplied by the number of orthogonal comparisons that are present in a particular set of comparisons. For instance, for Factor A there are $(a - 1)$ orthogonal comparisons, so the typical critical F value is multiplied by $(a - 1)$. The Scheffé test is simple and straightforward—one computes the comparison in the usual way but uses a new critical value F' that is more conservative and controls the overall Type I error rate. We reiterate that the Scheffé test can be used even when the comparisons being considered are not orthogonal. Brown and Forsythe (1974) discuss how to perform the Scheffé test in a manner that does not assume equal population variances; their approach is to correct the denominator degrees of freedom in the spirit of the Welch test and then compute the Scheffé test with the new denominator degrees of freedom.

11.2 Pairwise Comparisons in Factorial Designs

Consider a two-factor experiment in which Factor A has four levels and Factor B has two levels. Assume that the main effects for Factors A and B are statistically significant but the interaction is not statistically significant. The interpretation of the statistically significant main effect for the B factor is simple because Factor B has two levels—the means for levels B_1 and B_2 averaged over the four levels of Factor A—so the critical question is whether the two marginal means of Factor B differ significantly. But the significant main effect for Factor A indicates that there are significant differences among the four A means averaged over the two levels of Factor B, and this result alone is not very satisfying because it is an omnibus test over four marginal means of Factor A. We would like to know more about specific differences among the four marginal means of Factor A. We have argued throughout the book that comparisons help a researcher understand the pattern that was observed among the cell means. The researcher may be inclined to test all possible pairwise differences of the means of A. This will be a generalization of the material presented in a previous chapter on pairwise comparisons to all possible comparisons of marginal means. Here we extend the procedure for pairwise comparisons to factorial designs.

If we have a set of comparisons on the means for Factor A that are of theoretical or practical interest, the tests of significance for these comparisons will aid in interpreting the significant omnibus main effect of A. But if we have no planned or orthogonal comparisons to make, then we should consider making all possible pairwise comparisons on the means for Factor A. It is possible, for example, that the mean for A_1 differs significantly from the means for A_2, A_3, and A_4 but that the means for A_2, A_3, and A_4 do not differ among themselves. Simply reporting that the main effect for Factor A is statistically significant would not provide the specific information about which means are statistically different. If an unlimited number of comparisons will be tested, then the Scheffé test can be used. However, if the experimenter wishes to compare only the possible pairwise marginal means on Factor A

(for example, A_1 versus A_2, A_1 versus A_3, A_3 versus A_4, etc.), then the Scheffé test will be conservative because the Scheffé test corrects for an unlimited number of comparisons, not only the pairwise comparisons. Recall that pairwise comparisons are special-case comparisons with one coefficient equal to 1, another coefficient equal to -1, and the remaining coefficients equal to 0.

Whenever we obtain a significant omnibus F test for a factor having more than two levels (or an interaction involving at least one factor with more than two levels), we should make some effort to specify which treatments have contributed to the significant omnibus test. Planned comparisons provide one way that we can accomplish this goal of testing specific research questions on treatment means because of how they partition the sum of squares. We could also rely on pairwise comparisons with a statistical correction for multiple tests to deal with the multiplicity problem.

The question of which is the best procedure for performing pairwise comparisons in factorial designs is currently under debate. Here we present one method: the Tukey test. As we argued in Chapter 7 in the context of the one-way between-subjects design, it is not necessary to compute the omnibus test (nor does the omnibus test need to be statistically significant) in order to perform a Tukey test. The Tukey test requires that the "level" (or "family") of the pairwise tests be identified (that is, at the level of cell means or marginal means). For instance, in the example above where Factor A has four levels and Factor B has two levels, the investigator might be interested in performing all possible pairwise means between the four marginal means for the main effect for Factor A. Clearly, there is no need to perform pairwise tests for the two means on Factor B because Factor B has two levels and that result is identical to the main effect for Factor B, which involves only two marginal means. The investigator may also be interested in calculating all pairwise tests between the eight cell means that define the 4×2 interaction. The difference between an analysis of pairwise comparisons of the eight cell means versus the pairwise comparison of the four main effect means of Factor A boils down to whether or not one wishes to collapse the means of

Factor B. Comparing all possible pairwise means across the eight conditions allows for comparisons across the second factor (Factor B), whereas focusing on the marginal means of Factor A averages over the means of Factor B. For example, the Tukey test on pairs of cell means permits a test of those participants in condition B_1 who were also in condition A_1 compared to those participants in condition B_1 who were also in condition A_2. The marginal means (that is, main effects comparison) however compare all participants in condition A_1 to all participants in condition A_2 regardless of whether or not they were in condition B_1.

The Tukey test fixes the Type I error rate at α within each "family." That is, the Tukey test on the main effect for Factor A controls the Type I error rate at α for all the pairwise comparisons within the marginal means of Factor A. Similarly, the Tukey test on the eight individual cell means (the interaction effect) controls the Type I error rate at α for all the pairwise comparisons on the individual treatment means.

The Tukey test on the individual cell means (the interaction effect) is identical to the Tukey test presented in Chapter 7. This is because comparing all possible cell means in a factorial design is tantamount to comparing all possible treatment means, as we did in Chapter 7.

However, a Tukey test on the marginal means (for example, comparing the four marginal means of Factor A) requires a minor modification to the formula presented in Chapter 7. Recall that the formula for the Tukey test when all cells have equal sample sizes was given as

$$\frac{\bar{X}_i \quad \bar{X}_j}{\sqrt{MS_W/n}}$$

and the observed value is compared to the tabled value of the studentized range statistic in Table B.6. However, when comparing the marginal means in a factorial design, we need to use the total number of observations that went into the computation of the marginal means. Thus, rather than using n, the cell sample size, in the Tukey formula, we use s, the total number of observations in a given marginal mean. Thus, the Tukey test for main effects in a factorial design when each marginal mean is based on an equal number of observations is given as

$$\frac{\bar{X}_i - \bar{X}_j}{\sqrt{MS_W/s}}$$

where s is the number of observations that enter the marginal mean.

For example, in a 4×2 factorial design with 5 participants per cell (for a total of 40 participants in the study because there are eight cells in the design), each of the four marginal means of Factor A is based on 10 participants. That is, the 5 participants from one level of Factor B plus the 5 participants from the other level of Factor B equal a total of 10 participants for a given level of Factor A. So, in this example the number s to plug into the above formula would be 10. The MS_W is the usual within treatment error term from the source table. Further, when finding the critical value of q in Table B.6, use the total number of pairwise means within the family. In the ongoing example, when performing a Tukey test on the four marginal means of Factor A, use $k = 4$ in Table B.6, but when performing a Tukey test on the eight cell means of the interaction, use $k = 8$. This procedure will control the Type I error rate separately for the main effect on Factor A and for the interaction between Factor A and Factor B.

A note of caution when interpreting the Tukey test over the individual treatment means (the interaction) in the context of a factorial design: when comparing all possible cell means, one may be confounding factors on some of the tests. For instance, suppose one had a 2×2 factorial design where one factor involved incentives given to participants (high or low) and the other factor involved the amount of time given to study a word list (say, 1 minute versus 10 minutes). The Tukey test on the four cell means of this factorial design is statistically permissible but may not be easy to interpret. How would one interpret the difference between the means in the high incentive/1-minute group and the low incentive/10-minute group? These two means confound the two factors. Why did the cell means differ? Was the difference due to the incentive? Was it due to study time? Was it due to a combination of both factors? Usually, it is easier to interpret differences in cell means within one level of a given factor. For instance, we can interpret the pairwise comparison between the high incentive/1-minute group and the low incentive/1-minute group because those two conditions differ in their level of incentive but have the same 1-minute study time. The caution we raise is interpretational and relevant only when comparing cell means that cross at least two factors; the Tukey test still gives the appropriate Type I error correction despite the possibility that the interpretation of the differences may be difficult.

11.3 Unequal Sample Sizes in a Factorial Design

The formulas presented so far for calculating sums of squares in factorial experiments require that we have an equal number of observations in each treatment combination. For instance, the result that the sum of squares for a main effect is equal to the sum of the separate sum of squares from an orthogonal set of comparisons assumed equal sample sizes. Unequal sample sizes creates nonorthogonality across the comparisons, so the sum of squares no longer adds up the way one would expect with orthogonal comparisons. In other words, with unequal sample sizes a pie chart of sum of squares will not neatly decompose into nonoverlapping slices.

The ling of unequal cell sizes, especially in the context of a factorial design, has been hotly debated in the literature. Several methods for dealing with unequal sample sizes have emerged. The methods differ in how the main effects are treated. It turns out that the interaction effect is not a problem because the interaction effect is based on individual cells; so, regardless of which method one uses, the interaction effect remains the same. Also, the computation of the mean square error term, MS_W, is unaffected by unequal sample sizes. The different methods for handling unequal sample sizes concern themselves exclusively with how the main effects are computed in a factorial design (thus, in a one-way between-subjects design unequal sample sizes is not an inherent problem). Intuitively, unequal sample sizes create the following problem: because main effects involve collapsing over cells of another factor or factors (for instance, the main effect for Factor A requires one to collapse over the levels of Factor B), differing sample sizes might skew the resulting collapsed means because they have differing numbers of observations. Unequal sample sizes make the main effects nonindependent, and dealing with this nonindependence can be tricky in some situations.

It is useful to be clear about why unequal sample sizes occur in designs. Different reasons will lead to different solutions. Did the investigator intentionally plan unequal sample sizes in order to create a representative sample? For instance, the investigator may wish to treat gender as a factor (male/female), and in the population under investigation there are 30% males and 70% females. In this case, the investigator may intentionally create unequal sample sizes in order to match the 30%/70% distribution of gender. Or, did the unequal sample sizes occur by accident? For instance, the experimenter came up against the last day possible for using human subjects from the introductory psychology course and could get 18, 12, 16, and 21 participants in each of the four cells of a 2×2 design. In this latter case, the unequal sample sizes are more of an accident, or random event, than the intentionally planned strategy of having the sample represent the composition of the population.

It turns out that these two situations are handled differently. If unequal sample sizes occurred by accident, then the experimenter proceeds "business as usual" and may compute comparisons on main effects in the usual way. No special procedure is needed to handle the main effects in this situation, and each individual comparison can be tested in the usual way described in the preceding chapter. The experimenter can use the usual comparison weights (such as 1, -1, 1, -1 to test a main effect in a 2×2 factorial design). However, if the experimenter intentionally planned unequal sample sizes (say in order to represent the population), then the main effect analyses are more complicated because the comparison weights need to take into account the sample sizes. Performing factorial designs in the context of representative sampling is a complicated topic that is beyond the scope of this book. We refer the interested reader to a book on sampling by Cochran (1977) or Groves et al. (2004); the reader may also wish to examine a more technical book on ANOVA for details on the various methods of testing main effects in the presence of unequal sample sizes (for example, Kirk, 1995; Maxwell & Delaney, 1990).

Statistical packages sometimes differ in their default treatment of unequal sample sizes in a factorial design. If the unequal sample sizes are unintentional, then, as stated in the preceding paragraph, the analyst can test all individual comparisons using the standard MS_W. Some people and statistics programs refer to this method of testing all individual comparisons (all main effects and interactions) simultaneously as Type III sum of squares. This is to differentiate other procedures for handling nonorthogonal decomposition due to unequal sample sizes (you guessed correctly that there is a Type I and Type II, and there is even a Type IV—it sure would facilitate communication and understanding if more informative titles were used). Make sure you know what your statistics package does by default; you may need to change the automatic features of the program in order to get the analysis you want.

The experimenter should consider other reasons that could account for the final sample size. Some individuals assigned to a particular treatment may refuse to participate. In this case the refusal to participate can be related to a treatment, and this could cloud the interpretation of the results. That is, if participants assigned to treatment A_1 refuse to participate more often than participants assigned to treatment A_2, then generalizability is compromised. Or, if participants assigned to treatment A_1 (say, being shocked by a strong electrical current) drop out of the study at a higher rate than those assigned to treatment A_2 (say, being tickled on the nose by a soft feather), then the unequal sample sizes occurred because of the difference in the treatments that the participants experienced. Such a differential attrition rate could compromise the random assignment and the inferences that can be made from the study. Making causal inferences from studies that have differential rates of participation or attrition across treatments is a major area of research in statistics.

11.4 Individual Difference Factors

It is now commonplace in behavioral research to treat nonmanipulated variables as factors in an experimental design. For instance, some researchers treat gender (male versus female) as a factor. Some researchers treat people who come to the lab with either high or low self-esteem as a factor. These factors (the gender of the participant or a participant's level of self-esteem prior to the study) are not manipulated variables. Instead, they are individual difference factors, as we discussed in Chapter 1. From a statistical standpoint, it is fine to perform analysis of variance and all its calculations on individual difference factors. However, the investigator must be very careful not to draw causal inferences about the effects of such factors. Participants were not randomly assigned to the male or female level of the gender factor. Random assignment to treatments is a critical and necessary design feature (as many methodologists would argue) when making causal inferences. If you find a significant main effect for an individual difference factor like gender (or an interaction between an individual difference factor and any other factor), be careful not to discuss it with causal language. Rather, results involving individual difference factors should be described as they are observed ("females on average scored higher than males on the dependent variable ..."). Such results are associations because the factors were not manipulated and participants were not randomly assigned to the levels of the factor. Research on individual difference factors is important, both theoretically and practically. Our point is merely that the associations between the factors and the dependent variable remain associations and are not magically transformed into "causal factors" because the data analyst uses the statistical machinery of experimental design.

11.5 Control Variables

It is well known that a group of participants will vary with respect to almost any variable that we might care to measure (except in special circumstances, such as in animal research, where researchers can control the genetic composition of the subject pool and their environment conditions to minimize individual differences). Participants differ in their reaction times, their abilities to solve problems, to learn, to recall, to perceive, to stay alert during the study, and so on. In many experiments, there may be widespread individual differences. If random assignment treatment conditions are not possible (for example, there are individual difference factors), then there may be systematic variability in the dependent variable due to these variables. When random assignment is not possible, there is a statistical alternative available.

A factor may group participants into common clusters of individual variables. Such a factor is called a **blocking factor**. For instance, a block can be created by identifying a group of people who have one characteristic in common such as age, self-esteem, IQ, etc. The blocks are formed under the assumption that the units within each block will be more homogeneous in their response on some dependent variable of interest than units selected completely at random. By taking into account the differences existing among blocks in the analysis of variance, the experimenter anticipates that a smaller mean square within treatments will be obtained for the same number of observations than would have been the case if a standard design had been used. The design, in other words, is one in which the differences among blocks are eliminated from the estimate of experimental error.

For example, an investigator may be interested in the effect of exercise on weight loss, so randomly assigns participants to a workout condition or a control condition. However, it may be that the age of the participants is also related to the degree of weight loss. Simple random assignment to treatment is one way of dealing with the effect of age because randomization equates within limits the age distribution in the two conditions. But age still contributes to the variability in the scores and thus contributes to the mean square within (MS_W) term. One way to handle this is to create blocks, such as ages 18–21, 22–25, and 26–29, and include those blocks as levels of an age factor in the design. This will partition the sum of squares that can be attributable to age, thereby reducing the MS_W term and providing a more powerful statistical test (assuming, of course, that the reduction in sum of squares error is not offset by the reduction in degrees of freedom by the inclusion of the blocking factor).

Usually the factor playing the role of a "blocking variable" is not of primary concern to the researcher. Instead, the blocking factor is included to reduce the within-treatments mean square term (MS_W). This is justified because if the experimenter's hunch that the blocking factor really does account for some of the within treatments error is correct, then a more sensitive test will be given on the other variables of interest. On the other hand, if the investigator's hunch is wrong and the blocking factor is unrelated to the dependent variable so there is no systematic difference across its levels on the dependent variable, then the power of the test will be compromised (as compared to not including the blocking factor) because of the reduction in degrees of freedom without a corresponding reduction in the sum of squares within treatments.

One experimenter's blocking variable may be another experimenter's primary variable. For instance, one investigator may be interested in exploring gender differences on a sensory task. For this investigator, the factor gender with two levels is the main factor of interest. On the other hand, a different investigator might be interested in age differences on the same sensory task. He or she is not interested in the study of gender differences, but realizes that there may be gender differences (may even cite the work of the other researcher in his or her writeup of the study). In this example, the investigator interested in age may benefit by including the factor gender in his or her study because if there are any gender differences the inclusion of gender as a factor will reduce the MS_W (in comparison to not including gender as a factor) and thus will give a more powerful test for the age factor. The more powerful test will occur as long as the drop in the within-treatment sum of squares is not offset by the drop in degrees of freedom from including the extra factor gender (and all interactions with the factor gender).

In the experimental design literature there are many different designs that use control, or blocking, variables. The most commonly used design is to include control variables in the context of a factorial design. This makes matters simple because a control variable in a factorial design is treated just like any other factor in a factorial design. The computation proceeds as usual. There are other types of designs that are described in detail in most textbooks of this kind; we briefly mention a few of these designs here. It is our belief that for most applications the advantages of making the extra effort to perform a factorial design provides the investigator with useful information and is well worth the effort. However, sometimes one does not have the resources necessary to collect all the data needed to complete the cells of a factorial design. There is a class of designs that makes use of blocking factors but also makes the tradeoff of giving up some interaction terms in order to reduce the number of required observations.

One simple design that uses a control variable is called a **randomized-block design**. This basic design is most commonly introduced in its simple form. Suppose there are two factors. One factor is a control variable and the second factor is the one of primary interest to the experimenter. So far, this resembles a factorial design with two factors. However, suppose that only one observation per cell was obtained. Is it possible to do a factorial design with only one observation per cell? The answer is a qualified yes. In such a design with only one observation per cell it is impossible to find the two-way interaction term as well as the two main effects; it is possible however to compute the main effects for the two factors. In other words, the source table for a randomized-block design contains an MS_A term, an MS_B term, and an MS_W term, but not a mean square term corresponding to the interaction. The structural model for this randomized-block design is

$$Y_{ijk} = \mu + \alpha_i + \beta_j + \epsilon_{ijk} \tag{11.1}$$

representing the kth participant's score in condition i and condition j, a treatment effect α for Factor A, and a treatment effect β for Factor B. There is no interaction term $\alpha\beta$.

In all other respects, however, this design is identical to a two-way factorial design. The reason there is no interaction term is that with only one observation per cell there is no variability within a cell, hence no mean square interaction can be extracted that is unique from the MS_W. Sometimes the control variable in a randomized-block design is treated as a random effect (see the next section for a discussion of random-effects factors). In any event, this simple design enables the researcher to ask questions about the primary factor of interest in a situation where the MS_W has been reduced by the control factor. This can be accomplished with relatively minimal data because only one observation per cell is needed.

In Chapter 15 we review a different technique for statistically controlling the effects of a variable—the analysis of covariance (ANCOVA). The randomized-block design is related to the analysis of covariance in that the effect of one variable (or more variables in the case of ANCOVA) are controlled statistically. They differ in that the randomized-block design typically creates blocks, or bins, in order to create levels of a factor, as we did in the example above with the age variable. ANCOVA tends to leave variables in their original form (no blocking or binning of the variable). Both approaches have their advantages and disadvantages. The typical randomized-block design allows more accessible summary of means and does not constrain the modeling of the blocking factor to be linear; the ANCOVA in its simplest form requires that the covariate be kept in its original form and entered as a linear predictor. In more advanced forms of ANCOVA it is possible to deal with nonlinearity in the covariate. For a detailed discussion of dichotomization in the context of a factorial design see Maxwell and Delaney (1993).

Another design worth mentioning is the **latin-square design**. This design contains three factors in its most basic form: one factor is the main factor of interest, and the two other factors are typically treated as control factors. In its simplest form, a latin-square design contains one observation per cell so that no interactions (the two-way interactions nor the three-way interactions) can be tested. For instance, a latin-square design with three factors can test three main effects but nothing else. The structural model for this latin-square design is

$$Y_{ijkl} = \mu + \alpha_i + \beta_j + \gamma_k + \epsilon_{ijkl} \tag{11.2}$$

representing the lth participant's score in conditions i, j, and k, a treatment effect α for Factor A, a treatment effect β for Factor B, and a treatment effect γ for Factor C.

The main advantage of the latin-square design is that it reduces the number of treatment cells needed relative to a complete factorial design. For instance, suppose the three factors each had four levels. A complete factorial design would require $4 \times 4 \times 4 = 64$ treatments. If each treatment contained 10 participants, then 640 participants would be needed for the experiment. However, in the simplest latin-square design, only 16 treatments are needed for a total of 16 observations. Thus, by sacrificing the ability to test interactions, one can have fewer participants. If participants are expensive or time-consuming to test, and it is important to test the main effects of three factors, then a latin-square design may be used. The three-factor latin-square design illustrated here is also useful when the investigator wants to examine the main effect of the primary factor, controlling for the main effects of the other two factors. However, the latin-square design does not permit the study of interactions between the factors, which is one of the main advantages of factorial designs. So, both the randomized-block and the latin-square designs come at a price—no interactions are possible.

SPSS syntax for the randomized-block design and the latin-square design must clearly specify the omission of interactions. Here we show how to do this, using the MANOVA command. An example of a randomized-block design with factor A having three levels and factor B having 4 levels is computed in SPSS by

```
manova DV by A(1,3) B(1,4)
 /design = A, B.
```

Note that only the two main effects are listed in the design subcommand. Similarly, the latin-square design with three factors (A, B and C), each with four levels, is computed in SPSS by

```
manova DV by A(1,4) B(1,4) C(1,4)
 /design = A, B, C.
```

Again, only the three main effects appear in the design subcommand.

There has been much activity in developing new experimental designs that require fewer cells, but most of these designs come at the price of giving up higher-order interactions. One design, called the "fractional factorial design," has recently gained popularity in some behavioral areas. It allows one to study multiple main effects in a simple way with few treatment groups. For an application to social science, see Collins, Murphy, Nair, and Strecher (2005), and for a general review, see Wu and Hamada (2000). Collins et al. (2005) illustrate how an efficient design permits the evaluation of six factors, each with two levels, using only 16 cells. This is more efficient than the 64 cells that would be required for a full factorial (that is, 2^6). This literature has generated several useful statistical procedures, such as the concept of *d*-optimality, which provides a framework to make decisions about which combinations of conditions are most worthwhile to include in a study from the perspective of optimizing the design of the experiment (see Wu and Hamada, 2000, for technical details).

11.6 Random-Effect Factors

In our discussion of analysis of variance so far, we assumed that the levels of each factor were selected by the experimenter because they were the levels the experimenter was interested in. Generalizations based on tests of significance in these experiments were therefore confined to the particular levels of the factors and combinations of levels actually investigated. In other words, the levels of a factor were not considered to be a random selection from a larger population of possible levels. For example, one of the factors in an experiment may be electric shock with three levels. The experimenter obviously has a choice in selecting the levels or intensities of shock to be investigated. It is not likely, however, that the experimenter will decide to choose the three levels by random selection from a larger population of intensities. When the treatments, or levels of factor, are not randomly selected, the analysis of variance model is referred to as a **fixed-effect model**. Up until now we have been discussing the fixed-effect model for all the designs we have presented.

Assume that the levels of a factor have been randomly selected from some larger population. The analysis of variance model, in this instance, is referred to as a **random-effect model**. To illustrate suppose a researcher wanted to study three different marketing strategies at church bake sales. The investigator may randomly select several churches for inclusion in the study and could treat "church" as a control factor to help soak up some of the within-treatment error variance because any variability due to church would otherwise inflate the sum of squares within, that is, SS_W. Participants from a particular church are randomly assigned to each of the three treatments. This is repeated for each church that has been randomly selected for inclusion in the study. The levels of the factor church were not fixed in advance but were instead randomly selected from a larger population of churches. If the levels of some factors have been randomly selected and those of others have not (such as the three treatments in the church example), the analysis of variance model is referred to as a **mixed model**. Thus, the mixed model involves both fixed and random effects.

Stated differently, if all of the treatments, or levels, about which inferences are to be made are included in the experiment, then the treatments or levels may be regarded as fixed, and the fixed-effect model is appropriate for the analysis of variance. On the other hand, if generalizations and inferences about treatments, or levels, not included in the experiment are to be made, then the treatments investigated must be randomly selected from the population of interest. When this situation is the case, the treatments are regarded as random, and the analysis of variance must be altered slightly to take into account the random-effect factor.

Given that we have been discussing fixed-effect factors up until this point in the book, the reader can correctly infer that nothing needs to be done differently for fixed-effect factors. The statistical procedures we have reviewed in the previous chapters hold for fixed-effect factors. It turns out that if the factor in a one-way between-subjects design is random, it is also the case that everything we presented so far, computationally at least, is performed in exactly the same way. The only difference is that the random-effect factor is now interpreted with respect to the larger population of levels that could have been in the study rather than the specific levels that appeared in the study (as is the case with fixed-effect factors). The benefit of random-effect factors is that one can generalize the results to the population of possible levels.

Random-effect models permit the estimation of different variances, not only the usual variance of the error term ϵ. That is, in addition to the usual source of variability ϵ there is also a source of variability that is tied to the variability of the levels of the random factor. The term σ_ϵ^2 denotes the variance of the ϵ, and the term σ_{T}^2 denotes the variance of the random factor T. The label "variance component" is typically used to describe such variance terms. Further, one can normalize the variance component estimate for the random-effect factor to create an **intraclass correlation** estimate. The typical normalization is given by

$$\frac{\sigma_{\mathrm{T}}^2}{\sigma_{\mathrm{T}}^2 + \sigma_\epsilon^2} \tag{11.3}$$

The intraclass correlation can be interpreted as the proportion of variance attributable to the random-effect factor T compared to the total variance (that is, the denominator $\sigma_{\mathrm{T}}^2 + \sigma_\epsilon^2$). The actual estimation of such terms, including the intraclass correlation, is a major area of research, with several books written about the topic (for example, Wolter, 1985). The intraclass correlation can be quite informative in designs involving naturally occurring groups of subjects such as a collection of married couples. The social group such as couple is treated as one of the units of analysis and is treated as a random-effect if the groups were randomly selected (such as a study where married couples are randomly selected to participate in the study). This allows one to properly account for the variability due to couples and also provide a metric to assess interdependence, or the degree of similarity or dissimilarity within a couple (for example, Griffin & Gonzalez, 1995).

Unfortunately, things get a little complicated in factorial designs when random-effect factors are added to the design. Consider a 3×3 mixed model with the first factor a fixed effect and the second factor a random effect. Now there is a complication because the fixed-effect factor has an additional source of variance. Not only is the main effect for the fixed effect influenced by the usual within treatment error and treatment variability, it is also influenced by the random-effect factor. Recall that to compute the main effect for the fixed-effect factor one needs to collapse over the other factor, which in this case is a random-effect factor. Consequently, some of the main effect sum of squares will be influenced by the random-effect factor that is being collapsed in the computation of the main effect for the fixed-effect factor. The solution turns out to be simple. When testing the omnibus main effect for the fixed-effect factor, one uses the mean square interaction as the error term (not MS_W). The reason is that the interaction term has two sources of error—the usual MS_W term as well as the variability introduced by the random selection of the levels of the random-effect factor. Recall that in Chapter 6 we presented an intuitive explanation of the F ratio as having in the numerator a sum of terms we "care about" and terms we "don't care about," and in the denominator just the terms we "don't care about." It turns out that the mean square interaction term has all the terms we "don't care about," so taking a ratio of the mean square term for the main effect of the fixed-effect factor and the mean square interaction yields just the right F test.

Similarly, any test performed on the main effect fixed-effect factor (whether it be a planned comparison or a Tukey test on all possible marginal means for the fixed-effect factor) would use the mean square interaction as the error term rather than MS_W. The two other effects, the main effect for the random-effect factor and the two-way interaction, are tested using the usual MS_W as the error term. In sum, the analysis for a design where the levels of one factor are fixed and the levels of the second factor are random is treated in the usual way, with the mean square within term as the error term with one exception: the main effect of the fixed-effect is tested with the interaction term playing the role of the error term.

There are also complications when both factors are random in a two-way between-subjects design. The interaction term is still tested against the usual mean square within treatment, but each of the two main effect terms are tested using the mean square interaction as the error term. The reader interested in the general theory for determining which error term to use when random-effect factors are present should consult a more technical textbook such as Winer (1971). For an interesting discussion of how neglect of random-effect factors can produce incorrect statistical conclusions in empirical research, see Clark (1973).

11.7 Nested Factors

Consider again the example presented in Section 11.6 describing the sampling of churches in a study on marketing strategies. Assume there are three treatment conditions (that is, three different marketing strategies) treated as a fixed-effect one-way between-subjects design. Suppose that five churches were selected for each treatment (total of 15 churches), and that 40 participants were further selected from each church. So, there are a total of 600 participants (that is, 40 participants per each of the 15 churches) with 200 participants per treatment. We say that "churches" is a nested factor because the levels of "church" are not crossed with treatment in the form of

Table 11.1: Comparing crossed and nested factors.

	Treatment		
Church	1	2	3
1			
2			
3			
. . .			
15			

Representation of the church factor crossed with the treatment factor. The same 15 churches are represented in each treatment.

Treatment		
1	2	3
C1	C6	C11
C2	C7	C12
C3	C8	C13
C4	C9	C14
C5	C10	C15

Representation of the church factor nested within the treatment factor. A different set of five churches appears in each treatment, for a total of 15 churches.

a typical factorial design, but instead are nested in the sense that a given church appears in only one level of the treatment. That is, a particular church does not appear in each treatment (crossed) but appears in a single treatment (nested). See Table 11.1.

The careful reader will note that participants are also nested within church because each of the 40 participants comes from one church, and each participant is not crossed with church because the same participant does not appear in each of the 15 churches. Indeed, in all between-subjects designs there is an implicit factor representing participants that is nested within the treatments of the study. When there are an equal number of participants per treatment combination, this subjects factor has n levels, where n is the number of participants in a given treatment combination. Can you say whether this "subject" factor is typically treated as a random-effect or fixed-effect factor? Because we have usually sampled participants and are interested in generalizing to the population from which the participants were selected (rather than making statements just about the particular participants in the study), the subjects factor is typically treated as a random-effect factor.

Returning to the problem of testing the three marketing strategies, how should we test the three treatment means? There are two factors (treatment and church), but clearly the two factors are not crossed, so we cannot use the machinery of the factorial analysis of variance. The analysis should somehow consider the variability over churches because that contributes to the variability in the treatment means. We will present a very simple procedure for testing the treatment means of the fixed-effect marketing strategies factor that can be used in balanced designs. The trick is to compute the mean for each church, so there will be a total of 15 means. Now, perform the usual one-way between-subjects design on the 15 means (that is, the first set of 5 means corresponds to marketing strategy A_1, the second set of 5 means corresponds to the marketing strategy A_2, and the third set of 5 means corresponds to the marketing strategy A_3). Thus, one has three levels with 5 "observations" per level (where an observation here corresponds to a mean over 40 participants). This analysis correctly compares the three treatments with respect to the variability over churches. Note that the degrees of freedom for within treatments (that is, the error term) will be $12 = 15 - 3$ (number of "observations" minus treatments) in this analysis and will not be based on the 600 total participants in the study. The reasoning is that the effect of the treatment factor is based on the variability of the church factor.

The reader should note that the trick presented in the preceding paragraph is a special case of a nested design. It works when there are an equal number of participants at each level of the "nested" random-effect factor and there are an equal number of levels of the random-effect factor nested within the fixed-effect factor. As one deviates from this ideal situation (for example, has unequal numbers of participants per level of random effect, or plays around with which factors are fixed or random effect, or adds more complications such as additional factors), then one should consult an advanced book such as Winer (1971) in order to derive the correct mean square error terms.

The notion of nested factors becomes important in research involving groups of individuals. For instance, someone studying jury decision making will randomly select participants, randomly assign them into groups of 12, and then randomly assign the groups of 12 into different treatment combinations such as different trials or different types of evidence. If the treatments (say, types of evidence) correspond to a one-way fixed-effect factor, then the trick of computing means for each jury and then inputing the jury means into the one-way design will work. However, if the investigator is interested in comparing two or more dependent variables, then care must be taken when comparing jury means on one dependent variable to jury means on a second dependent variable. See the work of Griffin and Gonzalez (1995; also Gonzalez & Griffin, 2001, and Kenny et al., 2006) for a discussion of the underlying problems and suggested analyses in the framework of multivariate data with nested participants.

The general topic of nested factors is now a major area of research in social science statistics. The topic is called **multilevel modeling** because each layer of nesting can be conceptualized as a "level of analysis." For example, when we have a study about school children's academic performance, the children are nested in schools, schools are nested in school districts, and school districts may be nested within states. Suppose the investigator kept track of not only the child but also the school and the school district. In this case we have the opportunity for different levels of analysis such as at the level of the child, at the level of the school, at the level of the district, and at the level of the state. There may be different factors that impact academic performance at each level of analysis. Proper accounting of the different levels of influence is important in order to have clear statistical results. That is, if properties of school and school district influence academic performance in important ways, it is necessary to account for this in the statistical model. Multilevel modeling techniques permit simultaneous modeling of all the levels that are accounted for in the design. For a discussion of how to use multilevel modeling in experimental designs see Raudenbush (1993), and for a general discussion about the technique see Raudenbush and Bryk (2002). In the next chapter we present a natural way to view repeated-measures designs within a multilevel context.

11.7.1 Nested Designs in SPSS

There are several ways of using SPSS to estimate designs containing nested factors. We describe two commands that have the benefit of automatically computing the proper error term—the UNIANOVA and MIXED commands. The syntax for these two commands is given in turn. The design has factor B nested within levels of factor A and dependent variable DV. In our example, factor A is treated as a fixed effect and factor B is treated as a random effect. We use the SPSS convention that the notation B(A) is interpreted as "factor B nested within factor A."

```
UNIANOVA DV by A B
 /method = sstype(3)
 /intercept = include
 /random = B
 /design = A B(A).

MIXED DV by A B
 /print = solution testcov
 /fixed = A
 /method = REML
 /random = B(A).
```

For balanced designs both commands produce the same results, but for unbalanced designs the MIXED command is preferred because it is based on a more well-behaved likelihood method rather than the method-of-moment approach used by UNIANOVA. Both commands provide output that includes a test of significance for the fixed-effect A and a test of significance for the effect of factor B nested within factor A. Of course, in this subsection we highlight the syntax of the statistical test. We assume the reader has already computed the relevant descriptive statistics at each level of analysis, has checked the equality of variance assumption, has checked for outliers, etc., and has determined that the data satisfy the usual statistical assumptions. The inferential test described in this subsection assesses the goodness of fit of the statistical model and provides a p value in order to make a statistically based decision.

11.8 Homogeneity of Variance

The factorial design presented here assumes that the treatment combinations have the same population variances. In other words, the population variance in each cell (for example, treatment combinations $A_1B_1C_1$, $A_1B_1C_2$, $A_1B_2C_1$, etc.) are identical. Of course, because of sampling we do not expect the observed variance in each treatment combination to be identical. The assumption is adequately met when the observed variances are close enough that a common population assumption is plausible. In our opinion the best way to check this assumption is through boxplots, as we discussed in Chapter 5. The investigator should graph a separate boxplot for each treatment combination. For ease of comparison the separate boxplots could be put on the same page (and should have the same horizontal scale). For example, in a $2 \times 3 \times 4$ factorial design there are a total of 24 treatment combinations, so there would need to be 24 boxplots printed side by side.

The main feature to check in the boxplot is whether the widths of the boxes (that is, the interquartile range) differ dramatically and whether the whiskers (the lines coming out of the box) are systematically asymmetric. What constitutes a dramatic difference? It is not possible to give guidelines because it depends on many factors. One factor is the sample size. If there are a few observations per treatment combinations (say, 10 observations), then the investigator can be more tolerant of discrepant boxplots. However, if the sample sizes are relatively large (say, 100 observations per treatment combination), then the investigator should be less tolerant of discrepant boxplots. We do not recommend performing tests of significance on the observed treatment variances as a procedure for checking the equality of variance assumption. It seems silly to perform a test of significance in order to check whether it is permissible to perform another test of significance.

What should an investigator do when the equality of variance assumption appears violated in a between-subjects factorial design? Unfortunately, there isn't much an investigator can do. We recommend caution when using variance-stabilizing transformations (as discussed in Chapter 5) in factorial designs. Even though transformations may be sensible in a one-way between-subjects design, a transformation in the context of a factorial design can drastically change the interpretation of the results, in particular the interaction. For instance, sometimes we can make an interaction between two factors significant or nonsignificant simply by applying a transformation. Making sense of which transformation is appropriate in a factorial design is not easy because it depends on many issues such as the nature of the scale of the dependent variable and the nature of the theoretical models we are testing. Recall that factorial designs can be computed in the context of a one-way between-subjects design with proper use of comparisons. Our point is that the interpretation of some of those comparisons can be adversely affected by transformations. Also, nonparametric statistical procedures (discussed in a later chapter) may not help because they are not yet completely developed for between-subjects factorial designs.

If an investigator chooses to test all questions in a factorial design through single degree of freedom comparisons, then the Welch-like procedure introduced by Brown and Forsythe (1974) is applicable. The computation of this statistic is not easy, but the intuition is that the denominator degrees of freedom term is reduced in relation to how much the data violate the equality of variance assumption. As mentioned above, the Brown and Forsythe approach can also be applied in the context of the Scheffé procedure. This makes the Scheffé procedure quite versatile. An investigator can examine an orthogonal set of planned comparisons or can examine as many comparisons as he or she wants. We have another example of how single degree of freedom comparisons make statistical analyses simple!

11.9 Summary

This chapter presented many ideas related to the factorial design. Entire textbooks can be devoted (and have been) to each of the subsections in this chapter, and we simply mentioned the relevant issues. We list the major issues an investigator should keep in mind when testing a factorial design. Am I properly correcting the Type I error rate when performing many comparisons? Did I correctly handle unequal sample sizes? Did I make use of all the relevant control factors? Are some of my factors random effect? Are some of my factors of interest nested within other factors? Do my data violate the equality of variance assumption?

11.10 Questions and Problems

1. Describe the difference between a fixed-effect model and a random-effect model. Give an example from your own area of research where a factor may be treated as a random effect.

2. Using data presented in Question 2 from the preceding chapter, perform the (1, −1, 0) comparison over the main effect of Factor B, using the Scheffé criterion.

3. Using data presented in Question 2 from the preceding chapter, perform a Tukey test for all possible pairwise comparisons of the marginal means of Factor B.

4. Give an example of a plausible control variable that could be used in your own area of research. Explain why you think this control variable could help your design.

5. Skim a behavioral research journal, such as in social, personality, or clinical psychology, to look for an example of a published article using an individual difference variable as a factor in an analysis of variance. Read the article carefully. Point out cases in which the author(s) make causal inferences, and explain why these may not be appropriate.

6. Explain in your own words why the assumption of equality of variance is important in factorial designs.

7. Sketch by hand an example set of boxplots from a 2×2 design where the equality of variance is violated. Include relevant information to make the boxplot interpretable. [Hint: What feature of a study would make one more or less tolerant of discrepant boxplots?]

12

Within-Subjects Factors: One-Way and 2^k Factorial Designs

12.1 Introduction

This chapter introduces within-subjects factors in the context of one-way and 2^k factorial designs. More complicated designs, including designs with both between-subjects factors and within-subjects factors, will be discussed in the next chapter.

A **within-subjects factor** means that the same individual is measured more than once on the same variable. An example is the typical pre/post (or before/after) design where a measurement is taken before the treatment and the same measurement is made again after the treatment. Research questions involving change processes are important in many areas of behavioral research, including developmental and educational psychology. The question under investigation in such a design is whether there is a difference between the means for the two times. Does the mean change over the two times of measuring the dependent variable? A within-subjects factor is also known as a repeated-measures factor.

Within-subjects designs introduce new complications, as we will see throughout the next two chapters. In keeping with the emphasis of this book to focus primarily on single degree of freedom comparisons, we will concentrate on planned comparisons. Many of the statistical complications that arise in within-subjects designs are due to pooling the error term. Single degree of freedom comparisons avoid the need for some of the pooling; thus there is a need for fewer assumptions, which leads to relatively simple tests. Also, as we have argued throughout this book, comparisons are more focused than omnibus tests, so they offer an interpretational advantage because they test specific hypotheses on means.

12.2 Example: One-Way ANOVA with a Within-Subjects Factor

We illustrate comparisons on a within-subjects factor with a simple example having five participants, each measured three times on the same dependent variable. Each participant was asked at three points during the term to denote their satisfaction with their statistics instructor. The responses were on a nine-point Likert scale ranging from not at all satisfied (coded as 1) to very satisfied (coded as 9). The data are presented in Table 12.1. The means for the three times (denoted $T1$, $T2$, and $T3$) are 3, 5, and 6. Suppose the investigator wants to test the null hypothesis that the population mean for $T1$ is the same as the population mean for $T3$, that is, $\mu_1 = \mu_3$. The null hypothesis is that the population mean does not change over the first and third time points.

A direct way to test this hypothesis, that is, the (1, 0, -1) comparison, is to apply the comparison participant by participant to create a new score using the original data. That is, take the three scores for the first participant, and apply the coefficients to those three scores. So, participant 1 had the

Table 12.1: Data from a one-way within-subjects experiment.

	Treatments		
Participants	$T1$	$T2$	$T3$
1	5	6	6
2	4	4	8
3	3	5	4
4	2	7	7
5	1	3	5
Mean	3	5	6

three scores 5, 6 and 6 across the three times. Applying the comparison to these scores yields: $(1)5+(0)6+(-1)6 = -1$. Perform the same computation for the remaining four participants. This will produce a new column of numbers, which we call d (we also assign a subscript to d to identify the comparison). The d_1 scores for this first comparison are -1, -4, -1, -5, and -4. The mean of the d_1 scores is $-15/5 = -3$. Thus, on average the comparison weights lead to an average difference of 3 between the $T3$ and $T1$ scores. In some cases the investigator will find it more natural to reverse the sign (that is, apply the -1, 0, 1 comparison) so that the d variable is ordered in the same direction as the alternative hypothesis, but the statistical significance for a two-tailed test will be the same regardless of whether the sign of each coefficient is reversed.

The test of significance is very simple. We perform a one-sample t test on the new column of d_1 scores. The null hypothesis for this test is that the population mean of d_1 is equal to 0, that is, that there is no difference between the population mean of the $T1$ scores and the population mean of the $T3$ scores. Recall that a one-sample t test is given by the ratio

$$t = \frac{\bar{d_1}\sqrt{N}}{stdev(d_1)} \tag{12.1}$$

where $\bar{d_1}$ is the observed mean of the new scores created by applying the comparison directly to the original data, $stdev(d_1)$ is the standard deviation of the new scores, and N is the total sample size (that is, the number of participants). This computed ratio is compared to the tabled value for t (Appendix B, Table B.1) corresponding to the desired α level and $N - 1$ degrees of freedom.

In the present example, the mean of the d_1 scores is -3, the standard deviation is 1.87, and we have

$$t = \frac{-3\sqrt{5}}{1.87} = -3.59$$

The computed value exceeds the critical value of 2.776 from Table B.1 for 4 degrees of freedom and two-tailed $\alpha = 0.05$, so we reject the null hypothesis. The standard deviation of the d_1 scores is computed just like any standard deviation (see Chapter 3). Recall that the definitional formula for the standard deviation of the d_1 scores is the square root of the sum over all observations of the squared deviations between the d_1 scores and the mean of the d_1 scores, divided by the number of subjects minus 1 or, in symbols,

$$\sqrt{\frac{\sum^{N}(d_1 - \bar{d}_1)^2}{N-1}}$$

Applying this formulation to the present data yields a standard deviation for the d_1 scores equal to 1.87.

This test is easy to implement in statistical programs such as SPSS. Follow the simple recipe of creating weighted sums of each time where the weights are the coefficients of the comparisons, and perform a one-sample t test. The syntax and resulting output for the first comparison are presented in Table 12.2.

A confidence interval around the observed mean $\bar{d}_1$ can be constructed in the usual way by the formula

$$\bar{d}_1 \pm t se_{d_1}$$

where $\bar{d}_1$ is the observed mean of the comparison, t is the critical value from the t table with a two-tailed α and $N - 1$ degrees of freedom, and se_{d_1} is the standard error of d_1. The standard error of d_1 is given by the ratio

Table 12.2: Example illustrating SPSS syntax and output. The compute command creates the new variable d_1 according to the comparison weights (1, 0, −1).

SPSS Syntax

```
data list free / S  T1  T2  T3.

begin data
1  5   6   6
2  4   4   8
3  3   5   4
4  2   7   7
5  1   3   5
end data.

compute d1 = (1)*T1 + (0)*T2 + (-1)*T3.
execute.

T-TEST
  /TESTVAL=0
  /VARIABLES=d1 .
```

SPSS Output

Variable	N	Mean	Std. deviation	Std. error mean
D1	5	−3.00	1.8708	0.8366

t	df	p value	Mean difference	95% Conf. Int. Lower	95% Conf. Int. Upper
−3.586	4	0.023	−3.00	−5.323	−0.677

of the standard deviation of d_1 and the square root of the sample size, or symbolically the standard error of d_1, or se_{d_1}, is given by

$$se_{d_1} = \frac{stdev(d_1)}{\sqrt{N}}$$

For a general approach to confidence intervals in within-subjects designs see Loftus and Masson (1994).

Because there are three time periods in this example, there are two orthogonal comparisons that can be made. More generally, an orthogonal set of comparisons over a within-subjects factor will contain as many comparisons as there are number of time periods minus one. This is similar to what we saw in Chapter 8 for the one-way between-subjects design. We saw in that chapter that the number of comparisons in an orthogonal set consisted of the number of treatments minus one comparison. The difference is that now we are counting the number of time periods that observations were made rather than the number of levels in a between-subjects factor. Having tested the (1, 0, −1) comparison in the present example, we can also test the (1, −2, 1) comparison over the three time periods; these two comparisons are orthogonal. In order to test this second comparison, define a new variable, denoted d_2, that is equal to $(1)T1+(-2)T2+(1)T3$, or more simply $T1-2T2+T3$.[1] Creating this new variable for the data presented in Table 12.1, we have these five values for d_2: −1, 4, −3, −5, and 0. Now we can apply Equation 12.1 to compute the t test for the null hypothesis that the population mean of d_2 equals 0,

$$\begin{aligned} t &= \frac{\bar{d}_2\sqrt{N}}{stdev(d_2)} \\ &= \frac{-1\sqrt{5}}{3.39} \\ &= -0.66 \end{aligned}$$

[1]In SPSS this comparison is created by the compute command:
compute d2 = (1)T1 + (−2)T2 + (1)T3.
execute.

Again, the critical value from Table B.1 is 2.776 (for 4 degrees of freedom with a two-tailed $\alpha = 0.05$). The computed t fails to exceed the critical value, so we fail to reject the null hypothesis that the population mean of $\bar{d}_2$ equals 0.

We hope you can appreciate the simplicity of these two one-sample t tests. The comparison was applied directly to the observed scores, a new variable d was created, and a one-sample t test was performed with $N - 1$ degrees of freedom testing the null hypothesis that thc population mcan equals 0. This template will be applied in a more general way to factorial designs, as we will see shortly.

12.3 Trend Analysis on One-Way Within-Subjects Designs

Sometimes the primary objective is to study the trend, or trajectory, of the means over successive trials in an experiment. The means for a series of trials may reveal that the trend is either upward or downward; that is, the means may either increase or decrease with successive trials. Apparent trends can, of course, occur as a result of random variation. For example, the population mean over two times may remain stable, but the sample mean at time 1 may be less than the population mean and the sample mean at time 2 may be greater than the population mean. Thus, we seek a way to evaluate changes in sample means over time. From the experimenter's point of view, the important question is whether the upward or downward trend can be regarded as meeting the requirements of statistical significance or are consistent with random fluctuation from sampling. Similarly, the trend in the means, in addition to being downward or upward, as the case may be, may also show a bend, or some curvature. Again, if there is a bend or curvature in the trend of the means, we wish to be able to determine whether the curvature meets the requirements of statistical significance. The nature of the curvature can be informative about the underlying psychological process.

For example, if differences with successive time periods become smaller, this suggests an asymptote. Another possibility is that habituation or satiation may be occurring, such as the enjoyment of eating a second orange may be greater than the enjoyment of eating a sixth orange, which may be greater than the 12th orange.

The methods of analysis described in this section do not deal with the problem of finding an equation that will describe the trend of the trial means. The problem of curve fitting is important but beyond the scope of this book (curve fitting falls in the domain of nonlinear regression). Our concern in this section, however, is in providing the experimenter with a method for determining whether particular characteristics of the trend of the means (such as a linear trend) are statistically significant or whether they can be attributed to random variations. Before one undertakes the analyses and tests of significance described, it is good practice to plot the means for the successive trials for each time period. We advise researchers to present either a table of means or the plot of the means when reporting trend data.

Assessing trends in one-way within-subjects designs is relatively easy. One uses the polynomial comparisons in Table B.5 in Appendix B and tests the relevant linear, quadratic, cubic, etc., comparisons. The t test we will use is identical to that described in the previous section. Once a comparison is identified, then the comparison weights are used to create a new score d, and a one-sample t test is computed on the new score d. The key to examining trends is the choice of the particular set of orthogonal comparisons.

We illustrate using the same data from the preceding section (Table 12.1). The three means are plotted in Figure 12.1. In this example there were three time periods, so the set of orthogonal comparisons contains two comparisons. Using polynomial comparisons on the three time periods, the linear comparison would be (1, 0, −1) and the quadratic comparison would be (1, −2, 1). One creates two new scores, d_l for the linear comparison and d_q for the quadratic comparison, and performs one-sample t tests separately on d_l and d_q. Note that in the preceding section we tested exactly these two comparisons but did not interpret them as linear and quadratic

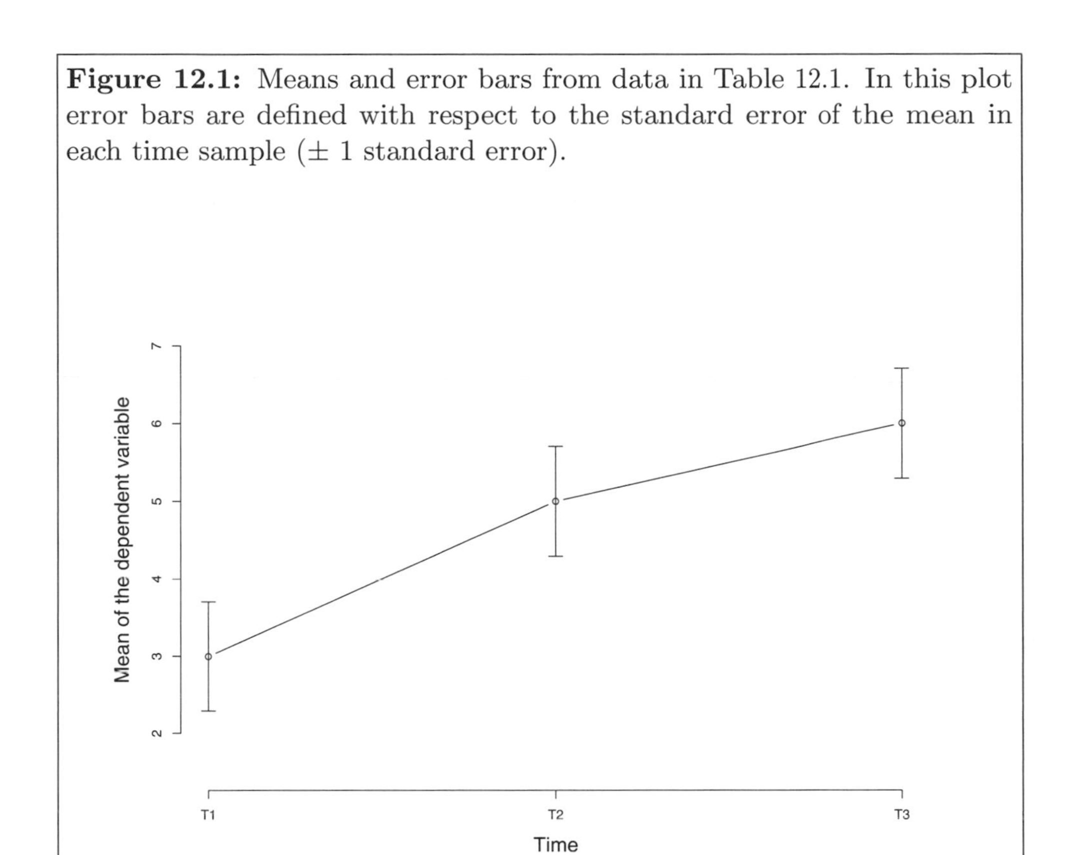

Figure 12.1: Means and error bars from data in Table 12.1. In this plot error bars are defined with respect to the standard error of the mean in each time sample ($\pm$ 1 standard error).

comparisons. The values of the computed t ratios (and the corresponding p values) will be the same.

For the present example with five participants, the test on d_l was statistically significant, but the test on d_q was not (using a two-tailed $\alpha = 0.05$). The visual pattern in Figure 12.1 suggests that the increase is not linear in the sense that there is an increase in the mean from $T1$ to $T2$ but little increase from $T2$ to $T3$. It appears that the satisfaction of these five students increased substantially from the beginning of the term until the middle and then increased a little more from the middle to the end of the term. It will be useful to perform an additional statistical test between the means of $T1$ and $T2$. This illustrates the need to interpret the results of a statistical test directly from the descriptive statistics (for example, the means). Trend analysis will be easier to interpret if the time intervals are equally spaced (for example, each of the measurements is taken 1 hour apart, or each taken 1 day apart, or each taken 1 month apart).

The lesson here is the same as always: never base your interpretations on the p value alone. In this case the linear trend was increasing, and that is what was predicted. But we can easily imagine a scenario where the linear trend is statistically significant with the identical p value but the linear trend is decreasing. Unless we looked at the descriptive statistics (in this case the pattern in the means), we would not know the direction of the linear trend. As always, look at your descriptive statistics when interpreting p values.

12.4 Assumptions and Effect Size Measures

The simplicity of the technique we present for testing comparisons on within-subjects designs occurs because there is no pooling across groups of the error term, and we are freed from making an equality of variance assumption over the time factor. We also do not have to make an assumption about the equality of covariances, as is needed in other techniques for within-subjects factors with three or more levels. A covariance is a measure of whether individual scores on two variables are on the same side of their respective means. The covariance is similar to a variance but is defined over two variables X and Y as follows:

$$\text{Cov}(X,Y) = \frac{\sum_{i=1}^{N}(X_i - \bar{X})(Y_i - \bar{Y})}{N-1}$$

The variance is a special case of the covariance—when the variable X is the same as variable Y, then the formula is identical to the variance. Consequently, a variance can be interpreted as the covariance of a variable with itself. In symbols, Cov(X,X)= Var(X).

The covariance can be interpreted by examining its numerator. For a given participant, the score on variable X can be either above the mean of variable X, below the mean of variable X, or at the mean of variable X. This leads to a difference on variable X that is either positive, negative or zero. Similarly, there is a difference from the mean of Y that is either positive, negative or zero on dependent variable Y. If the participant's scores are on the same side of the mean on both variables, the product will be positive (that is, a product of two positives or a product of two negatives). However, if the participant's scores are on one side of one variable's mean, but on a different side of the other variable's mean, then the product will be negative (that is, the product of one positive and one negative). If the participant's score is at the mean for either or both variables, then the product will be zero. The numerator for the covariance is the sum over all participants of all such products. Of course, the numerator takes into account the magnitude of the discrepancy as well as the sign of the discrepancy, but most of the intuition for what the covariance is doing comes from focusing on the sign of the difference.

Why is the covariance relevant in a within-subjects design? Let's consider the case of a within-subjects factor with four levels. In this design there are four observed variables: $T1$, $T2$, $T3$ and $T4$. Each of these variables has a variance. In addition, there are six covariances representing the six pairwise relations between the four variables. Using Cov to denote covariance, these six covariances are Cov($T1$, $T2$), Cov($T1$, $T3$), Cov($T1$, $T4$), Cov($T2$, $T3$), Cov($T2$, $T4$), Cov($T3$, $T4$). The covariance is symmetric in that Cov(X, Y) = Cov(Y, X).

If we want to merge (that is, to average or "pool over") the four time periods to create a single error term analogous to the MS_W term that we had in the between-subjects case, then we will have to make some assumptions on the four variances as well as the six covariances. One technique that was used many years ago was to assume that the four variances have the same population value and that the six covariances have the same population value (but that the value for the variance need not equal the value for the covariance). Under this assumption, one can average the four variances into a single number (denoted $\bar{V}$ to indicate a mean over variances) and average the six covariances into a single number (denoted $\bar{C}$). A pooled error term can then be defined as $MS_W = \bar{V} - \bar{C}$. But these assumptions are too restrictive. Why should the time periods have equal population variances? Why should the covariances between all the time periods be estimates of the same population value? It may be that variables closer in time (such as $T1$ and $T2$) have a greater covariance than variables further apart in time (such as $T1$ and $T4$). Data rarely satisfy these restrictive conditions, so techniques that require this level of pooling are not desirable.

The technique we present in this chapter does not pool variances and covariances to create a single error term for all comparisons. Instead, a new error term is automatically constructed for each comparison. We will not go into the details of what makes up a given error term because that is best described using matrix notation. We simply highlight that the error term consists of a weighted sum of variances and covariances, where the weights depend on the comparison values. Thus, each comparison can produce a different weighted sum of variances and covariances, leading to a uniquely constructed error term based on the coefficients of the comparison.

The technique presented here for analyzing within-subjects designs does make a normality assumption in order to perform the one-sample t test (that is, the population of the new variable d must be normally distributed for the p value calculation). However, with a large enough sample the one-sample t test is fairly robust to violations of normality, as we observed in Chapter 3. In our own experience, we have found that the one-sample t test performs quite well when the distribution is symmetric, but there has been some detailed work that challenges that conclusion (for example, Wilcox, 1993, 1995).

Unfortunately, there does not appear to be much consensus on the definition of an effect size measure for within-subjects factors. On the surface it would seem that the effect size index would be easy to define because we are performing a one-sample t test on the new variable d. However, there is a debate over how to handle the covariance(s) between the variables involved. Some authors have argued for measures that ignore the covariance(s) from the measure of effect size; other authors have argued for measures that take into account the covariance(s) in the measure of effect size. It is our opinion that little progress will be made on this debate until there is a theory of effect size; at present it appears, to us at least, that most effect size measures are ad hoc without a solid foundation.

We suggest the following simple ad hoc formula for the effect size of a comparison on a within-subjects factor:

$$\sqrt{\frac{t^2}{t^2 + (N - 1)}}$$

which is analogous to the formula presented for between-subjects designs. Care must be taken, though, in that this index no longer carries a "proportion of unique variance accounted for by the comparison" interpretation because we are not using a common error term across all comparisons. We suggest this index because it is on the same scale as the one for between-subjects designs.

12.5 2^k Factorial Designs: All Within-Subjects Factors

Recall that for 2^k between-subjects factorial designs (Chapter 9) we presented the method of converting a factorial design into a set of orthogonal comparisons where each comparison tested a particular feature of the factorial design (a main effect, a two-way interaction, a three-way interaction, etc.). The analogous method is presented here for 2^k within-subjects factorial designs.

Suppose that we have a 2^2 within-subjects factorial design. An example involves participants who are tested at the beginning of the academic year

Table 12.3: Hypothetical 2 × 2 design with two within-subjects factors. The placeholder "data" denotes observations in each of the cells.

	Academic year	
Test within session	Beginning	End
Pretest	Data	Data
Posttest	Data	Data

and again at the end of the academic year. At each of the two sessions in the lab, each participant is given a pretest and a posttest. Thus, each participant is observed four times: beginning academic year pretest, beginning academic year posttest, end academic year pretest, and end academic year posttest. See Table 12.3 for a tabular form of the design. We know that with four time periods three orthogonal comparisons are needed to code this factorial design. Maintaining the order of the four treatments listed above, the (1, 1, −1, −1) comparison tests for main effect of beginning versus end of the academic year, the (1, −1, 1, −1) tests for the main effect of pretest versus posttest, and the (1, −1, −1, 1) comparison tests for the interaction between academic year and pre/post.

The tests of these comparisons proceed in the manner described in the preceding two sections. Create a new variable that is a weighted sum of the original observations, where the comparison values serve as the weights, and perform a one-sample t test against the null hypothesis of zero. For instance, to test the interaction, one creates a new variable $d_3 = T1 - T2 - T3 + T4$, where the Ts correspond to the four time periods in the order described (that is, for the present example, begin/pre, begin/post, end/pre, and end/post). The variable d_3 represents the weighted sum of the four scores. One performs a one-sample t test (using Equation 12.1) on the new d_3 scores against the null hypothesis that the population mean of d_3 equals 0. Again, the degrees of freedom for each of these tests is $N-1$ (that is, the number of participants measured minus one).

Of course, different tests could be performed too, such as the difference between pretest and posttest during the beginning of the academic year assessment, and this would be tested with the (1, −1, 0, 0) comparison. This comparison is called a simple main effect. A positive feature of testing individual comparisons is that each comparison has its own error term. That is, every time a one-sample t test is performed, a new $stdev(d)$ is computed for that comparison. Thus, there isn't a single MS_W term (as in the case of the between-subjects ANOVA) because we have avoided pooling. This makes constructing pairwise tests analogous to the Tukey test rather difficult because there are different error terms for each pairwise comparison. There has been little research done on the problem of testing multiple comparisons and pairwise tests for within-subjects designs.

For the general 2^k within-subjects factorial design, one applies the same logic. Take the total number of time periods (that is, the value of 2^k) and define an orthogonal set of $(2^k - 1)$ comparisons that codes the factorial structure—that is, a main effects comparison for each variable, two-way interaction comparisons between every pair of variables, etc., as we showed for the between-subjects case in Chapter 9. For each comparison, create a new weighted difference score variable d, and perform a one-sample t test on that new variable to test its mean against the population value of zero. This procedure will automatically test all the main effects and interactions because in a 2^k factorial design all main effects and interactions have a single degree of freedom, and, thus, each test will be identical to the test of a comparison.

The investigator should examine the individual treatment means when interpreting the results of the main effects and interactions (by treatment mean we refer to the mean of a single time period, such as beginning academic year/posttest). Merely stating in a results section that the value of the new variable d is statistically significant is not informative. There are many patterns that would lead to the same value of d. So, for instance, to understand the meaning of the two-way interaction term in the example presented, one must interpret the pattern present in the four observed means (that is, the means in the four conditions: beginning academic year pretest, beginning academic year posttest, end academic year pretest, and end academic year posttest). We reiterate this important point: the statistical test is performed on the new weighted scores d, but the interpretation of the statistical test follows directly from the means in each time period.

We illustrate the need to interpret results, using the means at each time point with a simple example. Suppose the interaction comparison is computed for a 2^2 within-subjects factorial design and the mean of the new d scores (that is, $\bar{d}_i$) is 4. Think of all the ways that four treatment means weighted by the comparison (1, −1, −1, 1) could produce a new d variable having a mean of 4. We present two examples, and the reader is invited to create additional ones. In one example, the individual time periods have means of 4, 2, 2, and 4. In a different example, the means for the individual time periods could be 1, 2, 1, and 6. Both of these situations would produce a $\bar{d}_i$ equal to 4, but, as Figure 12.2 shows, they each carry a different interpretation of what happened in the study. A similar example could be constructed for other comparisons too (like comparisons involving main effect terms in a 2^k factorial design). The same caveat can be made for between-subjects designs, and our previous recommendation to graph the treatment means was an attempt to force interpretation of the treatment means. The same logic applies in within-subjects designs. As we have consistently argued, interpretation follows from the descriptive statistics such as means; the p values that emerge from the statistical tests provide information about the status of a hypothesis test. The meat comes from the interpretation of the means—the p value along with the critical value α provide a "stamp of approval" that the statistical significance of the result passes conventional standards.

12.6 Multiple Tests

Some readers may wonder about our procedure of performing so many one-sample t tests. Aren't we capitalizing on chance by making so many comparisons? Our first answer is that, in the case of the 2^k factorial design, the procedure we present is identical to the results of the usual main effect,

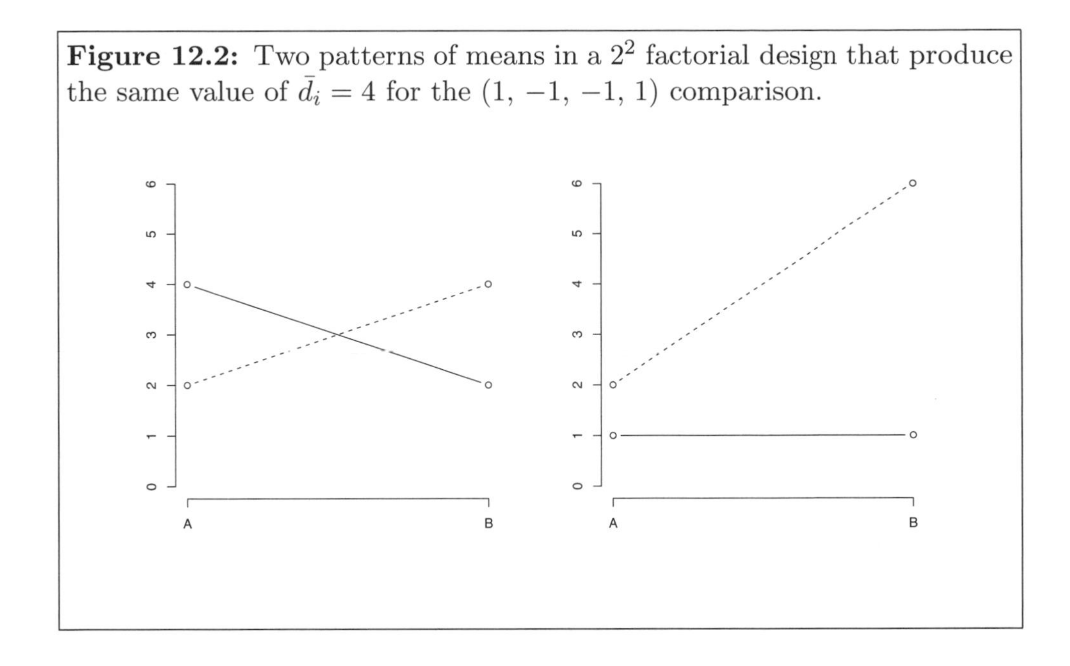

Figure 12.2: Two patterns of means in a 2^2 factorial design that produce the same value of $\bar{d}_i = 4$ for the (1, −1, −1, 1) comparison.

two-way interaction(s), three-way interaction(s), etc. So, the same number of tests as in the traditional within-subjects factorial design are performed. For instance, in the 2^2 case we perform three tests that are identical to the tests for the two main effects and interaction that would be part of the standard one-way ANOVA.

In the one-way within-subjects design we perform $K - 1$ comparisons (where K is the number of time periods), whereas the traditional method only performs one omnibus test over all K time periods followed by a series of additional tests such as pairwise tests. As we argued in the between-subjects chapters, if the investigator has a set of orthogonal comparisons, then he or she should test those comparisons directly—bypassing the omnibus test, which is usually not informative. Replication is the best way to deal with concerns about multiple tests and inflated Type I error rates (Greenwald et al., 1996). But, if replication is not easy and the investigator wants a quick and dirty correction for having performed multiple tests, then a Bonferroni procedure can be used. Simply divide the desired α level by the number of tests performed and base the critical t value on the reduced α level. Because we decided not to pool measures to create a single global error term, we do not have the luxury of using procedures analogous to the Tukey procedure that tests all possible pairwise comparisons using a pooled error term—a small price to pay, in our opinion, for the benefit of clear-cut one-sample t tests. In a later section we present the Scheffé test for within-subjects factors, which permits as many comparisons as the investigator wishes, regardless of whether the comparisons are orthogonal or nonorthogonal, planned or not planned.

12.7 Design Considerations with Within-Subjects Designs

There are pros and cons to within-subject factors. The pros include the following: (1) Each participant serves as his or her own control. All things being equal, this will increase the statistical power to detect differences between means. This increase in power occurs because changes are based on individual scores. (2) The within-subjects factor is useful for testing hypotheses about change over time, learning, and trends. (3) The within-subjects design requires fewer participants than a comparable between-subjects design for a given level of power. For instance, in a 2^2 factorial design there are four treatments. A between-subjects design requires n participants per treatment, or $4n$ participants in all; in a within-subjects design n participants are needed because each participant is measured four times. The denominator degrees of freedom are not identical, so this comparison on sample size is not fair (recall that a between-subjects design has $4n-T$ degrees for freedom for the denominator, but the within-subjects design has $n-1$ degrees of freedom, where n is the number of participants in a treatment or time period).

The cons involve methodological issues and interpretational concerns. (1) There may be practice and fatigue effects that result from measuring the participant repeatedly. Participants might improve over time not because of the treatment but because the repeated trials provide an opportunity for practice or familiarity with the experimental task. Similarly, participants may perform worse because they get tired or bored with the task and do not pay attention. Thus, one must be careful when interpreting a trend in the means over time periods. (2) As participants become familiar with the task, they may develop their own hypotheses about the study, and these hypotheses may influence their behavior on subsequent measurement. This may be a concern because participants are in all treatments and can detect variations across treatments. (3) For within-subjects factors to increase power, there must be a sufficiently high correlation across time periods. Otherwise, the loss of degrees of freedom (what in the preceding paragraph was deemed a virtue because fewer participants were needed relative to a between-subjects factor) becomes detrimental to the test of statistical significance. (4) Issues of missing data present problems with how within-subjects designs are usually implemented in most statistics programs and the simple comparison method presented here. In order to compute the analysis of variance or the comparison value, it is necessary to have all participants provide data on all time points. Most programs completely drop a participant who has at least one missing data point. More recent approaches to within-subjects designs permit missing data.

Statisticians such as Rubin (2005) have argued that causal inference requires a design where one estimates the difference between being in a treatment and not being in a treatment. In practice this cannot be accomplished easily because one cannot place a participant in one treatment, remove that person from the treatment wiping out all effects, and magically place that participant in another treatment. These statisticians point out that additional assumptions are necessary in order to make causal inferences. This is an active area of current research, and we hope that new understanding will emerge (though we acknowledge that the problem of causal inference has plagued philosophers and scientists for centuries).

The first two concerns listed above can be dealt with by introducing another factor into the design. Consider a between-subjects factor where one group of participants is measured K times and receives the relevant manipulations over the time periods, whereas a second group of participants is also measured K times but does not receive the relevant manipulations. The interaction between time and this new between-subjects factor helps disentangle general effects over time from those that can be attributable to the treatment. Any practice or fatigue effects would be present in both groups, however one group of participants received the treatment, so any difference in the trend over time relative to the control group can be attributed to the treatment. In the next chapter we discuss how to perform an analysis of variance when there are both between-subjects and within-subjects factors in the same design. For a more detailed discussion of the methodological issues surrounding within-subjects designs the interested reader should consult Greenwald (1976).

12.8 Scheffé Test for Within-Subjects Factors

We discussed the Scheffé test in Chapter 8 for the case of between-subjects factors. Recall that the Scheffé test allows for testing as many comparisons as needed, controlling the Type I error rate for the family of comparisons. For instance, in a two-way between-subjects factorial design one can perform as many comparisons as one can dream up on the first factor, on the second factor, and on the interaction. The Scheffé test involves computing the comparison in the usual manner, but the computed t ratio is compared to a new critical value to establish whether or not the null hypothesis has been rejected. The new critical value is computed for each "family" of comparisons (that is, separately for each omnibus main effect and omnibus interaction); the Scheffé test works its "magic" through the new critical value. The comparisons on the first factor (collapsing the second factor) make up one family of comparisons. Similarly, the comparisons on the second factor (collapsing the first factor) and the comparisons for the interaction each make up two additional families of comparisons, respectively. The Scheffé test controls the Type I error rate within each family of comparisons.

The logic of the Scheffé test can be extended to within-subjects factors. One computes the t ratio on the comparison in the usual way for a within-subjects factor but constructs a new critical value to compare the computed t ratio. The formula for the new critical value is

$$t' = \sqrt{\frac{(n-1)(a-1)F_{\alpha,df_{(a-1)},df_w}}{n-a+1}} \tag{12.2}$$

This formula looks cumbersome, but by now this formula should make some sense. The term $F_{\alpha,df_{(a-1)},df_w}$ is the value from the F table corresponding to the following three variables: (1) the desired α level, (2) the degrees of freedom for the "family" $df_{(a-1)}$ where a is the number of levels of the repeated-measures factor, and (3) the error degrees of freedom df_w. For example, in a one-way within-subjects factor with 25 participants each measured five times (that is, the one-way within-subjects factor has five levels) we have the following: $df_{(a-1)}$ = number of levels minus one $= 5 - 1 = 4$ and df_w = number of participants minus number of levels of factor A plus one $= 25 - 5 + 1 = 21$. Looking at Table B.2.1 in Appendix B using $\alpha = 0.05$ we find a tabled F equal to 2.84. The other terms in Equation 12.2 can be interpreted as correction factors that adjust the tabled value of F in relation to the number of participants in the study and the number of time periods.

For the example just presented the new critical value is $t' = \sqrt{\frac{24*4*2.84}{21}} = 3.60$, hence any comparison over the five levels of the within-subjects factor having an observed t value exceeding 3.60 will be statistically significant by the generalized Scheffé criterion. Again, the observed t ratio is computed in the manner presented earlier in this chapter. That is, the comparison weights are used to create a new variable, and a one-sample t test is computed on the new variable. The only difference when applying the Scheffé criterion is that one uses the Scheffé t' as the critical value rather than the usual tabled t critical value.

12.9 SPSS Syntax

There are several ways of performing a within-subjects design in SPSS. We present several ways to implement such designs. Consider the case of two within-subjects factors, each with two levels, so this is a 2^2 within-subjects factorial design. One implementation in SPSS uses the MANOVA command and specifies the comparisons for the main effects and interactions in the form of comparisons directly in the syntax:

```
MANOVA t1 t2 t3 t4
   /WSFACTOR time(4)
   /CONTRAST(time) = SPECIAL( 1 1 1 1
                              1 1 -1 -1
                              1 -1 1 -1
                              1 -1 -1 1)
   /PRINT= parameters(estim)
   /WSDESIGN time.
```

The data are entered so that each subject's four time points are in four different columns (t1 to t4 to indicate the four time periods). If there are N subjects, there will be N rows and 4 columns to the dataset. The four columns are labeled t1 to t4. The syntax for the MANOVA command specifies the four observed variables t1–t4. The MANOVA syntax also defines a name for the factor called time (using the WSFACTOR subcommand). Time is not a variable or column in the data set but a name given to the collection of four time observations t1 to t4. The CONTRAST subcommand defines four row contrasts: the first is the unit vector, and this is essentially a placeholder for the grand mean effect; the second is the main effect comparing t1 and t2 to t3 and t4 (the comparison 1, 1, -1, -1); the third is the main effect comparing t1 and t3 to t2 and t4 (the comparison 1, -1, 1, -1); and the fourth comparison is the interaction comparison, which is the product of the coefficients for the two previous main effects. The WSDESIGN subcommand specifies the variable time, the name we gave the observed variables t1–t4, as the main effect factor. There are other ways of coding this design, including specifying two factors, but we thought that treating the design as a one-way within-subjects design and specifying the main effect and interaction comparisons directly would be a clear way of conveying one way that repeated-measures analysis of variance can be conducted in SPSS.

An equivalent but alternative way to test the two main effects and interaction is to create the comparison scores directly, as we described in the previous sections. For each subject, apply the coefficients to their individual scores, and compute a one-sample t test on each comparison. The SPSS syntax to accomplish this is given by

```
compute cont1 = t1 + t2 - t3 - t4.
compute cont2 = t1 - t2 + t3 - t4.
compute cont3 = t1 - t2 - t3 + t4.

t-test
 /testval=0
 /variables cont1 cont2 cont3.
```

This syntax first applies the coefficients to each of the four time observations t1–t4. This creates three new variables, which we call cont1, cont2, and cont3 to correspond to the three resulting contrasts. We then compute one-sample t tests to each column of data cont1 to cont3 to get the t tests for each of the three comparisons. This syntax produces identical results to the MANOVA syntax above that defines the contrasts directly in the CONTRAST subcommand.

12.10 Multilevel Approach to Within-Subjects Designs

A different way to represent a within-subject design is to nest time within the subjects factor, where subjects is treated as a random-effects factor. The usual structural model for the within-subjects design models each observation as an additive sum of the following terms:

$$X_{ij} = \mu + \alpha_i + \pi_j + \epsilon_{ij} \tag{12.3}$$

where μ is the usual grand mean, α corresponds to the main effect of time (there is one α for each level of time), π represents the subject factor that is treated as a random effect because subjects are randomly selected from a population (there are as many π's as there are participants), and ϵ is the usual error term (there are as many ϵ terms as there are number of participants times number of observations over time). In the parenthetical statements in the preceding sentence it should be noted that there are some constraints imposed on the computed values of the α's, π's and ϵ's, so technically there are not as many unique values as there are number of levels (α's) and number of subjects (π's and ϵ's). Both π and ϵ are random variables assumed to follow a normal distribution with mean 0 and variances σ_π and σ_ϵ, respectively. However, in within-subjects designs the variance σ_π represents a square matrix with as many rows and columns as there are times, with variances in the diagonal and covariances in the off-diagonal. The reason that the σ_ϵ term is not a matrix is because subjects are assumed to be independent from one another (so all the covariances between subjects are 0) and are assumed to have the same error distribution, which is the homogeneity of variance assumption.

There is another way to specify the within-subjects model. The multilevel approach takes the time factor as nested within the subject factor and writes a two-level structural model. The first level models the observed data as a function of a subject effect β_j, time and the residual, as in

$$X_{ij} = \beta_j + \alpha_i + \epsilon_{ij} \qquad (12.4)$$

There is also a second structural model for the β_j, which is written as

$$\beta_j = \mu + \pi_j \qquad (12.5)$$

where π is a random effect parameter assumed to be sampled from a normal distribution with mean 0 and variance matrix σ_π. Note that if one substitutes the linear definition of β (Equation 12.5) into the first-level Equation 12.4, one gets the identical structural model for the within-subjects design presented in Equation 12.3. So, the multilevel model is not a different model as much as a different approach to handling repeated-measures. The advantage of the multilevel approach comes in its generalization to other models and a common framework for handling many different kinds of designs under one umbrella. For a review of multilevel models see Raudenbush and Bryk (2002). For a specific comparison of ANOVA and multilevel models see Raudenbush (1993).

One of the key benefits of the multilevel approach to repeated-measures is that it handles missing data in an elegant way. Unlike ANOVA, which discards the entire subject from the data set if there is at least one missing time point for that participant, the multilevel model makes use of all available data for each subject.

The SPSS implementation of this approach uses the command MIXED. The syntax to produce the identical output, as in the preceding section, is

```
MIXED data  BY time
  /FIXED = time
  /REPEATED = time | SUBJECT(subject) COVTYPE(UN)
  /TEST  'main effect 1'  time 1 1 -1 -1
  /TEST  'main effect 2'  time 1 -1 1 -1
  /TEST  'interaction'    time 1 -1 -1 1.
```

This syntax requires that data be entered in a different format than typically done for repeated-measures designs. Rather than entering each subject's data in a single row, every observation is entered in a separate row and new variables are included that code which subject and which time for each observation. So if there are 12 participants and each participant provided four observations, then the data set will have 48 = 12 * 4 rows, a column of numbers 1 through 12 to indicate to which subject each observation belongs, and a column of numbers 1 through 4 to indicate to which time each observation belongs. This is a strange way to implement a repeated-measures design for those who are well versed in the ways of repeated measures, where multiple times for the same person are entered in the same row. It is good to get in the habit of sorting the data file by subject because some programs require this sorting and yield inappropriate results if data aren't sorted.

The subcommand FIXED instructs SPSS to use time as a fixed effect, and the REPEATED subcommand sets up the structure where the time scores are nested within the subject factor. The setup for the REPEATED subcommand is quite general to allow several different types of correlational structures. The syntax above specifies the covariance to be unstructured. This is the typical assumption that corresponds to the single degree of freedom tests that we present in this book; there are different types of covariance structures that are possible. We refer readers to the manual of their statistics package, which in the case of SPSS and most major programs tends to have good documentation on the different covariance structures for the time factor that are possible within their commands for testing multilevel models.

The MIXED syntax specifies each comparison in separate TEST subcommands. For balanced designs this MIXED syntax produces the same output as the MANOVA command and the set of one-sample t tests we introduced in the preceding chapter. There are several different ways of implementing this design within the MIXED command that yield the identical results, and we refer the reader to the online documentation for the MIXED command (see Peugh & Enders, 2005).

12.11 Summary

Within-subjects designs are simple as long as one focuses on testing comparisons. In that special case, the test of statistical significance for a comparison reduces to a simple one-sample t test. A complete 2^k within-subjects factorial design can be tested using a sequence of one-sample t tests, as we have shown in this chapter. The result of these one-sample t tests are identical to the output of most statistical packages currently on the market (such as SPSS). Thus, we are not presenting a new technique but are describing an existing technique in a simple way. This simplicity occurs because we did not pool the time periods to find a single error term to use for all comparisons (as we did in the case of the between-subjects design when computing MS_W). Multiplicity of tests is not a serious problem, especially if replication is possible, but if the investigator is concerned about having made multiple tests, then Bonferroni and Scheffé procedures are possible.

12.12 Questions and Problems

1. A study consists of four observations from each of 10 participants. The observations were scores from weekly pop quizzes in vocabulary that were given in a sixth-grade class. The dependent variable was the total number of correctly defined words (a perfect score on each quiz is 5).

	Quiz			
Subject	1	2	3	4
1	5	3	4	3
2	1	3	3	2
3	1	3	5	5
4	4	4	4	4
5	2	1	4	5
6	3	3	4	5
7	1	1	0	1
8	4	5	5	4
9	5	5	5	5
10	2	3	3	5

(a) Plot the means of these four quizzes over time.

(b) Test whether there is a linear, a quadratic, or a cubic trend in these data. Use a two-tailed $\alpha = 0.05$. Be sure to base your interpretation of the results on the means of the four quizzes.

(c) In the preceding question you tested three orthogonal comparisons. Use the Bonferroni procedure on these three comparisons. Test the comparison against the Scheffé criterion.

(d) Test whether the $(1, 1, -1, -1)$ comparison, that is, the first two means for the quizzes versus the last means for the two quizzes, is statistically significant, using a two-tailed $\alpha = 0.05$.

2. Pick a domain of empirical research that interests you. Look in a relevant journal that publishes empirical research, and find one example of a within-subjects design. Discuss the pros and cons of adopting a within-subjects test in that research domain. Focus on methodological issues such as the participant serving as his or her own control, possible confounds such as fatigue effects, etc. If possible, identify safeguards the authors used in their design to minimize the negative features of within-subjects factors, and, if possible, make suggestions for how the design can be altered to avoid such confounds.

13

Within-Subjects Factors: General Designs

13.1 Introduction

In this chapter we review general issues surrounding within-subjects factors and present more general within-subjects designs including factorial designs that include both between-subjects and within-subjects factors. We conclude this chapter with a brief discussion of omnibus tests for within-subjects designs.

13.2 General Within-Subjects Factorial Design

In this section we extend the technique presented in the preceding chapter for 2^k within-subjects factorial designs to factorial designs with within-subjects factors having arbitrary number of levels. Recall our discussion in Chapter 10 of how to define sets of orthogonal comparisons that code main effects and interactions. We combine the logic in Chapter 10 for defining comparisons on the main effects and creating interaction comparisons by multiplying the pairs of comparisons from different main effects with the logic presented in Chapter 12 for testing comparisons on within-subjects factors in such a way that each comparison has a unique error term.

To illustrate, we use a study of memory. Suppose that three types of word lists are created, each list having 25 words matched on length of characters. One list consists of low-frequency words, the second list consists of mid-frequency words, and the third list consists of high-frequency words (there are several published sources that rank words on frequency of use in the English language). Each word in a list was presented on a computer screen for 1 second (with a 3-second lag between words). The participant studied one word list, performed a 15-minute distracter task, and then completed a test of recall. The participant followed this procedure (study, 15-minute distracter, recall test) three times, once for each list.

The order of presentation of the list was randomized so that the results would generalize and not be tied to a specific order in the study. There were six possible orders for presenting the list in one session (using L for low-frequency, M for mid-frequency, and H for high-frequency). The orders were:

	Temporal position within session		
	1	2	3
Order 1:	L	M	H
Order 2:	L	H	M
Order 3:	M	L	H
Order 4:	M	H	L
Order 5:	H	L	M
Order 6:	H	M	L

Thus in one session the participant had three different scores: the total number of correctly recalled words for each list. Assume the same eight participants came to the lab three times to complete a total of three sessions.

Table 13.1: Example of a 3 × 3 within-subjects factorial design. The study included eight participants who studied three different word lists (low, medium, and high frequency) in three different positions (first, second, and third). Data are the total number of words correctly recalled (highest score possible in a cell was 50). The correlation between any two columns in this table is relatively high. Note that data from the same participant are entered in the same row.

Participant	L1	L2	L3	M1	M2	M3	H1	H2	H3
1	9	9	10	21	18	19	38	41	41
2	12	12	12	22	20	22	42	43	39
3	6	7	7	16	15	14	34	36	35
4	12	15	13	24	23	21	42	43	44
5	17	16	18	27	27	29	48	47	49
6	14	12	11	23	22	23	41	42	42
7	6	5	4	14	17	15	37	35	35
8	6	7	7	17	15	16	36	38	36

The researcher is interested in whether there are differences across the three lists L, M, and H. The time and list factors create a 3 × 3 within-subjects factorial design. The first factor is frequency of words and the second factor is list. In all, there are nine possible conditions (L first, L second, L third, M first, M second, etc.). The data for each participant are presented in Table 13.1. We ignore the fact that each list appeared twice in each position and simply add the scores from both times to create a possible range between 0 and 50. We discuss the issue of collapsing the two administrations of each list within each position at the end of this section.

The mean score for each of the nine cells is presented below. In addition, we show the marginal means for both list and position.

	List L	List M	List H	Marginal means
Position 1	10.25	20.50	39.75	23.50
Position 2	10.37	19.65	40.65	23.56
Position 3	10.25	19.87	40.12	23.41
Marginal means	10.29	20.00	40.17	

Visual inspection of the means suggests that the marginal means for list differed from one another but that the marginal means for position were fairly comparable, suggesting a possible main effect for list but not for position. Also, there does not appear to be an interaction between position and list as is evident both by looking at the pattern in the nine cell means and also Figure 13.1. Of course, such eyeball assessments of differences between means would be better informed by also including confidence intervals.

Suppose we wanted to test the (−1, 0, 1) and (1, −2, 1) comparisons on the list factor. That is, we want to test the research questions: Did the mean for List L differ from the mean for List H? Did the average of Lists L and H differ from List M? This involves creating comparisons with nine weights (one weight for each of the $9 = 3 \times 3$ cells in the design). Using the order presented in Table 13.2, the first comparison is (−1, −1, −1, 0, 0, 0, 1, 1, 1), and the second comparison is (1, 1, 1, −2, −2, −2, 1, 1, 1). These two comparisons are orthogonal; they are presented in the first two rows of Table 13.2, which shows a complete set of eight orthogonal comparisons.

The comparisons can be tested by creating a new variable d, much like we did in the preceding chapter. For each participant create a new score that is the weighted sum of the nine observed scores, using the comparison values as the weights. For instance, for the first participant in Table 13.1, the first comparison yields $d_1 = (-1) \times 9 + (-1) \times 9 + (-1) \times 10 + (0) \times 21 + (0) \times 18 + (0) \times 19 + (1) \times 38 + (1) \times 41 + (1) \times 41 = 92$. We compute the value of d for each participant in the same manner. The eight d scores for each participant on this first comparison are presented in column 1 of Table 13.3. Once the d scores are in hand, then one performs a one-sample t test on those d scores, and this test corresponds to the test of significance for (−1, 0, 1) comparison over the list factor. The t test for the first comparison is statistically significant, as is the t test for the second comparison.

Table 13.2 presents a complete set of orthogonal comparisons for this 3×3 design. It was constructed as follows. The first two comparisons were discussed in the preceding paragraph and these correspond to two possible comparisons on the list factor. Assume that we also wanted to test the same

Figure 13.1: The plot of the individual cell means in the 3×3 within-subjects factorial design.

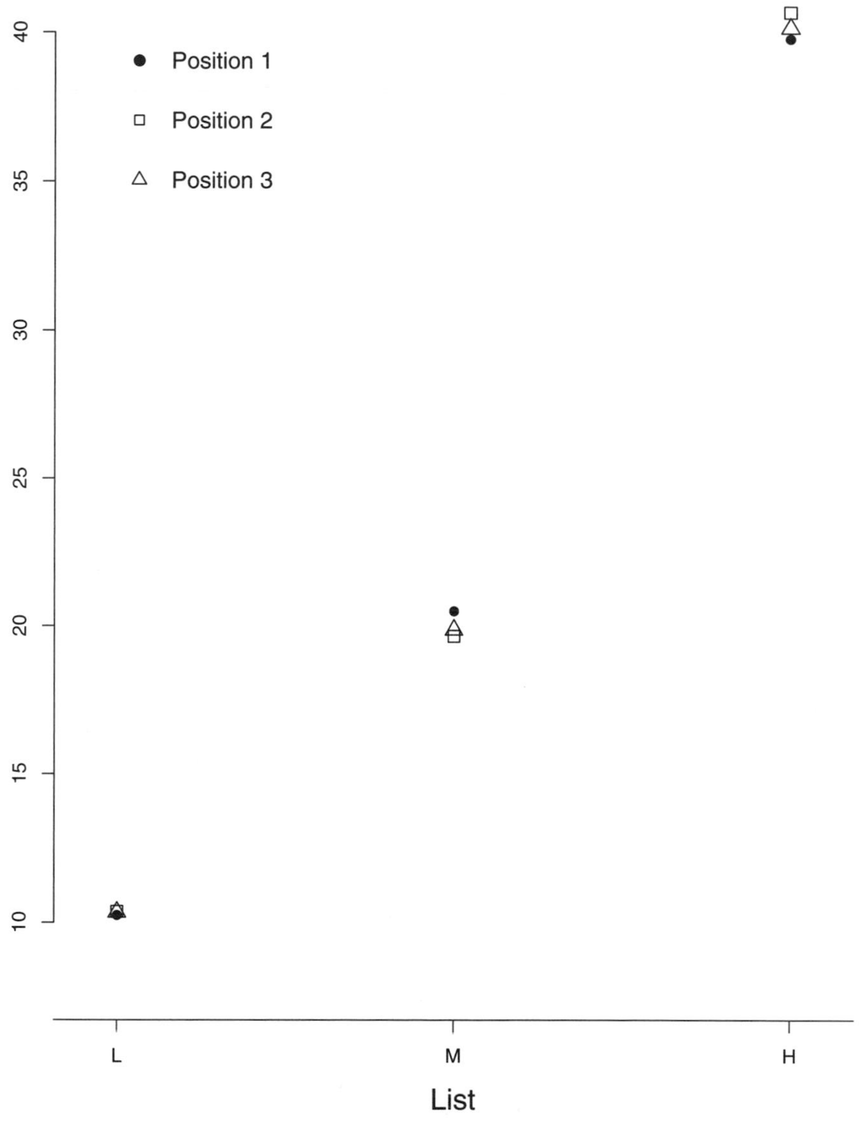

Table 13.2: A set of orthogonal comparisons that code a 3 × 3 design. Each of the eight comparisons corresponds to a different row in the table, and the columns correspond to the nine cells in the design.

	Treatments								
Comparison	L1	L2	L3	M1	M2	M3	H1	H2	H3
1	−1	−1	−1	0	0	0	1	1	1
2	−1	−1	−1	2	2	2	−1	−1	−1
3	−1	0	1	−1	0	1	−1	0	1
4	−1	2	−1	−1	2	−1	−1	2	−1
5	−1	0	1	0	0	0	1	0	−1
6	−1	2	−1	0	0	0	1	−2	1
7	−1	0	1	2	0	−2	−1	0	1
8	−1	2	−1	2	−4	2	−1	2	−1

Table 13.3: New d scores created by applying the eight comparisons in each row of Table 13.2 to the observed data in Table 13.1.

	Comparison number							
Participant	d_1	d_2	d_3	d_4	d_5	d_6	d_7	d_8
1	92	−32	2	−2	−2	−4	8	10
2	88	−32	−3	1	3	−5	−3	13
3	85	−35	0	4	0	−2	6	4
4	89	−33	0	6	−1	5	9	3
5	93	−29	4	−8	0	0	−2	−2
6	88	−26	−2	−2	−4	−2	−2	4
7	92	−30	−3	3	0	2	−6	−12
8	90	−34	0	2	1	−3	3	11

two comparisons on the Position Factor (columns 3 and 4 in Table 13.2). We use the analogous coefficients on the Position Factor. Thus, the first four columns of Table 13.2 give main effect comparisons for each of the two factors. The remaining four interaction comparisons were created by multiplying every possible main effect comparison on one factor by a main effect comparison on the other factor (for example, the first list comparison multiplied by the first position comparison, etc.). This is the same procedure that was described in Chapter 10 for the case of between-subjects designs. Table 13.3 gives the eight sets of d scores created by applying the eight

Table 13.4: One-sample t tests for each of the eight comparisons in Table 13.2. Each t test has number of participants minus $1 = 8 - 1 = 7$ degrees of freedom.

Comparison	Computed t value
1	94.97
2	−30.34
3	−0.29
4	0.32
5	−0.51
6	−0.96
7	0.82
8	1.35

comparisons listed in Table 13.2. Table 13.4 gives the one-sample t tests on each of the eight new d variables. We see that only the first two comparisons (that is, those coding the main effect for the list factor) are statistically significant using $\alpha = 0.05$.

If we wanted to perform the Scheffé test on these comparisons, we would use Equation 12.2. Because there are 8 participants and each factor has 3 levels, the new critical value corresponding to the main effect comparisons is $t' = \sqrt{\frac{(8-1)(3-1)5.14}{8-3+1}} = 3.46$. Thus, any computed t ratio on a comparison involving a main effect that exceeds 3.46 is statistically significant according to the generalized Scheffé criterion. We see that two of the main effect comparisons (comparisons #1 and #2) are statistical significant by the Scheffé criterion. Similarly, the Scheffé critical value for the two-way interaction comparison is $t' = \sqrt{\frac{(8-1)(3-1)(3-1)6.39}{8-((3-1)(3-1))}} = 6.69$; the critical F corresponding to 4 degrees of freedom for the numerator (that is, $(3-1)(3-1) = 4$) and 4 degrees of freedom for the denominator (that is, $n - (3-1)(3-1) = 4$) is 6.39. By comparing the observed t values in Table 13.4 to the relevant t', we see that none of the four interaction comparisons is statistically significant according to the Scheffé criterion.

We remind the reader that the comparisons listed in Table 13.2 represent one possible set of eight orthogonal comparisons. Any other set of orthogonal comparisons will do. For instance, the researcher may be interested in testing polynomial comparisons to assess trends over time (such as linear versus quadratic). The important point is that the data analyst should select comparisons that map onto the research questions that he or she wants to test. Usually, the strategy of focusing on the main effect comparisons first and then creating the interaction comparisons by multiplying the main effect comparisons provides a satisfactory set of comparisons. However, for some complicated research questions it may be best to focus on the interaction comparisons one wants to test and then work "backwards" to find the remaining comparisons. Recall that the Scheffé test does not require orthogonality; it can be performed even when there are nonorthogonal comparisons.

The example presented in this section can be made more complicated. For instance, one might consider the possibility of treating words as a random factor nested within the list factor. In our example we simplified matters by assuming that words are a fixed effect factor and so generalizations will be limited to just these stimuli. One might also consider the possibility of distinguishing the two times that each word list appears in each position. For instance, list L appears first both in order 1 and in order 2. The formulation of the present design can easily be extended to handle this new factor, time, by treating the design as a $3 \times 3 \times 2$ factorial design with list, position, and time as the three factors. This design requires 18 "treatments" (that is, $3 \times 3 \times 2$), and indeed the table in the text showing the six possible orders contains 18 treatments. We chose to analyze the present example as a 3×3 factorial design because the time factor doesn't make much sense to us. Appearing first versus appearing second is silly when the lists are presented a total of six times (once each session). If the reader disagrees with how we treated the time factor, he or she can easily treat the current example as a $3 \times 3 \times 2$ factorial design. We performed orthogonal comparisons (that is, number of cells in the 3×3 design minus 1), whereas including time as a factor requires 17 orthogonal comparisons (that is, number of cells in the $3 \times 3 \times 2$ design minus 1). The reader should convince him- or herself that the eight comparisons we test are a subset of the 17, and the tests of significance are identical for those 8 comparisons regardless of whether they were tested in the context of the 3×3 design or the $3 \times 3 \times 2$ design. The remaining nine comparisons $(17 - 8)$ are the main effect for time (one comparison), the interaction between time and position (two comparisons), the interaction between time and list (two comparisons), and the three-way interaction between time, position and list (four comparisons). Thus, it isn't that our simplification to a 3×3 design gave incorrect results; rather, our simplification provided a smaller subset of comparisons. The larger set of comparisons is justified if the researcher wants those questions answered.

There are also additional designs one may consider. For instance, one may want to vary how the six orders are presented to the participant. In the example, all eight participants saw the six orders (that is, the six sessions) in the same order. If order was thought to be theoretically important, then one may consider running different groups of participants such that each group is given a particular order. For example, one group of participants might receive the six orders in the manner presented in the text, and another group of participants might receive Order 2, Order 6, Order 1, Order 3, Order 5, and Order 4 for each consecutive session. If the experimenter has patience (and enough resources), many different orders can be tested. The complexity of the research design should be driven by the particular research question that the analyst is attempting to answer.

13.3 Designs Containing Both Within-Subjects and Between-Subjects Factors

The merger of within-subjects factors and between-subjects factors into the same experimental design requires a simple combination of the ideas we have, up until now, discussed separately for between- and within-subjects factors. Except for one small bookkeeping feature the logic of designs that have both repeated-measures and between-subjects factors is as follows: (1)

create a new set of d scores corresponding to the comparisons you want to test on the within-subjects factor, (2) create a new variable that is the sum of all the within-subjects measurements (that is, for each participant, add every score obtained into a single score, and do that for each participant), and (3) perform either one-sample t tests or between-subjects tests on these new variables. Through this procedure we can test research hypotheses about repeated-measures factors, between-subjects factors, and interactions between the two. An exception, of course, is that this method does not provide an easy way to compute general omnibus tests involving more than one comparison.

We illustrate with a 2×2 factorial design where one factor is between-subjects and the other factor is within-subjects. In other words this design consists of two groups of participants who were measured twice. We assume the investigator wants to test the following three research questions: Collapsing both data collection times, did the means of the two groups differ? Collapsing the two groups, did the means for the two data collection times differ? Was there an interaction between group and time (that is, did the difference in the two time periods differ across the two groups)?

Data for this example are presented in Table 13.5. The marginal means for the two levels of the between-subjects factor A are 4.5 and 5.7. The marginal means for the two times (that is, B1 and B2) are 3.7 and 6.5. The four cell means are presented in the following table:

	B1	B2
A1	3	6
A2	4.4	7

Next we create two new variables. One variable is the sum of the two times, and the other variable is the difference between the two times. We list the two variables in Table 13.6. Our strategy will be to perform two separate analyses, one on each of these two variables. First, perform a one-way between-subjects ANOVA (in this example, this happens to be identical to a two-sample t test because there are two groups) on the sum of the two

Table 13.5: Data from an experiment in which five participants are randomly assigned to each level of between-subjects factor A and each participant is tested under all levels of the repeated-measures factor B.

		Within-subjects factor	
	Participant	$B1$	$B2$
	1	1	3
	2	1	5
$A1$	3	2	6
	4	6	8
	5	5	8
	6	3	5
	7	2	4
$A2$	8	4	7
	9	5	8
	10	8	11

times d_1 (column 1 in Table 13.6). In other words, one test compares the means of the d_1 scores for the participants in each level of the between-subjects factor A. This simple test is equivalent to the main effect of factor A because the analysis has collapsed the within-subjects factor B. Second, perform a one-way between-subjects ANOVA on the difference of the two times d_2 (column 2 in Table 13.6). That is, compare the mean of the d_2 scores for the participants in each level of the between-subjects factor A. This tests the interaction between factor A and within-subjects factor B, and uses the correct error term for that comparison. Finally, we need to test the remaining main effect, the difference between the two levels of the within-subjects factor B, collapsing factor A. This tests whether the mean difference between the two levels of B is statistically significant, ignoring that there are different groups of participants. This last test is tricky—one cannot compute a one-sample t test on variable d_2 and interpret that test as a main effect because that would be based on an incorrect number of degrees of freedom. The test for the main effect of within-subjects factor B needs to include in its bookkeeping the degrees of freedom loss because there are groups of participants. The answer is found in the second step, where

Table 13.6: Two new variables created from the data in Table 13.5: the sum of and difference between, respectively, the two levels of the repeated-measures factor B.

Participant	Sum d_1	Difference d_2
1	4	2
2	6	4
3	8	4
4	14	2
5	13	3
6	8	2
7	6	2
8	11	3
9	13	3
10	19	7

the one-way between-subjects ANOVA on variable d_2 was computed. Some statistical packages provide tests for the grand mean (sometimes labeled "CONSTANT" in the output). This happens to be one of the few cases in social science research where the test for the grand mean is interpretable. The test for the grand mean of variable d_2 (in this example, the difference between the two times) against the null hypothesis of zero in the context of a one-way between-subjects ANOVA is identical to the main effect for the within-subjects factor B. That is, the grand mean of variable d_2 is testing whether there is a difference between the means of levels $B1$ and $B2$, and the test is based on the correct number of degrees of freedom. Of course, a diligent analyst can perform the standard one-sample t test and then manually correct the result for the appropriate degrees of freedom (a shortcut we ourselves have followed at times) instead of looking at the test for the grand mean.

This strategy of using comparisons can be generalized to any factorial design containing both between- and within-subjects factors. First, write the comparisons you want to perform on the within-subjects factor(s), and create new variables that use those comparisons as weights. These comparisons may be polynomial comparisons that assess trends, or they may involve other comparisons between the means corresponding to levels of the within-subjects factors. Second, create a new variable that is the sum of all within-subjects measurements. Third, list the comparisons you want to perform on the between-subjects factor(s). Fourth, test the between-subjects comparisons on each of the new variables you created, making use of the test for the grand mean on each variable (this corresponds to the within-subject comparison, collapsing the between-subjects factors).

This strategy may seem cumbersomc, but once the strategy is mastered the data analyst will quickly see the benefits. Every test boils down to a specific comparison applied to either a within-subjects factor, a between-subjects factor, or both (in the case of interactions that mix between- and within-subjects factors). The interpretation of these tests is straightforward because they are based on specific comparisons. Further, we avoid the mess of omnibus tests that require additional assumptions in the traditional framework or concepts from linear algebra in the multivariate framework.

13.3.1 Structural Model

We will now build the structural model for a two-way ANOVA where one factor is between-subjects and the other factor is within-subjects. You can think of this design as being composed of two structural models—one for the between-subjects part and one for the within-subjects part. The reason for needing two parts is that subjects are nested with respect to the between-factor but subjects are crossed with respect to the within-factor. Thus, we need to treat the two subparts of the design (the between-subpart and the within-subpart) differently.

The structural model for a between-subjects one-way ANOVA is

$$Y = \mu + \alpha + \epsilon_b \tag{13.1}$$

and the structural model for the one-way within-subjects ANOVA is

$$Y = \mu + \beta + \pi + \epsilon_w \tag{13.2}$$

Here β is the fixed time factor (we are using β instead of α to avoid confusion in the next paragraph). We subscript the error terms in each structural model to keep track of whether it comes from the between-subjects or the within-subjects model.

We combine the two structural models. An ANOVA with both between- and within-factors is simply a concatenation of a between-subjects ANOVA on one hand and a within-subjects ANOVA on the other. Literally sum the two structural models above (Equations 13.1 and 13.2). Be sure not to count the grand mean twice, and also include an interaction between α (a between-subjects factor) and β (a within-subjects factor). This results in the following structural model:

$$Y = \mu + \alpha + \beta + \alpha\beta + \pi + \epsilon_w + \epsilon_b \tag{13.3}$$

Be sure you understand how all these terms emerge. You can see why we distinguished the two error terms in each structural model, so that when the models are added together the different error components are listed separately.

Most computer programs print out only five lines in the source table rather than the six that you would expect from the structural model. The five lines that are typically printed tend to be organized in this manner:

1 α
2 ϵ_b

3 β
4 $\alpha\beta$
5 ϵ_w

The first line is the between-subjects factor, and the second line is the error term that is used to test the between-subjects factor. The third and fourth lines are the main effect for time and the interaction between the two factors, respectively. The last line is the error term used for both β and $\alpha\beta$.

13.3.2 SPSS Syntax

We present SPSS syntax for designs that include one factor that is between-subjects and a second factor that is within-subjects. We assume that each participant provides two scores ($t1$ and $t2$) and there are two groups of participants (experimental and control). The data file will have three columns: one column codes group membership (experimental/control), one column codes the scores from the first test ($t1$), and the third column codes the scores from the second test ($t2$). We first present the SPSS syntax for the method that uses sum and difference scores, then we present SPSS for the method that specifies between- and within-subjects factors directly in the syntax. Both versions yield identical results for the comparisons involving main effect and the interaction. This example is a 2×2 design.

```
COMPUTE timesum = t2 + t1.
COMPUTE timediff  = t2 - t1.
EXECUTE.

MANOVA timesum BY group(1,2)
  /DESIGN group.

MANOVA timediff BY group(1,2)
  /DESIGN constant group.
```

This next version of SPSS syntax yields identical results. It codes the factors as within- and between-subjects directly in the syntax.

```
MANOVA t1 t2 BY group(1,2)
```

```
  /WSFACTOR time(2)
  /WSDESIGN time
  /DESIGN group.
```

As before, the word "time" is not the name of a specific variable in the data file but a name given to the within-subject factor that consists of the two columns representing the two observations ($t1$ and $t2$). The WSFACTOR subcommand both defines the within-subjects variable name and specifies the number of levels (in this case two because there are two times at which observations were made). The WSDESIGN subcommand and the DESIGN subcommand define the structural model for both the within- and between-subjects factors, respectively. The interaction between within- and between-subjects factors is computed automatically by the MANOVA command. It is also possible to specify individual comparisons using the CONTRAST(special) subcommand (as illustrated in Chapter 12), defined for each within- and between-subjects factor.

The MIXED command in SPSS can also estimate designs involving within- and between-subjects factors. The data file needs to be arranged in long format, with each observation on a different row. The MIXED command with the estimation procedure REML, and with the option for an unrestricted covariance matrix, is identical to the way repeated measures has been presented in this book. The SPSS syntax for reading data in long format (data from Table 13.5), sorting the data by subject and time, and computing the ANOVA, is given below. The syntax for specific comparisons is quite complicated in the MIXED command, and the reader is referred to the SPSS documentation.

```
data list free / group sub dv time.
begin data
1  1  1  1
1  2  1  1
1  3  2  1
1  4  6  1
```

```
1  5  5  1
2  6  3  1
2  7  2  1
2  8  4  1
2  9  5  1
2  10 8  1
1  1  3  2
1  2  5  2
1  3  6  2
1  4  8  2
1  5  8  2
2  6  5  2
2  7  4  2
2  8  7  2
2  9  8  2
2  10 11 2
end data.

SORT CASES BY sub(A) time(A).

mixed dv by group time
 /fixed = group time group*time |  SSTYPE(3)
 /method= REML
 /print = solution
 /repeated = time | subject(sub) covtype(un)
 /emmeans=tables(group*time) compare(time).
```

13.4 Omnibus Tests

Our focus throughout the book has been on reducing ANOVA designs down to their elementary comparisons and analyzing individual comparisons. We have argued against the use of omnibus tests, making the point that when an omnibus test is statistically significant little information is imparted to the data analyst. The data analyst usually wants to know something more detailed about the pattern in the means than "There is a difference between some combination of means somewhere in this study." Some behavioral researchers believe that one should perform an omnibus test first in order to "protect" against a Type I error rate. As we have argued throughout these chapters, there are four basic ways to safeguard against the multiplicity problem. Our favorite strategy is to replicate the study (after all, replication is in the spirit of scientific inquiry). However, replication is not always feasible (such as when conducting a 30-year longitudinal study). Other options include the Bonferroni correction, the use of the Scheffé criterion, or, in the special case of between-subjects pairwise comparisons, the Tukey test. Thus, as we have seen in previous chapters, the omnibus test is not needed for purposes of protection against the Type I error rate, and some researchers have suggested that using the omnibus test as a criterion may not accurately correct the Type I error rate.

All statistical packages for data analysis on the market today include omnibus tests in the computer output. We briefly discuss a couple of the tests that appear in such output for within-subjects factors so that the reader can become familiar with the terminology and the meaning of such analyses.

A common omnibus test reported in computer output involves a measure that is known as ϵ, or "epsilon." Recall from Chapter 12 that in order to perform an omnibus test on a within-subjects factor one can pool the covariance matrix to construct a single mean square within treatments term. However, for this pooling procedure to be sensible some additional assumptions need to be in place. We have suggested in previous chapters that data rarely satisfy these restrictive additional assumptions. Statisticians have, over the years, found ways of assessing the degree of deviation from the assumption on the covariance matrix, and then have used that assessment in a procedure to adjust the degrees of freedom. The index known as ϵ is one such measure of assumption violation. Recall that an analogous procedure was mentioned in Chapter 5. Welch (1938, 1947) developed a generalization of the traditional two-sample t test that does not require pooling; instead, the method penalizes the degrees of freedom in relation to how severely the assumption of homogeneity of variance is violated. The more severe the violation, the greater the reduction in degrees of freedom, and hence a more conservative test.

Greenhouse and Geisser (1959) and Huynh and Feldt (1970) both developed procedures that adjust the degrees of freedom of the statistical test in relation to how much the assumption on the covariance matrix is violated. The adjustment is based on the measure ϵ. The measure ϵ is an index of how well the assumptions on the covariance matrix are met. The measure ϵ can range between $1/(a-1)$ and 1, with 1 being the "best" possible score, meaning the assumptions are met (a represents the number of levels of the within-subjects factor).

There are also more general ways of defining omnibus tests on within-subjects factors that involve multivariate statistical techniques. Under some applications these multivariate techniques may be useful because they can be used to find comparison weights on the levels of the within-subjects factor that maximally discriminate between groups (that is, the between-subjects factors). We will not discuss these techniques here because a serious discussion involves a good understanding of multivariate techniques. Interested readers should consult books on multivariate analysis (for example, Tabachnik & Fidell, 1983). The multivariate approach also can be used to define an omnibus test in the spirit of the usual analysis of variance. This omnibus procedure involves doing complicated operations on the matrix of d values corresponding to an orthogonal set of comparisons (such as the matrix of d values presented in Table 13.3). Such extensions are covered in Maxwell and Delaney (1990).

13.5 Summary

This chapter covered much ground, but we kept matters relatively simple by focusing on comparisons and bypassing the discussion of omnibus tests as much as possible. We presented the factorial design where all factors are within-subjects factors and a general factorial design that can include both within-subjects factors and between-subjects factors. We hope that the reader appreciates the relative simplicity of testing comparisons on within-subjects factors (create a new score d and apply the one-sample t test). Of course, when there are many cells such as in a large factorial design this approach can be tedious because of the computation needed. However, by keeping track of the research questions and their corresponding comparisons, the data analyst can systematically test complex factorial designs using these techniques.

13.6 Questions and Problems

1. We have two factors: A and B. Factor A has two levels ($a = 2$), Factor B has six levels ($b = 6$), and there are $s = 10$ participants assigned to each level of A. Thus, A is between-subjects and B is within-subjects. The data are presented on the next page.

Groups	Subjects	$B1$	$B2$	$B3$	$B4$	$B5$	$B6$	Sum
	1	9	3	2	5	10	12	41
	2	10	7	3	11	11	11	53
	3	5	6	10	5	8	15	49
	4	4	11	9	13	5	11	53
	5	8	11	5	13	9	7	53
$A1$	6	6	3	4	10	10	12	45
	7	6	2	6	8	9	14	45
	8	4	4	6	10	14	6	44
	9	4	10	4	11	5	15	49
	10	1	6	4	5	11	13	40
	1	8	9	13	9	18	14	71
	2	5	10	9	7	10	12	53
	3	3	4	8	10	11	19	55
	4	5	8	13	11	15	18	70
	5	1	4	13	14	17	15	64
$A2$	6	4	8	6	12	9	20	59
	7	10	7	14	13	9	16	69
	8	10	9	9	16	9	11	64
	9	2	9	9	14	12	19	65
	10	9	10	8	12	16	11	66

(a) Create a plot of the 12 cell means (each of the six levels of factor B crossed with the two levels of factor A).

(b) Create a list of the polynomial comparisons corresponding to the linear, quadratic, etc., for the six levels of factor B.

(c) Create new variables for each of the polynomial comparisons. Also create a new variable corresponding to the sum of the six levels of factor B.

(d) Perform tests for each polynomial comparison ignoring group membership, for each polynomial comparison crossed group membership, and the main effect for group membership. [Hint: This will lead to a total of 11 tests. Be sure to categorize each of the 11 comparisons as coding only the marginal means of factor A, only the marginal means of factor B, or a specific comparison in the interaction of factors A and B.]

2. Discuss the merits of focusing on individual comparisons as opposed to omnibus tests, especially in the context of designs having within-subjects factors.

14

Contrasts on Binomial Data: Between-Subject Designs

14.1 Introduction

Binary data occur often in behavioral research. The dependent variable might be coded as yes or no, true or false, pass or fail, divorced or not divorced, donated to charity or did not donate, etc. Even though the analysis of binomial data may appear straightforward and simple, there are numerous techniques available. Upton (1982) reviewed 22 different statistical tests that can be applied to a special case of a binary variable with a second variable that is also binary (such as a variable coding two groups). The combination of a binary dependent variable and a binary grouping code is sometimes called a **2 × 2 contingency table**. Which test out of the many available should a data analyst use? There appears to be little consensus among behavioral researchers as to which technique to use. This chapter offers guidelines for making this decision in the context of testing planned contrasts on proportions that follow the binomial distribution.

The second study with the farmer from Whidbey Island (Chapter 2), where the farmer was given pairs of cans and asked to choose which one in the pair contains water, is related to the type of data we will discuss in this chapter. In that example, the farmer was given five pairs, so the number of correct choices can range between 0 and 5. At each choice, the experimenter observes a binary response (the farmer correctly chooses the can with water or incorrectly chooses). We showed in subsequent chapters how the observations can be approximated by a normal distribution. In this chapter however we illustrate the analysis of binary data where each research participant produces just one binary observation. For instance, if our study included 20 farmers and each farmer was given a single pair of cans, we would observe a score of 0 or 1 for each farmer. We can compute the proportion of farmers who correctly selected the can containing water. Note that this number represents a "between-participant" average (that is, proportion of individuals) as compared to the "within-participant" number of correctly selected cans (number of correct trials within the same individual) presented in Chapter 2. In this chapter we focus on comparisons of proportions, where each proportion is based on a set of research participants each producing a single binary observation (correct versus incorrect, smoker versus not smoker, condom user versus not condom user, etc). There can be any number of research conditions or treatments in the study, where each treatment produces a single proportion. The research question addressed throughout this chapter involves testing comparisons over a set of proportions from an experimental design.

As we have seen throughout this book, behavioral researchers typically design experiments with two or more treatments. As an example, consider an experiment by Langer and Abelson (1972) that studied compliance to favors. Behavior was coded in terms of whether or not the participant complied with the request. Two factors were manipulated: the legitimacy of the favor and the orientation of the appeal (that is, either victim- or target-oriented), leading to a 2×2 factorial design because there are two factors, each with two levels. The summary statistic in each condition was the proportion of participants complying to the favor. Langer and Abelson were interested in testing whether the interaction between the two factors was statistically significant. As we have seen in previous chapters, in the special case of a 2×2 factorial design, the interaction can be tested with a single degree of freedom contrast. Each of the two main effect contrasts also has one degree of freedom. The research question posed by Langer and Abelson corresponded to a specific interaction contrast.

We will place several tests into a common framework so that the tests may be compared. These tests include a test on raw proportions, a test on proportions that have been transformed by the arcsin, a test on proportions that have been transformed by the logit, and a test on proportions that have been transformed by the probit.

14.2 Preliminaries

In this chapter we will work with experimental designs that have binary data as observations. Each condition in the design has one observed proportion $\hat{\mathrm{p}}_i$ and a corresponding population proportion p_i. This is analogous to the cell sample mean $\overline{X}$ and the corresponding cell population mean μ. Within a cell in a factorial design, the responses are assumed to be a series of independent, identically distributed Bernoulli trials (that is, responses are binary). Independence is justified within the context of between-subjects experimental designs being addressed in this chapter.

Statistical theory shows that the observed proportion $\hat{\mathrm{p}}_i$ is an unbiased estimate of the population proportion p_i (for example, Hogg & Craig, 1978). It is also a maximum likelihood estimate, which implies that order-preserving transformations of $\hat{\mathrm{p}}_i$ are maximum likelihood estimates as well (Cox & Hinkley, 1974; Hogg & Craig, 1978). Thus, the arcsin, the logit, and the probit transformations discussed in later sections are also maximum likelihood estimates of the transformed population proportion. Note that lowercase p refers to a proportion, which is different than the italicized p (as used in p value) that we have been using throughout the book to indicate the result of a hypothesis test.

The standard error of the proportion p_i, given by

$$\sqrt{\frac{\mathrm{p}_i(1-\mathrm{p}_i)}{\mathrm{n}_i}}, \tag{14.1}$$

depends on the value of the cell population proportion p_i.[1] Note that the standard error reaches its maximum when $\mathrm{p}_i = 0.5$ (for fixed n). As shown later in this chapter, unlike the proportion the standard error of $\hat{\mathrm{p}}_i$ is not invariant under order-preserving transformation.

Contrasts will be used to test hypotheses about linear combinations of the population proportions (or linear combinations of the transformed population proportions). See Table 14.1 for the basic setup used throughout this chapter. Comparison coefficients will be denoted by λ_i, and the coefficients are chosen so that they sum to zero. Comparisons carry the relevant "design information" making it possible to express a factorial design in terms of an orthogonal set of contrasts.

A note about terminology. When referring to a treatment in an experimental design, we refer to the proportion. For instance, the Langer and Abelson (1972) study had four cells arranged in a 2 × 2 factorial design. The descriptive statistic in each cell was the proportion of participants who complied. However, when referring to a contingency table, we take the standard approach and refer to a table that has the frequency of participants giving each possible response in each cell. So, the Langer and Abelson study

[1] A problem with the standard error of $\hat{\mathrm{p}}$ defined in Equation 14.1 is that in research settings one rarely knows the population proportion p. Replacing p in Equation 14.1 with the observed sample proportion $\hat{\mathrm{p}}$ (which is a consistent estimate of p) yields a biased estimate of the standard error. An unbiased estimate of the variance is given by

$$\frac{\hat{\mathrm{p}}(1-\hat{\mathrm{p}})}{\mathrm{n}-1}$$

(Cochran, 1963; Edwards, 1964). Further research is needed to understand the impact of this bias, especially in the case of small sample sizes. For an initial attempt at understanding this problem through simulation see Berengut and Petkau (1979).

Table 14.1: The general framework for testing linear combinations of k proportions.

group	Cell proportion	Cell sample size	Contrast weight
1	$\hat{p}_1$	n_1	λ_1
2	$\hat{p}_2$	n_2	λ_2
$\vdots$	$\vdots$	$\vdots$	$\vdots$
k	$\hat{p}_k$	n_k	λ_k

Note: The framework assumes that there are at least two groups in the study (that is, $k \geq 2$).

produces a 4×2 contingency table (four groups with a participant giving one of two responses).

Four of the techniques reviewed in this chapter can be formulated as Wald statistics (Wald, 1943–see also Agresti, 1990). For the case of comparisons on independent proportions, the Wald statistic follows a normal distribution. The comparison on the observed proportions is denoted $\sum \lambda_i \hat{p}_i$. The Wald statistic for the null hypothesis on the linear combination

$$H_0: \quad \sum_i \lambda_i p_i = 0$$

is given by the following test statistic:

$$z = \frac{\sum \lambda_i \hat{p}_i}{\sqrt{\sum \lambda_i^2 \hat{\text{var}}(p_i)}} \tag{14.2}$$

where z is asymptotically normally distributed with mean equal to 0 and standard deviation equal to 1. The term $\hat{\text{var}}(p_i)$ in Equation 14.2 is the estimated variance associated with the observed proportion p_i. The definition of $\hat{\text{var}}(p_i)$ will vary across the tests reviewed in this chapter, depending on, among other things, the particular transformation of the observed proportion that is used.[2] A 95% confidence interval version of the Wald test is given by

$$\sum \lambda_i \hat{p}_i \quad \pm \quad 1.96\sqrt{\sum \lambda_i^2 \hat{var}(p_i)}$$

More generally, the observed proportion $\hat{p}_i$ may be replaced with $f(\hat{p}_i)$, where f is an order-preserving transformation.

14.3 Four Examples of Wald Tests

We exploit Equation 14.2 to express four well-known tests on proportions. The benefit of writing tests in a common form is that similarities and differences between the tests become readily apparent.

14.3.1 Comparisons on Raw Proportions

This test is a direct application of Equations 14.1 and 14.2 to observed proportions. The test statistic is given by

$$z_s \quad = \quad \frac{\sum \lambda_i \hat{p}_i}{\sqrt{\sum \lambda_i^2 \frac{\hat{p}_i(1-\hat{p}_i)}{n_i}}} \tag{14.3}$$

where the subscript "s" refers to separate variance, z_s is asymptotically distributed as a normal distribution with mean equal to 0 and standard deviation equal to 1, and $\hat{var}(p_i)$ is taken to be the estimated variance $\hat{p}_i(1-\hat{p}_i)/n_i$. The null hypothesis is the same as that for Equation 14.2.

The simple case for two proportions was given by Goodman (1964). An identical formulation was given in Rosenthal and Rosnow (1985); more general forms appeared in McCullagh and Nelder (1989) and Woodward,

[2]Throughout this chapter the null hypothesis will be $\sum_i \lambda_i f(p)_i = 0$. But, more generally, the null hypothesis on the linear combination could be set equal to some real value κ. The numerator in Equation 14.2 would then be $\sum \lambda_i \hat{p}_i \; - \; \kappa$.

Bonett, and Brecht (1990). The logic underlying the separate variance version of the test has been used to construct confidence intervals around the difference between two proportions at least since 1911 (Robbins, 1977; Yule, 1911).

14.3.2 The Arcsin Transformation

The arcsin transformation (sometimes denoted $\sin^{-1}$) of the observed proportion p_i is defined, in radian units, as

$$\phi(p_i) = 2 \arcsin \sqrt{p_i} \qquad (14.4)$$

for $0 < p_i < 1$ (Cochran, 1940; Fisher, 1954).[3] Radian units will be used in the present discussion, though analogous results can be found for degrees. The multiplicative constant 2 is used so that the range of $\phi(p)$ is $0 < \phi(p) < \pi$. A slight modification to the endpoints (that is, replacing observed proportions of 0 with $\frac{1}{4n_i}$ and observed proportions of 1 with $1 - \frac{1}{4n_i}$) provides an estimate with better asymptotic properties. The asymptotic variance of the arcsin transformation is $\frac{1}{n_i}$ (Bartlett, 1936; Bishop, Fienberg, & Holland, 1975).

One advantage of the arcsin transformation is that it yields a variance that is (approximately) inversely related to the sample size. If the sample sizes are equal, then the variance of each transformed observed proportion $\phi(\hat{p}_i)$ is equal to the sample size, hence the variances are equal. It is this property that leads to this transformation being called "variance-stabilizing." Another advantage of the arcsin transformation is that the variance does not need to be estimated from the cell proportions because the variance is a function of the sample size, thus eliminating the problem

[3]Mosteller and Tukey (1949) and Freeman and Tukey (1950) proposed a variant of the arcsin transformation having an exact variance very close to $\frac{1}{n}$. Curtiss (1943), Anscombe (1948), and Ghurye (1949) also presented modifications of the arcsin transformation.

of a biased standard error discussed earlier. Once you know the sample size, you immediately have an estimate for the variance of the proportion.

The arcsin test on k groups has the basic form of the Wald test from Equation 14.2

$$z_{\mathrm{a}} = \frac{\sum \lambda_i \phi(\hat{\mathrm{p}}_i)}{\sqrt{\sum \lambda_i^2 \frac{1}{\mathrm{n}_i}}} \tag{14.5}$$

where n_i refers to the sample size in each cell and the subscript "a" denotes the arcsin. This form was presented in Cohen (1967) and is a generalization of the test used by Langer and Abelson (1972). The null hypothesis tested by z_{a} is

$$\sum \lambda_i \phi(\mathrm{p}_i) = 0 \tag{14.6}$$

The null hypothesis is expressed as a linear combination of proportions on the arcsin scale.

One concern when using the arcsin transformation is the problem of unequal sample sizes (Cochran, 1943). If the sample sizes in each cell differ, then the variances are not stabilized across cells because the variances will be inversely related to the sample sizes. When the samples sizes do not differ greatly, then the harmonic mean of the cell sizes can be used instead of the individual cell sizes (analogous to the unweighted-means technique in the analysis of variance). The harmonic mean is defined as

$$\mathrm{n}_h = \frac{k}{\sum \frac{1}{\mathrm{n}_i}}$$

where k is the number of groups. When the sample sizes are equal, then the harmonic mean will equal the cell sample size, that is, $\mathrm{n}_i = \mathrm{n}_h$. The effect of unequal samples sizes on z_{a} and the use of the harmonic sample size deserve additional research attention.

Variance-stabilizing transformations appear to be limited to between-subjects designs. Holland (1973) discussed the difficulties of finding covariance-stabilizing transformations—a necessary requirement to generalize this technique to within-subjects designs.

14.3.3 The Logit Transformation

Logistic regression is a special case of the more general log-linear model. For discussions of logistic regression and its generalizations see Agresti (1990), Cox and Snell (1989), McCullagh and Nelder (1989), and Wickens (1989). For the special case of testing contrasts in between-subjects experimental designs, logistic regression is relatively simple. When a complete orthogonal set of contrasts is tested, the predicted scores of the model are the logits of the cell proportions (Grizzle, 1961).

Logistic regression uses the logit transformation on the proportion. The logit of the observed proportion p_i is given by

$$\log\left(\frac{\hat{p}_i}{1 - \hat{p}_i}\right) = \log(\text{odds}_i) \tag{14.7}$$

One justification for taking the logarithm of the odds is to achieve symmetry around the proportion value of 0.50. To eliminate the problem that the logit is undefined at the endpoints $p = 0$ and $p = 1$, constants may be added to the numerator and denominator (Agresti, 1990; Cox & Snell, 1989; Gart, Pettigrew, & Thomas, 1985); however, this practice produces bias, which is not well understood. The asymptotic variance of the logit for the ith proportion is

$$\frac{1}{n_i p_i (1 - p_i)}$$

(Bishop et al., 1975). The variance of the logit reaches its maximum when the proportion p is near the endpoints of 0 or 1 (for fixed n).

Comparisons in logistic regression take the form of Equation 14.2

$$z_l = \frac{\sum \lambda_i \text{logit}(\hat{p}_i)}{\sqrt{\sum \lambda_i^2 \frac{1}{n_i \hat{p}_i (1 - \hat{p}_i)}}} \tag{14.8}$$

The subscript "l" denotes the logit. The null hypothesis tested by z_l is

$$\sum \lambda_i \text{logit}(\text{p}_i) \quad = \quad 0, \qquad (14.9)$$

which is on the logit scale.

Logistic regression has been generalized to more complicated situations including covariates, polychotomous variables (that is, dependent variables having more than two response categories), and within-subjects designs (McCullagh & Nelder, 1989). In more general models, however, the parameters are estimated under different constraints and are not simply the logits of the cell proportions (Rindskopf, 1990; Wickens, 1989).

14.3.4 The Probit Transformation

Probit analysis (Finney, 1971) involves transforming the observed proportions into Z scores corresponding to the inverse of the cumulative normal distribution. For instance, a proportion of 0.5 is transformed to a Z score = 0; a proportion of 0.95 is transformed to a Z score = 1.96. The comparison over observed proportions is then applied on the Z transformation of the proportions.

The asymptotic variance of a proportion that has been transformed by the probit is

$$\hat{\text{var}}(\text{p}_i) \quad = \quad \frac{\text{p}_i(1-\text{p}_i)}{\text{n}_i \text{g}(\hat{z}_i)^2}, \qquad (14.10)$$

where $\text{g}(z_i)$ denotes the ordinate of the normal curve evaluated at the point corresponding to the inverse, cumulative normal of p_i. So, for example, a $z = 1$ corresponds to an ordinate value of 0.242 because for the special case of a normal distribution with mean equal to 0 and variance equal to 1 the ordinate is given by the formula

$$g(z) \quad = \quad \frac{\exp(\frac{-z^2}{2})}{\sqrt{2\pi}} \qquad (14.11)$$

For completeness we note that an observed proportion of 0.8413 would be converted to a $z = 1$, and then the ordinate of $z = 1$, which is $g(1) = 0.242$, would enter the formula for the variance of a proportion that has been transformed by the probit.

With the approximation to the standard error of the probit in hand, the Wald test in Equation 14.2 becomes

$$z_Z = \frac{\sum \lambda_i Z(\hat{\mathrm{p}}_i)}{\sqrt{\sum \lambda_i^2 \frac{\hat{\mathrm{p}}_i(1-\hat{\mathrm{p}}_i)}{\mathrm{n}_i \mathrm{g}(z_i)^2}}} \tag{14.12}$$

letting $Z(\hat{\mathrm{p}}_i)$ denote the probit transformation of the cell proportion p_i. The subscript "z" denotes the Z score of the probit transformation. The null hypothesis tested by z_{z} is

$$\sum \lambda_i Z(\mathrm{p}_i) = 0 \tag{14.13}$$

which is distributed as a normal distribution with mean equal to 0 and standard deviation equal to 1. An observed z_{z} that exceeds 1.96 in absolute value will reject the null hypothesis, using a two-tailed criterion at $\alpha = 0.05$.

14.4 Other Statistical Tests for Comparisons on Proportions

This section reviews other procedures that have appeared in the literature. The tests are placed in this section because they are not expressible in terms of the Wald formulation of Equation 14.2.

14.4.1 Contrast on Raw Proportions (Pooled Error Term)

We consider first the special case of two treatments where the observed data are binary. The observed difference between the proportion observed in group 1 and the proportion observed in group 2 is divided by the estimated standard error of the difference, yielding

$$z_{\mathrm{p}} = \frac{\hat{\mathrm{p}}_1 - \hat{\mathrm{p}}_2}{\sqrt{\hat{\mathrm{p}}_{\mathrm{pooled}}(1 - \hat{\mathrm{p}}_{\mathrm{pooled}})}\sqrt{\frac{1}{\mathrm{n}_1} + \frac{1}{\mathrm{n}_2}}} \tag{14.14}$$

where $\hat{\mathrm{p}}_{\mathrm{pooled}}$ is the pooled estimate of p across the two groups (the subscript denotes "pooled"). This ratio converges to normal distribution with mean = 0 and standard deviation = 1 (a "unit normal") as the sample sizes approach infinity.

The familiar χ^2 test on a 2×2 contingency table and the pooled test of Equation 14.14 are equivalent (for example, Fleiss, 1981)—both test the identical null hypothesis. Further, for the special case of two groups, both the pooled variance and the separate variance tests approach the unit normal as sample sizes become large. Cressie (1978) compares the two tests on the basis of how quickly they approach normality. The results are complicated because they depend on the population proportions, the sample sizes, and the nature of the alternative hypothesis. The interested reader is referred to Cressie's article.

Some authors (such as Fleiss, 1981) have suggested the following generalization of Equation 14.14:

$$z_{\mathrm{p}} = \frac{\sum \lambda_i \hat{\mathrm{p}}_i}{\sqrt{\sum \lambda_i^2 \frac{\hat{\mathrm{p}}_{\mathrm{pooled}}(1-\hat{\mathrm{p}}_{\mathrm{pooled}})}{\mathrm{n}_i}}} \tag{14.15}$$

where $\hat{\mathrm{p}}_{\mathrm{pooled}}$ is the pooled estimate of p across all k groups.

When the sample sizes across the k conditions are equal, then the z_{p} test is equivalent to methods that test contrasts by collapsing the contingency table into smaller 2×2 tables (Castellan, 1965). For instance, to test the (3, −1, −1, −1) contrast on a 4 (group) × 2 (response) contingency table one can perform the standard χ^2 test on the 2×2 contingency table formed by collapsing the last three groups (that is, groups assigned a contrast value of -1) into a single "meta-group." A numerical example of this collapsing procedure will be presented below.

The z_{p} test (and methods equivalent to it such as collapsing contingency tables) should be used with caution because the pooling procedure loses the information of the individual cell variances. For example, consider a design having four cells each with equal sample sizes. The observed proportions were 0.9, 0.1, 0.4, and 0.4; the investigator tests the contrast (1, 1, -1, -1). The pooled version test of this contrast is identical to collapsing the first two groups and the second two groups, yielding the two observed proportions 0.5 and 0.4, and then testing the two collapsed proportions with the contrast (1, -1). In this case, the pooling procedure leads to a test that has an inappropriate error term because the variances of the collapsed cells do not represent the variances of the original cells. For this reason the z_{p} test should be limited to cases where the null hypothesis is that all population proportions are equal. Note that when this null hypothesis is true, the four Wald tests highlighted at the beginning of this review are asymptotically equivalent to one another and to Equation 14.15. Another difference between the pooled and the separate variance forms is that the z_{p} test uses all cells to estimate the pooled variance when testing a contrast whereas the z_{s} test uses only the proportions that are given nonzero contrast values to estimate the variance. Because the cell variance is related to the cell mean for the binomial, it does not make much sense to include cells receiving contrast values of zero in the calculation of the error term. These points make z_{s} preferable to z_{p} for testing hypotheses about linear contrasts.

14.4.2 Student's *t* Test and the ANOVA

The standard two-sample t test on a binary response variable is included in this review because of the frequent reference to the t test's "robustness" to violations of normality (for example, Cox & Snell, 1989; Rosenthal & Rosnow, 1985). Simulations suggest that, remarkably, such a t test performs well for proportions that are not too extreme, that is, $0.2 \leq \mathrm{p} \leq 0.8$, and when the cells have equal sample sizes (for example, Lunney, 1970).

There are similarities between the test using a pooled error term (z_{p}) and the t test on binary data—both involve pooling data between the two groups to estimate the standard error of the difference between the two proportions. However, the pooling procedure for the t test is justified on the assumption that the mean and variance are independent—in the case of proportions this assumption is not generally true. Aside from the obvious difference of using different reference distributions to calculate the critical region, there is a subtle difference in the pooling procedures. The pooling procedure in the z_{p} test is defined on the observed proportion itself, that is, the number of total successes across the two groups divided by the number of total subjects. Thus the z_{p} test pools the two proportions to arrive at an estimate of $\mathrm{p}_{\mathrm{pooled}}$ to use in the calculation of the pooled variance. By comparison, the pooling procedure in the t test is defined using the estimated variances of each group; in other words, the estimated variances, not the proportions, are pooled in the t test.

The relation between the standard t test applied to binary data and the z_{p} test for two binomial proportions (Equation 14.14) is given by

$$t = z_{\mathrm{p}}\sqrt{\frac{\mathrm{N}-2}{\mathrm{N}-z_{\mathrm{p}}^2}}$$

where N refers to the total number of subjects in both groups and t is distributed as Student's t with N − 2 degrees of freedom (Cochran, 1954; D'Agostino, 1971, 1972). The two tests are asymptotically equivalent (that is, as N approaches infinity). Unfortunately, in the case of the t test the equality of variance assumption may be violated when the two observed proportions differ and are not symmetric around 0.50. This could be problematic in situations with extreme proportions or unequal sample sizes where the robustness of the t test breaks down.

The generalization of the t test to contrasts on proportions involves the between-subjects analysis of variance applied directly to the binary data.

Denote the estimate of the pooled standard error as $\sqrt{\text{MSE}}$, where MSE is the mean square error from the ANOVA source table. The critical test statistic is t (with N $-$ k degrees of freedom where k is the number of groups) rather than the standard normal (Rodger, 1969; Woodward et al., 1990). Of course, the t distribution asymptotically approaches the normal distribution as the sample size gets large. For completeness, we give the test for a contrast on binary data in ANOVA:

$$t = \frac{\sum \lambda_i \hat{\text{p}}_i}{\sqrt{\sum \lambda_i^2 \frac{\text{MSE}}{\text{n}_i}}} \tag{14.16}$$

with N $-$ k degrees of freedom.

The ANOVA model uses a structural model having parameter estimates that are identical to the separate variance test, but the test of significance differs from the separate variance test because (1) the two tests use different reference distributions, and (2) the two tests calculate the error term differently (Aldrich & Nelson, 1984; McCullagh & Nelder, 1989).

14.4.3 Comments on the *t* Test and the ANOVA

A major problem with performing an ANOVA on binary data is that the assumption of homogeneity of variance may be violated, making the pooling procedure suspect. One suggestion that has been made to alleviate the problem of heteroscedasticity of variance is to perform a two-stage weighted least squares regression. One first performs a linear regression of the binary response variable on the grouping variable. For the case of two groups the test of significance of the slope of this first regression is equivalent to the two-sample t test on the binary data. The purpose of this first regression is not to test a hypothesis but to calculate the predicted scores (that is, the $\hat{\text{Y}}$'s that follow from the regression equation). For the second step, predicted scores from the first regression are used to estimate weights that enter into a second weighted least squares regression (see Neter, Wasserman, & Kutner,

1996, for a discussion and some examples). A limitation of this technique is that the predicted scores from the first regression may not fall in the [0, 1] interval of the proportion. This creates difficulties when interpreting the weights used in the second step.

A second method to alleviate the problem of heteroscedasticity is to develop separate variance t tests for binomial data. For example, one might use a Welch-type procedure that adjusts the degrees of freedom to deal with the problem of heterogeneous variances (Welch, 1938). To our knowledge such possibilities have not been extensively explored with binomial data. This is a natural domain for further research.

14.5 Numerical Examples

This section illustrates the tests reviewed here using the Langer and Abelson (1972) study investigating compliance to favors. Langer and Abelson examined compliance rates for requests by varying the legitimacy (legitimate versus illegitimate) of the favor and the orientation of the appeal (that is, either victim-oriented or target-oriented). The dependent variable was whether or not the subject complied with the request. Thus, they had a 2×2 factorial design with a binary response measure (see Table 14.2). Langer and Abelson assumed that within each cell the number of subjects who complied with the request followed a binomial distribution. They tested whether the interaction between legitimacy and orientation was statistically significant, using the contrast (1, −1, −1, 1).

Langer and Abelson analyzed their data with the arcsin transformation. The experiment had equal sample sizes in each cell ($\mathrm{n} = 20$), so the test statistic for the contrast (Equation 14.5) simplifies to

$$z_{\mathrm{a}} = \frac{\sqrt{\mathrm{n}} \sum_i \lambda_i \phi(\hat{\mathrm{p}}_i)}{2}$$

where n is the cell sample size, λ is the contrast, and $\phi(\hat{\mathrm{p}})$ is the vector of transformed cell proportions. The null hypothesis that Langer and Abelson tested was

$$\begin{aligned}\phi(\mathrm{p}_{\mathrm{leg.victim}}) + \phi(\mathrm{p}_{\mathrm{illeg.target}}) &= \\ \phi(\mathrm{p}_{\mathrm{leg.target}}) + \phi(\mathrm{p}_{\mathrm{illleg.victim}})\end{aligned}$$

For the data in Table 14.2 the test statistic is

$$\begin{aligned}z_{\mathrm{a}} &= \frac{\sqrt{20}[\phi(.70) - \phi(.30) - \phi(.35) + \phi(.50)]}{2} \\ &= \frac{\sqrt{20}[1.982 - 1.159 - 1.266 + 1.571]}{2} \\ &= 2.52.\end{aligned}$$

The observed z_{a} exceeds the critical value of 1.96 (two-tailed $\alpha = 0.05$).

Table 14.3 shows the results for six tests considered in this review. The information supplied in Table 14.2 is sufficient to calculate the six tests. The MSE was 0.2362 with 76 degrees of freedom. For these data the six statistics yield comparable statistical results.

The (1, −1, −1, 1) contrast tested above is interpreted as the interaction contrast. Because there are four cells, a total of three orthogonal contrasts may be tested. Two natural contrasts to test are the main effect for legitimacy (rows) and the main effect for orientation (columns); these contrasts are (1, 1, −1, −1) and (1, −1, 1, −1), respectively. We leave these comparisons as exercises at the end of the chapter.

14.5.1 Collapsing the *k* × 2 Contingency Table

One practice that some researchers have advocated for testing contrasts on $k \times 2$ contingency tables is collapsing the table and performing test statistics on the reduced table (for example, Castellan, 1965; Kastenbaum, 1960). When sample sizes are equal, this procedure is equivalent to the pooled

Table 14.2: Data from Langer and Abelson, 1972 (Study 1; $n = 20$). The first table shows raw proportions of subjects who complied with the favor, the second shows proportions that have been transformed by the arcsin (Equation 14.4), the third shows proportions that have been transformed by the logit (Equation 14.7), and the fourth shows proportions that have been transformed by the probit (inverse cumulative normal).

Raw Proportions

	Orientation		
Legitimacy	Victim	Target	Marginal mean
Legitimate	0.700	0.300	0.500
Illegitimate	0.350	0.500	0.425
Marginal mean	0.525	0.400	Grand mean = 0.462

Arcsin Transformation

Legitimacy	Victim	Target	Marginal mean
Legitimate	1.98	1.16	1.57
Illegitimate	1.27	1.57	1.42
Marginal mean	1.62	1.36	Grand mean = 1.49

Logit Transformation

Legitimacy	Victim	Target	Marginal mean
Legitimate	0.85	−0.85	0.00
Illegitimate	−0.62	0.00	−0.31
Marginal mean	0.11	−0.42	Grand mean = −0.15

Probit Transformation

Legitimacy	Victim	Target	Marginal mean
Legitimate	0.52	−0.52	0.00
Illegitimate	−0.38	0.00	−0.19
Marginal mean	0.07	−0.26	Grand mean = −0.096

Interaction contrast

Legitimacy	Victim	Target
Legitimate	1	−1
Illegitimate	−1	1

Table 14.3: The (1, −1, −1, 1) contrast on data from Langer and Abelson (1972) for six tests considered in this review.

Statistic	Value
z_p (pooled variance)	2.47
z_s (separate variance)	2.60
t (ANOVA on binary data)	2.53
z_a (arcsin)	2.52
z_l (logit)	2.44
z_p (probit)	2.48

variance test z_p. For example, the interaction contrast tested by Langer and Abelson can be formulated in terms of a reduced 2 × 2 contingency table. The 24 "successes" out of 40 for the two groups coded +1 by the contrast are compared to the 13 "successes" out of 40 for the two groups coded −1. A standard χ^2 test on this reduced table yields the equivalent result as the interaction contrast tested with z_p (that is, $\sqrt{\chi^2} = 2.47 = z_p$).

As another example, consider a test of the contrast (3, −1, −1, −1), that is, the legitimate/victim cell is assigned the value 3 and the three remaining cells are assigned the value −1. This test can be formulated in terms of a 2 × 2 contingency table where 14 out of 20 "successes" in the legitimate/victim cell are compared to the 23 out of 60 "successes" in the remaining three cells. This reduced contingency table is shown in Table 14.4. The χ^2 test in Table 14.4 is 6.051. The square root of this value is equal to the (3,−1,−1,−1) contrast applied to the four original proportions using the pooled binomial test ($z_p = 2.46$). The values of all six tests using this contrast are also shown in Table 14.4.

The reader should note that the equivalence between the test on the reduced contingency table and the z_p test occurs when the cells have equal sample size. Further, the equivalence between the standard χ^2 statistic on the reduced 2 × 2 contingency table and z_p only holds when all contrast values are nonzero. Castellan (1965) presented a modification for collapsing a contingency table for the situation where some weights in the contrast are zero.

Table 14.4: Illustrating the (3, −1, −1, −1) contrast on the Langer and Abelson (1972) data.

Condition	Number who did perform favor	Number who did not perform favor	Row total
Legitimate/victim cell	14	6	20
All other cells	23	37	60
Column total	37	43	80

For comparison, the contrast (3, −1, −1, −1) is applied to the four cell proportions, using each of six test statistics reviewed in this paper.

z_{p} (pooled variance)	2.46
z_{s} (separate variance)	2.65
t (ANOVA on binary data)	2.52
z_{a} (arcsin)	2.52
z_{l} (logit)	2.40
z_{p} (probit)	2.45

The equivalence between the test statistic on the collapsed contingency table and z_{p} (for equal sample sizes) highlights two critical aspects of the z_{p} test. First, the pooling procedure uses all cells to estimate the standard error of the contrast even when some cells receive a contrast weight of zero and do not enter into the particular linear combination.[4] As stated earlier, the mean and variance are not independent for binomial data, so it may not be appropriate to include cells that receive a contrast value of zero when estimating the error term of a contrast. Second, the pooling procedure used in z_{p} has the property that in some situations the estimated error term is incorrect. An example will illustrate this problem. If the population proportion vector is (0.9, 0.1, 0.9, 0.1) and one tests the contrast (1, −1, −1, 1) using z_{p}, then the pooling procedure makes this test equivalent to testing the difference between two proportions that each have a population value of 0.50. The variance of proportions near the endpoints (for example, 0.1 and

[4] The same issue applies to MSE when using the standard ANOVA on binary data.

0.9) differs from the variance of proportions near the middle of the scale (for example, 0.5). Thus the pooled error term may not adequately represent the cell variances.

14.6 How Do These Tests Differ and What Do They Test?

Which of these tests should one choose? The tests are asymptotically equivalent under the very special situation that the population proportions are identical, which is a stronger constraint than the usual linear constraint imposed by contrasts. However, in general the null hypothesis and the structural model differ across these tests, and these differences have implications for the interpretation. We argue that the choice should be based on consideration of the scale of the null hypothesis. There are other considerations (such as the nature of the structural model implied by each of the tests) that we do not consider in this book.

What null hypothesis is being tested? The separate variance z_s procedure tests hypotheses on raw proportions, the z_a procedure tests hypotheses on the arcsin scale, logistic regression tests hypotheses on the log odds scale, and probit analysis tests hypotheses on the "Z score" scale. Thus, these four tests differ on the scale the proportions are compared (that is, see Equations 14.2, 14.6, 14.9, and 14.13, respectively).

The question of whether or not to use a nonlinear transformation boils down to the research question: On what scale is the hypothesis based? Does the investigator want to test a hypothesis about a linear combination of proportions, a linear combination of angles, a linear combination of log odds, or a linear combination of Z scores? Under most applications in behavioral research, the natural scale will be the raw proportion, but the scaling decision should be guided, in part, by the particular theory the researcher is testing. Consider the null hypothesis for the special case of two proportions. If the researcher believes that a difference of 0.10 between two proportions means the same regardless of the location on the unit interval, the separate variance test is appropriate (for example, population proportions of 0.05 versus 0.15, or 0.50 versus 0.60, or 0.80 versus 0.90 all lead to a difference of 0.10; of course, the three situations have different variances).

Emerson (1991) makes an analogous argument when he considers treatments to reduce smoking behavior and asks whether a difference in the percentage of smokers between 30% and 29% in one condition is comparable to a difference in percentage between 2% and 1% in another condition. When there are equal sample sizes across the conditions, these differences imply that both pairs of conditions have the same difference in frequency. Yet, at the same time, the latter pair appears more dramatic because "it represents cutting in half the number of smokers" (p. 389).

Consider a more complicated experimental design consisting of four treatment groups, each group generating a proportion of participants who performed a particular behavior. The population proportions in the four groups are 0.45, 0.05, 0.25, and 0.25. The researcher wants to test the (1, 1, −1, −1) contrast. On the raw proportion scale this linear combination of proportions leads to a population value of zero, but on the logit scale the linear combination for the population is −0.948. In this case the logit test can be statistically significant, given sufficient power. This example was adapted from Steiger (1980), who argued that nonlinear transformations (in his case, the arcsin) should be avoided. Note that it is possible to have the reverse situation too: proportions on the logit scale (or the arcsin scale, or the probit scale) can lead to a linear combination equal to zero, but the linear combination on the raw proportions can lead to a nonzero value. For example, the (1, 1, −1, −1) contrast on the proportion vector (0.45, 0.05, 0.172, 0.172) is zero on the logit scale but nonzero on the raw scale. One implication of this observation is that the choice of test cannot be settled exclusively by simulation of effective Type I and Type II error rates because any single test can be made to outperform the others by careful selection of the population proportions.

In most applications researchers are likely to be interested in comparing cell proportions, and so the separate variance test will be most appropriate (see also Steiger, 1980). There are occasions, however, where a nonlinear transformation of the proportions will be appropriate. Consider this argument by D'Agostino (1971, p. 330), albeit in a slightly different context:

> [A] reason for considering a transformation is that it may aid in removing effects due solely to the original scale (here the sample proportions, p, in each cell). For example, consider a problem of item analysis for a set of exam questions. In an I × J classification say the data are the proportions of correct responses to a given question where the columns refer to classification of individuals by I.Q. (low, middle, high) and the rows are the different exam questions. For the easy questions almost all will get the correct answer (so p will be close to one). With the very hard questions almost all will get a wrong answer (so p will be close to zero). If an advantageous effect due to I.Q. does exist it will be reflected only for the remaining middle range of questions. This implies that in the original scale row and column effects cannot be strictly additive over the full range. An interaction exists which may be solely a consequence of the scale.

Thus, D'Agostino suggests that sometimes it might make good sense to consider a nonlinear transformation of the observed proportions.

Unfortunately, general guidelines for when to use a specific transformation cannot be given. The issue is problem-dependent. A cautious strategy is to recommend the separate variance test as the default procedure. A researcher may decide to deviate from the default procedure if the problem at hand suggests that converting the raw proportions to angles, logits or Z scores (or some appropriate transformation f—see Table 14.5) provides a more "natural" scale on which to test hypotheses. If the researcher is primarily interested in testing differences in proportions across cells of an experimental design, as is usually done, then the separate variance test provides the most natural scale. This recommendation runs counter to an apparent recent trend in the behavioral literature to automatically use logistic regression whenever one wants to compare proportions in an experimental design.

14.6.1 Structural Model

We now turn to a comparison of the structural models underlying the four tests. In particular, we compare the separate variance test (z_s) and the test using the logit(z_l). The z_s test models data as additive on the probability scale, whereas the z_l test models data as multiplicative on the odds scale (equivalently, a model that is additive on the log odds scale).

In between-subjects experimental designs the separate variance test models the cell proportions as a linear combination of the treatment effects. For instance, the Langer and Abelson (1972) study described in the introduction denotes the legitimacy effect by α, the orientation effect by β, and the interaction as $\alpha\beta$. The data for the study are listed in Table 14.2. The separate variance test models the cell proportions as (with μ representing the "grand mean," or intercept, depending on the parameterization)

$$\hat{p} = \mu + \alpha + \beta + \alpha\beta \tag{14.17}$$

Numerical estimates are found in the usual way (for example, the parameter β for column 1 is estimated by the difference between the proportion for the first column marginal and the grand mean—see Rosenthal & Rosnow, 1985, for a general discussion of these parameters). If group membership was not known, then the predicted proportion would be the grand mean, or 0.4625 (37 out of 80 subjects). The experimental factors, as indexed by α, β, and $\alpha\beta$, serve to improve the prediction, and each term is additive on the proportion scale. The inclusion of all terms leads to a perfect prediction of the cell proportions.

Table 14.5: Six tests reviewed in this chapter. The null hypothesis that the linear combination $\sum \lambda_i f(p_i)$ equals 0 is tested in each case. The term $\hat{p}_P$ denotes the pooled proportion $\hat{p}_{pooled}$.

Procedure	Transformation f	Estimated standard error	Test statistic
Pooled variance	Identity	$\frac{\hat{p}_P(1-\hat{p}_P)}{n_i}$	$z_P = \frac{\sum \lambda_i \hat{p}_i}{\sqrt{\sum \lambda_i^2 \frac{\hat{p}_P(1-\hat{p}_P)}{n_i}}}$
Separate variance	Identity	$\frac{\hat{p}_i(1-\hat{p}_i)}{n_i}$	$z_S = \frac{\sum \lambda_i \hat{p}_i}{\sqrt{\sum \lambda_i^2 \frac{\hat{p}_i(1-\hat{p}_i)}{n_i}}}$
ANOVA	Identity	$\frac{MSE}{n_i}$	$t = \frac{\sum \lambda_i \hat{p}_i}{\sqrt{\sum \lambda_i^2 \frac{MSE}{n_i}}}$
Variance stabilizing	Arcsin	$\frac{1}{n_i}$	$z_a = \frac{\sum \lambda_i \phi(\hat{p}_i)}{\sqrt{\sum \lambda_i^2 \frac{1}{n_i}}}$
Logistic regression	Logit	$\frac{1}{n_i \hat{p}_i(1-\hat{p}_i)}$	$z_l = \frac{\sum \lambda_i \text{logit}(\hat{p}_i)}{\sqrt{\sum \lambda_i^2 \frac{1}{n_i \hat{p}_i(1-\hat{p}_i)}}}$
Probit	Probit	$\frac{p_i(1-p_i)}{N_i g(z_i)^2}$	$z_p = \frac{\sum \lambda_i \text{probit}(\hat{p}_i)}{\sqrt{\sum \lambda_i^2 \frac{p_i(1-p_i)}{N_i g(z_i)^2}}}$

The model underlying the logit test is multiplicative on the odds scale rather than additive on the probability scale, as in the separate variance test. Thus the z_l test models the odds as

$$\widehat{\left(\frac{\hat{p}}{1-\hat{p}}\right)} = e^{\mu+\alpha+\beta+\alpha\beta}$$
$$= e^{\mu}e^{\alpha}e^{\beta}e^{\alpha\beta}$$
$$= \mu'\alpha'\beta'(\alpha\beta)' \qquad (14.18)$$

where the prime denotes a reparametrization. This multiplicative property leads naturally to statements of independence between factors in an ANOVA design (see Wickens, 1989). It is easy to estimate these parameters on the (additive) log odds scale—follow the identical procedure used for the raw proportions but use the logit transformation instead. For the Langer and Abelson data the grand log odds was -0.1548 (that is, the average of the four logits in Table 14.2). One can exponentiate the log odds to convert this estimate to the odds scale, or $e^{-0.1548} = 0.8565$. Thus the "grand mean" odds was 0.8568 to 1 (or approximately 5:6) that subjects across all conditions complied with the request. The grand mean on the log odds scale corresponds to the log of the geometric mean of the four cell odds.

As an illustration, the β on the logit scale for compliance in the victim orientation condition is 0.26, that is, marginal logit 0.11 minus grand mean logit (-0.15). This main effect can be expressed in terms of odds by exponentiating the treatment effect, that is, $e^{0.26} = 1.3$. Thus, the effect of the victim orientation is to change, or increase, the grand mean odds by a factor of 1.3. Similarly, the main effect odds for compliance in the target level changes, or decreases, the grand odds by a factor of 0.77 (that is, $e^{-0.26}$).

Both models are saturated in the sense that all degrees of freedom are used. Therefore, when all terms in the model are included (for example, in a two-way factorial design the terms are $\mu, \alpha, \beta, \alpha\beta$), then both models give identical predictions. For instance, applying the full model containing all terms to the victim orientation/legitimate cell in the Langer and Abelson

(1972) study leads to a prediction of 0.7, which is the observed proportion when using the identity model, whereas it leads to a prediction of 0.85, which is the observed logit when using the logit model. However, the individual parameters in one model are not the inverse function of the parameters in the other model. This difference in the individual parameter values is due to the underlying difference that one model is additive on the probability scale whereas the other is multiplicative on the odds scale. Structural models analogous to Equations 14.17 and 14.18 can be constructed for the arcsin and the probit frameworks.

Logistic regression and the separate variance test (z_S) can be extended to general linear models that allow the use of covariates and the testing of repeated measures designs (Guthrie, 1981; McCullagh & Nelder, 1989). Both are special cases of the **generalized linear model**[5] and they differ in what is called the **link function**, which specifies the relationship between the predicted values and their corresponding expected values from the error distribution. Logistic regression specifies the logit as the link function, whereas the separate variance test uses the identity as the link function; both approaches model error with the binomial distribution (see Agresti, 1990; Bishop et al., 1975; Grizzle, Starmer, & Koch, 1969).

14.7 Summary

The tests are summarized in Table 14.5 in a Wald-like form. The recommendation is for the use of the separate variance test because it is on the proportion scale. In some contexts it may be justified to use a nonlinear

[5]The generalized linear model includes as a special case the well-known general linear model, which includes ANOVA and multiple regression. The generalized linear model extends the general linear model by allowing error distributions other than the normal and allowing a nonlinear transformation to relate predicted values to expected values. There are other useful features such as the ability to adjust for violations of the expected variance by varying a parameter called dispersion. For a complete discussion of the generalized linear model see McCullagh and Nelder (1989).

transformation such as the logit when testing a contrast, but researchers should interpret the parameters on the scale f rather than in terms of raw proportions. How these tests perform under very small samples (that is, n < 20) and under a wide range of parameter values (sample sizes, population values, number of cells) is a topic for subsequent research. An additional domain of study involves the issue of multiple comparisons and corrections thereof (for example, Goodman, 1964; Knoke, 1976; Levy, 1975; Marascuilo, 1966; Rodger, 1969). Further, researchers may consider using alternative theoretical frameworks to the Wald test; for example, the likelihood ratio test provides a more direct way to assess these different statistical tests.

14.8 Questions and Problems

1. Compute the two main effect comparisons (1, 1, −1, −1) and (1, −1, 1, −1) on the Langer and Abelson data from Table 14.2, using the separate variance test, the arcsin test, the logistic test and the probit test.

2. Analyze the Langer and Abelson data in Table 14.2, using a logistic regression procedure in a statistical package such as SPSS. Because the design is a 2 $\times$ 2 experimental design, there will be three predictors—two main effects and one interaction. Verify that the three tests printed in the output correspond to the hand computations presented in this chapter (the interaction test) and Question 1.

15

Debriefing

15.1 Introduction

The goal of this book is to show that the comparison (weights that are applied to treatment means) is the building block of the analysis of variance. Our hope is that the reader will find the value in framing research hypotheses in terms of comparisons and appreciate how the standard tools of analysis of variance (such as the omnibus test) follow from a relevant set of comparisons. We avoided clutter by not discussing different situations that require special attention. Instead, we focused on basic between-subjects and within-subjects designs common in behavioral research, and focused on comparisons on means from different treatments or different time periods.

We began with the example of the farmer from Whidbey Island (Chapter 2), which illustrated many subtle issues about data analysis, among them the importance of knowing the probability of an experimental outcome. We soon found that the computation of the relevant probabilities became very tedious, so we introduced a shortcut using the normal distribution (Chapter 3) and then extended the shortcut to the more typical case where the population variances are unknown (Chapter 4). In Chapter 5, we discussed the assumptions underlying that approximation and ways to deal with violations of the equality of variance assumption. In Chapter 6, we introduced the notion of decomposing the total sum of squares into a portion that is between-treatment means and another portion that is within-treatment means; the latter served a critical role in the rest of the book because of its relation to MS_W. The comparison of pairwise treatment means, the more general test of a comparison over a set of treatment means, and the correction for unplanned comparisons was discussed in Chapters 7 and 8. The rest of the book used those building blocks to cover 2^k between-subjects designs, general between-subjects designs, within-subjects factors, and designs containing both between-subjects and within-subjects factors.

The relatively simple approach used in this book comes at a price. We glossed over the mathematical foundations (for example, expected mean squares, quasi-F) and skipped several special situations (for example, latin square designs, random effects, different tests one could do in the case of a between-subjects factorial design with unequal sample sizes). We believe the trade-off makes sense because one gains clarity, simplicity, and, we hope, intuition about what the various tests are doing. The special cases and detail can come later as the reader gains experience and confidence in his or her statistical knowledge. There are many more advanced techniques to learn. We opted to focus on a few simple ideas in order to lay the foundation for the reader's more advanced learning that will follow in more advanced courses and additional reading.

15.2 Descriptive Statistics and Plotting Data

We didn't spend much time discussing the issue of examining one's data. It is very important that the data analyst examine data closely. Does the mean provide an adequate measure of central tendency for each treatment? That is, does the mean give the value corresponding to roughly "the middle" of the distribution in each treatment? Are there outliers, or is the distribution skewed, thereby distorting the value of the mean? One useful tool in detecting these potential problems is the boxplot, which we described in Chapter 5. Each treatment (or time period) should be plotted on a separate boxplot, but to facilitate comparison across treatments it helps to force the vertical scales to be identical across boxplots. Some computer programs allow different boxplots on the same page with the same vertical scale. The key questions to address when examining boxplots include (1) Are the data symmetric? (2) Are there outliers that could distort the sample mean? and (3) Is the variability within each group comparable across the different treatments?

The data analyst should also ask whether the mean is the most appropriate measure, given the research question that is being asked. Some research questions are framed in terms of distributions being "shifted." For instance, the prediction in a two-group study might be that "the two distributions are different." This type of research question calls for a different type of test than we have reviewed in this book. Clearly, the mean by itself is not adequate in assessing whether two distributions are different. Special procedures have been developed that permit these general questions, and they are reviewed by Cliff (1996). We note that the standard nonparametric tests based on ranks are not testing the questions that many researchers think they are testing, but we defer that discussion until later in this chapter.

15.3 Presenting Your Results

This book covered the problem of making statistical inferences, and many pages were devoted to computing the appropriate statistical test. Do not lose sight of what is most important in data analysis: describing your data. Be very clear about what the means in each treatment were (and, even better, supply the standard error for each mean so the reader can understand the variability around each treatment mean). The p value or the result of a statistical decision (such as the proclamation that the null hypothesis was rejected) are not the most important aspects of data analysis. Do not be misled by all the pages devoted in this book to statistical tests, or the amount of time you spent on homework and in lectures computing t's, F's, and p values. Your results section should highlight the descriptive statistics, such as the mean. Your results section should have sentences like "The participants assigned to the high-contact condition performed faster (mean = 2 seconds) than participants assigned to the low-contact condition (mean = 3.5 seconds), $t(28) = 2.15$, $p = 0.04$." and not sentences like "The statistical test between the two treatments was significant, using the two-tailed α criterion with $t(28) = 2.15$ and $p = 0.04$. The means were 2 seconds and 3.5 seconds." The former version highlights the observation, whereas the latter highlights the results of the statistical test.

Highlight the observed means in your results section. Use statistical tests and p values merely to punctuate your sentences. Recall the example of the farmer from Whidbey Island presented in Chapter 2. There the statistical test was merely a backdrop providing a model of chance performance. The result was not the probabilistic model surrounding the farmer's behavior in the experiment but the number of cans correctly identified as containing water.

Also, when reporting the p value of a statistical result, it is useful to report the actual p value rather than merely stating the decision of whether the test was or was not statistically significant. For example, it is better to report something like $p = 0.036$ than to report the inequality $p < 0.05$. Some theoretical underpinnings of statistics emphasize the statistical decision, and so have emphasized the reporting of whether the observed p value passes the α criterion. Other theoretical underpinnings, however, emphasize the observed result. Greenwald et al. (1996) present a justification for reporting observed p values from the standpoint of a replicability index. As new approaches to statistical inference gain popularity, such as the Bayesian approach, we will see new conventions for reporting statistical results as well as alternative measures to p values for assessing model fit and testing contrasts over sample means.

15.4 Nonparametric Statistical Tests

The reader will note that we did not discuss nonparametric statistics such as the Mann–Whitney U test or the Kruskal–Wallace test. The reason is that there is nothing special about most of the nonparametric procedures (that is, those tests based on ranks). It turns out that nonparametric tests based on ranks are parametric tests in disguise. That is, they are actually related to tests presented in this book but on data that have been transformed to ranks. We illustrate by sketching an example with two groups such that there are no ties in the data. Suppose the data analyst ranked all the data (1 being the highest score and N being the lowest score) where the ranking is done without concern for which group the data came from (that is, where the ranking for all observations in both groups is done simultaneously). Then the data analyst performed the classic two-sample t test (Chapter 4) on these ranks. It turns out that the two-sample t test on the ranks is equivalent to the Mann–Whitney U. By "equivalent," we mean both asymptotic equivalence (as the sample size gets large the p values of the two tests converge) and decision equivalence (either both tests will reject the null hypothesis or neither test will reject the null hypothesis). For proofs of this and related results, see Conover (1971). Note that the p value for the Mann–Whitney test computed by most computer programs is based on the z distribution rather than the t distribution, so the p values will differ slightly. However, the observed p values will be on the same side of the criterion α, so the conclusions about the rejection of the null hypothesis will be the same. If there are tied ranks, then a special correction needs to be applied to the Mann–Whitney test (the two-sample t test has the correction automatically built in). Because they are identical, the Mann–Whitney U test makes the same assumptions as the two-sample t test on the ranked data (for example, the population variance of the ranks is equal in both groups). Thus, the usual nonparametric tests are not assumption-free as many people incorrectly believe. They are equivalent to usual parametric tests on ranks.

There has been much activity in the past few years in developing new versions of statistical tests that require fewer assumptions. This research is reviewed by Cliff (1996).

15.5 Nonexperimental Controls

It is not always possible to conduct a well-controlled experiment in the sense that the researcher can control all factors relevant in the study. For instance, imagine that in a study on memory participants are in the lab one day to learn words and they return a week later for a test of recognition memory. The experimenter does not know whether the participants practiced or rehearsed the words during the week. Maybe some participants had final exams during that week, and others did not (the memory of those who had final exams might differ from those who did not). Of course, random assignment to treatment conditions helps eliminate these uncontrolled factors as alternative explanations. The assumption is that random assignment distributes all these differences across the treatment conditions. Thus, it is very important to make sure that the experimenter randomly assigns participants to different treatments. Random assignment differs from random selection, which is about how participants were sampled from a larger population to be included in the study. Random assignment deals with how participants are placed into a particular treatment once the participants are selected to be in the experiment. Note that, by definition, individual difference factors do not involve random assignment (for example, usually one cannot randomly assign a person to be either male or female).

One can still perform analyses in such situations as long as additional information is available and one is careful about the kinds of inferences that are made from the results. There are two kinds of situations. If the experimenter points to a factor that is a possible confound, then that factor can be included in the analysis itself. For instance, in the memory study described above, if I suspect that students who had final exams during the week between the two lab sessions would remember differently than those students who did not take final exams, then I could have "final exams" as a second factor in the analysis of variance. Now, for ethical and practical reasons I probably can't randomly assign participants to the different levels of final exam, but I can account for the additional noise (that is, additive noise) contributed by the "final exam factor" by including it in the analysis of variance. In this situation the final exam factor plays the role of a blocking factor (Chapter 11). With knowledge of which students had final exams and which didn't, I can attempt to control statistically what I could not control experimentally. Of course, this requires that I have enough advanced knowledge so I know to record in advance whether or not the participant took a final exam.

But sometimes it isn't possible to define simple cubbyholes such as "took final exams" versus "did not take final exams." In such cases one can take a measurement of the variable(s) suspected of influencing the experimental results and then perform a special type of analysis of variance. For example, suppose you wanted to study the effects of different breathing exercises on the subjective experience of pain. You have that participant immerse his or her arm in a large bucket of ice water for as long as he or she can hold it in (try this sometime—you'll be surprised how quickly the pain sets in). The manipulation involves different breathing exercises for relaxation performed while the participant's arm is in the ice water (fast breathing, slow breathing, a control condition where no breathing instruction is given). The dependent variable is how long each participant leaves his or her arm in the ice bucket. This is a one-way between-subjects design because participants are randomly assigned to one of the three treatments that can be analyzed with the techniques presented in Chapters 6, 7, and 8.

Now consider a variant of the breathing study where, instead of inducing pain by having participants immerse their arms in ice water, you do a field study of pregnant women and the amount of pain they report during labor. You are interested in examining whether the different breathing exercises lead to different perceptions of pain during labor. Things are not so easy now. For ethical reasons, the researcher may not be able to randomly assign women to each treatment condition. Instead, women are given some instruction about the different breathing techniques and make a choice as to which breathing technique they would like to use during labor (or possibly choose none). Suppose that the choice of breathing technique was related to some other factor such as tolerance for pain or anxiety about labor. We're making this up, but suppose that women who are more anxious about labor (say "first-time moms") tend to choose the fast breathing technique but women who are less anxious about labor tend to choose the slow breathing technique. If the researcher had thought about this possibility in advance, and included a measure of anxiety about labor, then that measure could serve as "covariate" in the analysis of variance (a covariate is a measure that is used as a blocking factor). That is, the additive effects on reported pain from the measured variable can be removed from the analysis so that the effects of breathing pattern can be studied. The difference between a blocking factor and a covariate is subtle. A blocking factor has categories whereas a covariate is continuous. A blocking factor is sometimes created by categorizing a continuous covariate (as we did with age when illustrating blocking factors in between-subjects designs). The current view among methodologists is that it is best to leave variables in their original form (that is, don't dichotomize variables) unless there is a problem such as skewness in which some type of categorization may help provide a more meaningful measure.

Analyses involving covariates introduce their own complications and special assumptions; they are best described in the context of multiple regression, which we do not cover in this book. See Edwards (1979) for a discussion of the analysis of covariance from a regression perspective; the ANOVA perspective that characterizes this book does not make discussion of analysis of covariance transparent. Care must be taken when using covariates in the context of regression analyses. For instance, the researcher should make sure to test the typical assumption that there is no interaction between the covariate and the factors of interest. It is important to center (that is, subtract the mean of the covariate) the covariate in the event that factors are interpreted in the context of an interaction between the factors and the covariate. The concept of centering is covered in any good regression book. We found that not all statistical packages properly center the covariate when the automatic tests of interaction are performed, which makes interpreting the other effects (that is, the factors of interest) difficult. Also, the analyst should ensure that, in addition to the covariate's not having an interactive effect with the other factors, there should be a good distribution of participants with similar values on the covariate across the different treatment conditions. For example, if age is a covariate and there are three treatment conditions, it would be a problem if all the participants less than 25 are in one condition, all the participants between 25 and 40 are in a second condition, and all the participants older than 40 are in a third condition. There should be some overlap across the conditions in age (for instance, there should be 40-year-olds in all three conditions) in order for the adjustment performed by the analysis of covariance to make any sense.

There are also new developments in the statistical literature that provide extensions to the usual idea of adding a covariate to a statistical design. The general techniques fall under the title "propensity score" methods. We refer the interested reader to Rubin (1997). The basic intuition of the propensity score approach, without getting into the specific details (and we admit that we gloss over much of the detail), is that one performs two analyses—one analysis computes propensities and the other uses the propensities as a blocking factor. Let's suppose that there are two groups for which one wants to compare treatment means and several covariates that one would like to control. One first runs a logistic regression using the grouping code (in this case there are two groups, so the grouping code is binary) as the dependent variable and the covariates as predictors. This regression results in predicted probabilities of group membership based on the covariates. Then there is a second analysis that uses the actual dependent variable as the data to model, the actual grouping code as the factor, and the propensity score as a blocking factor. The propensity score serves as a basis to perform a type of matching with subgroups of participants who have similar propensities based on their covariate profile. The analyst can easily examine whether there is sufficient overlap in the propensities across the two groups (as we described in the preceding paragraph with analysis of covariance). One of the nice features of this approach is that one doesn't eliminate the effect of the covariates, but rather one examines the effect the covariates have on the key effects. Thus, there is potential to learn more from one's data. There is a growing literature relating to propensities scores, and there is a strong statistical foundation underlying the technique.

We return to the example of the breathing study. Imagine that you observed the result that the participants in both the fast and slow breathing conditions reported lesser pain during labor than the participants who did not choose to use a breathing exercise. Can you say that the breathing technique itself lowers the perception of pain? It turns out that one cannot make that conclusion without much more data and possibly some experimental manipulation. There are many different issues that need to be accounted for in this field study. However, it turns out that we may not be able to make a causal inference about breathing in the ice immersion study sketched above either. Suppose that participants who did a breathing exercise were able to hold their arms in the ice water longer than participants who did not partake in a breathing exercise. Can we be sure it was breathing that did the trick? Maybe what was really going on was that breathing simply served to distract the participants from the pain (those participants who did not receive a breathing instruction were not able to distract themselves, so experienced and reported more pain). Of course, this alternative explanation can be tested by including a different condition in the study that operationalizes "distraction" in a different way than breathing such as counting backwards by 7 starting at 1,000. Or maybe the reason the breathing treatment led to differences in the perception of pain is because it relaxed the participant. In this case, it isn't breathing or distraction that increases the tolerance for pain, but relaxation. A new experimental condition could test the relaxation hypothesis. As new conditions are added, it is possible to conceive of new alternative explanations, and the research process continues.

To us, this illustrates the beauty of the scientific method, and we hope that the reader does not get discouraged or become pessimistic. The beauty of the method is that slowly, through a process of empirical study, we rule out alternative explanations and arrive at a better understanding of the phenomenon under investigation. Studies lead to new alternative explanations and new studies to test them.

The techniques presented in this book provide but a few tools that are applied at the level of the single study to determine whether the pattern observed can be rejected from chance alone under the null hypothesis of no difference across treatment means. That is the fundamental point from the farmer example in Chapter 2. The rest of the book merely elaborated that point in a few different types of applications such as between-subjects factorial designs. Causal inference, clarity about what our experimental manipulations are actually testing, and understanding emerge from a series of well-controlled studies.

We hope you enjoyed reading this book. To paraphrase Allen Edwards, who wrote predecessors to this book.... May all your statistical tests be statistically significant, and from time to time may you have a significant hypothesis to test.

15.6 Questions and Problems

1. Think about a blocking variable or a covariate that may be useful in an area of research that interests you. Explain how this variable may impact relevant variables and why it would be important to account for the effects of this variable through statistical means (either as a blocking factor or a covariate).

2. Select a journal article in an area of research that interests you that uses an analysis of variance to analyze the data. Critique the presentation of the results. Did the authors check all assumptions? Did they emphasize descriptive statistics or the inferential tests in their write-up? Did they consider alternative explanations? Did they consider relevant blocking variables or covariates? Do any of these issues limit the conclusions drawn in the article? Possible journals include *Developmental Psychology*, *Journal of Experimental Psychology*, and *Journal of Personality and Social Psychology*.

Appendix A
The Method of Least Squares

A.1 Introduction

In the analysis of variance designs we have discussed, we assumed that a given observation is a sum of a number of additive components plus a component that represents a random error. The additive components are unknown constants or parameters, whereas the random error is a variable. In this appendix we show how to derive some of the terms of these structural models using the method of least squares. This provides some foundational structure to an important concept in this book. Additional foundations for other details such as sampling distributions are not provided in this introductory text (for example, showing how the ratio of mean square terms follows an F distribution).

In the following pages in order to keep notation as simple as possible we take the liberty of abusing notation and do not distinguish the value of a counter from the total number of treatment conditions relevant to that counter (for example, we use i to be both a counter for the treatments as well as the total number of treatments). Meaning should follow from the context.

We illustrate with the basic between-subjects design in Chapter 6. Recall that we assumed

$$X_{ij} = \mu + \alpha_i + \epsilon_{ij}$$

where μ is a constant for all ij observations, α_i is a constant for those observations obtained for the ith treatment, and ϵ_{ij} is a random error associated with each of the ij observations. We will show the least squares foundation to this basic model.

The method of least squares is a technique for using the obtained data of an experiment to provide estimates of the unknown parameters or constants in such a way as to minimize the sum of the squared errors. In the sections that follow, we state without proof the elementary rules of the differential calculus that we need in order to apply the method of least squares to problems in the analysis of variance. The review is not complete, but it is sufficient for our purpose. The student who has no knowledge of differential calculus can verify the rules by consulting any elementary calculus text. We find that the method of least squares is easier to understand than other, perhaps more powerful and general, methods that could be presented, such as the method of maximum likelihood or Bayesian methods.

A.2 The Derivative of a Power Function

Let $y = x^n$. Then, y is a function of the power of x. The derivative of this function, written dy/dx, is

$$\frac{dy}{dx} = nx^{n-1} \tag{1}$$

(1) states the general rule for the derivative of $y = x^n$. We multiply x by its original exponent and decrease the original exponent by 1.

A.3 The Derivative of a Power Function with an Added Constant

Let $y = a + x^n$, where a is an added constant. Then, the derivative of this function with respect to x is

$$\frac{dy}{dx} = nx^{n-1} \tag{2}$$

(2) defines the general rule that an added constant does not appear in the derivative.

A.4 The Derivative of a Power Function with a Multiplicative Coefficient

If $y = ax^n$, where a is a constant and a coefficient of x, then the derivative of this function with respect to x is

$$\frac{dy}{dx} = anx^{n-1} \tag{3}$$

(3) defines the general rule that if a is a coefficient of the term with respect to which we are differentiating, it appears as a coefficient of that term in the derivative.

A.5 The Method of Least Squares

Suppose that we have error $\epsilon = X - a$ and that we have a sample of n values of X. At this point, the value of a is not known. We will use the method of least squares to define a. We let Q be the sum of squared errors

$$Q = \sum \epsilon^2 = \sum (X - a)^2 \tag{4}$$

and we want to find the value of a that will minimize Q. To do so, using the method of least squares, we differentiate Q with respect to a and set the derivative equal to zero. If we expand the right side of (4) and sum, we have

$$Q = \sum X^2 - 2a \sum X + na^2$$

and using the rules we have stated, we have[1]

$$\frac{dQ}{da} = -2\sum X + 2na$$

Setting the above derivative equal to zero, we have

$$-2\sum X + 2na = 0$$

or

$$2na = 2\sum X$$

Then, dividing both sides of the above expression by $2n$, we obtain

$$a = \bar{X}$$

Thus, we have shown through the method of least squares that the mean $\bar{X}$ is the value of a for which Q will be a minimum. In other words, we have just shown that the mean $\bar{X}$ produces the smallest possible sum of squared differences from a set of observations, that is,

$$\sum(X - \bar{X})^2 < \sum(X - a)^2$$

for any value of $a \neq \bar{X}$. This provides one justification for the use of the mean in experimental designs—the value that minimizes the sum of squared differences.

[1] For any given set of observations, $\sum X$ and $\sum X^2$ are constants, but a may conceivably take any one of an infinite number of values. The least squares solution for a, however, requires that we find the *single* value of a that will minimize Q for a set of n observations. If we were to let a take a different value for each possible value of X, then we would have the trivial solution that Q is minimized when $a = X$ for each value of X.

A.6 Another Method for Finding the Derivative of $\sum(X - a)^2$

We show another method of finding the derivative of $Q = \sum(X - a)^2$ that is useful. Again, we let

$$\epsilon = X - a \tag{5}$$

and

$$Q = \sum \epsilon^2 = \sum (X - a)^2 \tag{6}$$

Differentiating (5) with respect to a, we have

$$\frac{d\epsilon}{da} = -1$$

and if we differentiate (6) with respect to ϵ, we obtain

$$\frac{dQ}{d\epsilon} = 2 \sum \epsilon$$

Then, we have the general rule that

$$\frac{dQ}{da} = \frac{d\epsilon}{da} \times \frac{dQ}{d\epsilon} \tag{7}$$

and substituting in (7), we obtain

$$\begin{aligned} \frac{dQ}{da} &= -2 \sum \epsilon = -2 \sum (X - a) \\ &= -2 \sum X + 2na \end{aligned}$$

as before.

A.7 Least Squares Estimates α_i and μ for a Between-Subjects Design with One Factor

If we have a between-subjects design with one factor, with j treatments and n observations for each treatment (as we had in Chapter 6), then we have assumed the simple additive model

$$X_{ij} = \mu + \alpha_j + \epsilon_{ij}$$

where μ is common to all n observations in the ij combinations, α_i is a treatment effect that is common to the n observations receiving a given treatment, and ϵ_{ij} is a random error associated with each of the ij observations. Then, the error is expressed as

$$\epsilon_{ij} = X_{ij} - \alpha_i - \mu \tag{8}$$

We let

$$Q = \sum \epsilon_{ij}^2 = \sum (X_{ij} - \alpha_j - \mu)^2 \tag{9}$$

and we want to find the estimates of μ and the values of α_i that will minimize Q. In our discussion of the between-subjects design, we took $\bar{X}_{..}$ as an estimate of μ and $\bar{X}_{i.} - \bar{X}_{..}$ as an estimate of α_i. We now show that these estimates are the least squares estimates.

Differentiating (8) with respect to μ and (9) with respect to ϵ, we have -1 and $2\sum\epsilon$ as the respective derivatives. Then, multiplying the two derivatives, we obtain

$$\frac{dQ}{d\mu} = -2\sum\epsilon$$

or

$$\frac{dQ}{d\mu} = -2\left(\sum X_{..} - n\sum\alpha_i - in\mu\right)$$

In the analysis of variance model for the between-subjects design, we have shown that without any loss in generality we can let $\sum \alpha_i = 0$. Then, with $\sum \alpha_i = 0$, if we set the derivative equal to zero and solve for μ, we have

$$2in\mu = 2 \sum X_{..}$$

or

$$\mu = \bar{X}..$$

and $\bar{X}_{..}$ is the least squares estimate of μ.

If we differentiate (8) with respect to α_i and (9) with respect to ϵ and multiply the two derivatives, we have

$$\frac{dQ}{d\alpha_i} = -2 \sum \epsilon$$

But in this case, we have only n observation equations in which a given α_i appears, so that[2]

$$\frac{dQ}{d\alpha_i} = -2 \left(\sum X_i. - n\alpha_i - n\mu \right)$$

Setting this derivative equal to zero and solving for α_i, we have

$$2n\alpha_i = 2 \sum X_i. - 2n\mu$$

or

$$\alpha_i = \bar{X}_i. - \mu$$

Substituting in the above expression with the least squares estimate of α_i, we obtain

[2]If an observation equation for ϵ_{ij}^2 contains the α_i with respect to which we are differentiating, then its derivative is $-2(X_{ij} - \alpha_i - \mu)$. If the observation equation does not contain α_i, then its derivative is zero.

$$\alpha_i = \bar{X}_{i}. \;-\; \bar{X}..$$

and $\bar{X}_{k.} - \bar{X}_{..}$ is the least squares estimate of α_i.

If we now substitute the least squares estimates of μ and α_i in (9), we have

$$\sum_{1}^{ij} \epsilon_{ij}^2 = \sum_{1}^{i}\sum_{1}^{j} [X_{ij} \;-\; (\bar{X}_{i}. \;-\; \bar{X}..) \;-\; \bar{X}..]^2$$

$$= \sum_{1}^{i}\sum_{1}^{j} (\bar{X}_{ij} \;-\; \bar{X}_{i}.)^2 \qquad (10)$$

and (10) is the sum of squares within treatments.

Appendix B
Statistical Tables

B.1 *t* Table

This table displays the t value corresponding to degrees of freedom (*row*) and a particular α criterion (*column*). For example, for 1,000 degrees of freedom, the value 1.9623 corresponds to the tail area of 0.025. For the tail on the other side of the distribution, simply take the negative of the table entry. To illustrate, a 95% confidence interval takes 0.025 of the area on each side of the distribution. With 1,000 degrees of freedom, the t values of -1.9623 and 1.9623 are used to construct the 95% confidence interval. The t distribution approaches the standardized normal distribution as the degrees of freedom approach infinity (Table 4.2). See Chapter 4 for more details.

			α		
Degrees of freedom	0.25	0.1	0.05	0.025	0.001
1	1.0000	3.0777	6.3138	12.7062	318.3088
2	0.8165	1.8856	2.9200	4.3027	22.3271
3	0.7649	1.6377	2.3534	3.1824	10.2145
4	0.7407	1.5332	2.1318	2.7764	7.1732
5	0.7267	1.4759	2.0150	2.5706	5.8934
6	0.7176	1.4398	1.9432	2.4469	5.2076
7	0.7111	1.4149	1.8946	2.3646	4.7853
8	0.7064	1.3968	1.8595	2.3060	4.5008
9	0.7027	1.3830	1.8331	2.2622	4.2968
10	0.6998	1.3722	1.8125	2.2281	4.1437
11	0.6974	1.3634	1.7959	2.2010	4.0247
12	0.6955	1.3562	1.7823	2.1788	3.9296
13	0.6938	1.3502	1.7709	2.1604	3.8520
14	0.6924	1.3450	1.7613	2.1448	3.7874
15	0.6912	1.3406	1.7531	2.1314	3.7328
16	0.6901	1.3368	1.7459	2.1199	3.6862
17	0.6892	1.3334	1.7396	2.1098	3.6458
18	0.6884	1.3304	1.7341	2.1009	3.6105
19	0.6876	1.3277	1.7291	2.0930	3.5794
20	0.6870	1.3253	1.7247	2.0860	3.5518
21	0.6864	1.3232	1.7207	2.0796	3.5272
22	0.6858	1.3212	1.7171	2.0739	3.5050
23	0.6853	1.3195	1.7139	2.0687	3.4850
24	0.6848	1.3178	1.7109	2.0639	3.4668
25	0.6844	1.3163	1.7081	2.0595	3.4502
26	0.6840	1.3150	1.7056	2.0555	3.4350
27	0.6837	1.3137	1.7033	2.0518	3.4210
28	0.6834	1.3125	1.7011	2.0484	3.4082
29	0.6830	1.3114	1.6991	2.0452	3.3962
30	0.6828	1.3104	1.6973	2.0423	3.3852
60	0.6786	1.2958	1.6706	2.0003	3.2317
120	0.6765	1.2886	1.6577	1.9799	3.1595
300	0.6753	1.2844	1.6499	1.9679	3.1176
1000	0.6747	1.2824	1.6464	1.9623	3.0984

B.2 *F* Tables

The F distribution is used in this book to test the ratio of two variances and the ratio of two mean squared terms, such as MS_B and MS_W. Two F tables—corresponding to $\alpha = 0.05$ and $\alpha = 0.01$—are presented below. See Chapter 6 for more details about the F distribution. The abbreviations *df* and *den* represent degrees of freedom and denominator, respectively.

B.2.1 $\alpha = 0.05$

df	Degrees of freedom (numerator)								
(*den*)	1	2	3	4	5	6	7	8	inf.
3	10.1280	9.5521	9.2766	9.1172	9.0135	8.9406	8.8867	8.8452	8.5264
4	7.7086	6.9443	6.5914	6.3882	6.2561	6.1631	6.0942	6.0410	5.6281
5	6.6079	5.7861	5.4095	5.1922	5.0503	4.9503	4.8759	4.8183	4.3650
6	5.9874	5.1433	4.7571	4.5337	4.3874	4.2839	4.2067	4.1468	3.6689
7	5.5914	4.7374	4.3468	4.1203	3.9715	3.8660	3.7870	3.7257	3.2298
8	5.3177	4.4590	4.0662	3.8379	3.6875	3.5806	3.5005	3.4381	2.9276
9	5.1174	4.2565	3.8625	3.6331	3.4817	3.3738	3.2927	3.2296	2.7067
10	4.9646	4.1028	3.7083	3.4780	3.3258	3.2172	3.1355	3.0717	2.5379
11	4.8443	3.9823	3.5874	3.3567	3.2039	3.0946	3.0123	2.9480	2.4045
12	4.7472	3.8853	3.4903	3.2592	3.1059	2.9961	2.9134	2.8486	2.2962
13	4.6672	3.8056	3.4105	3.1791	3.0254	2.9153	2.8321	2.7669	2.2064
14	4.6001	3.7389	3.3439	3.1122	2.9582	2.8477	2.7642	2.6987	2.1307
15	4.5431	3.6823	3.2874	3.0556	2.9013	2.7905	2.7066	2.6408	2.0658
16	4.4940	3.6337	3.2389	3.0069	2.8524	2.7413	2.6572	2.5911	2.0096
17	4.4513	3.5915	3.1968	2.9647	2.8100	2.6987	2.6143	2.5480	1.9604
18	4.4139	3.5546	3.1599	2.9277	2.7729	2.6613	2.5767	2.5102	1.9168
19	4.3807	3.5219	3.1274	2.8951	2.7401	2.6283	2.5435	2.4768	1.8780
20	4.3512	3.4928	3.0984	2.8661	2.7109	2.5990	2.5140	2.4471	1.8432
25	4.2417	3.3852	2.9912	2.7587	2.6030	2.4904	2.4047	2.3371	1.7110
30	4.1709	3.3158	2.9223	2.6896	2.5336	2.4205	2.3343	2.2662	1.6223
35	4.1213	3.2674	2.8742	2.6415	2.4851	2.3718	2.2852	2.2167	1.5580
40	4.0847	3.2317	2.8387	2.6060	2.4495	2.3359	2.2490	2.1802	1.5089
45	4.0566	3.2043	2.8115	2.5787	2.4221	2.3083	2.2212	2.1521	1.4700
50	4.0343	3.1826	2.7900	2.5572	2.4004	2.2864	2.1992	2.1299	1.4383
55	4.0162	3.1650	2.7725	2.5397	2.3828	2.2687	2.1813	2.1119	1.4118
60	4.0012	3.1504	2.7581	2.5252	2.3683	2.2541	2.1665	2.0970	1.3893

B.2.2 $\alpha = 0.01$

df (*den*)	1	2	3	4	5	6	7	8	inf.
	Degrees of freedom (numerator)								
3	34.1162	30.8165	29.4567	28.7099	28.2371	27.9107	27.6717	27.4892	26.1252
4	21.1977	18.0000	16.6944	15.9770	15.5219	15.2069	14.9758	14.7989	13.4631
5	16.2582	13.2739	12.0600	11.3919	10.9670	10.6723	10.4555	10.2893	9.0204
6	13.7450	10.9248	9.7795	9.1483	8.7459	8.4661	8.2600	8.1017	6.8800
7	12.2464	9.5466	8.4513	7.8466	7.4604	7.1914	6.9928	6.8400	5.6495
8	11.2586	8.6491	7.5910	7.0061	6.6318	6.3707	6.1776	6.0289	4.8588
9	10.5614	8.0215	6.9919	6.4221	6.0569	5.8018	5.6129	5.4671	4.3105
10	10.0443	7.5594	6.5523	5.9943	5.6363	5.3858	5.2001	5.0567	3.9090
11	9.6460	7.2057	6.2167	5.6683	5.3160	5.0692	4.8861	4.7445	3.6024
12	9.3302	6.9266	5.9525	5.4120	5.0643	4.8206	4.6395	4.4994	3.3608
13	9.0738	6.7010	5.7394	5.2053	4.8616	4.6204	4.4410	4.3021	3.1654
14	8.8616	6.5149	5.5639	5.0354	4.6950	4.4558	4.2779	4.1399	3.0040
15	8.6831	6.3589	5.4170	4.8932	4.5556	4.3183	4.1415	4.0045	2.8684
16	8.5310	6.2262	5.2922	4.7726	4.4374	4.2016	4.0259	3.8896	2.7528
17	8.3997	6.1121	5.1850	4.6690	4.3359	4.1015	3.9267	3.7910	2.6530
18	8.2854	6.0129	5.0919	4.5790	4.2479	4.0146	3.8406	3.7054	2.5660
19	8.1849	5.9259	5.0103	4.5003	4.1708	3.9386	3.7653	3.6305	2.4893
20	8.0960	5.8489	4.9382	4.4307	4.1027	3.8714	3.6987	3.5644	2.4212
25	7.7698	5.5680	4.6755	4.1774	3.8550	3.6272	3.4568	3.3239	2.1694
30	7.5625	5.3903	4.5097	4.0179	3.6990	3.4735	3.3045	3.1726	2.0062
35	7.4191	5.2679	4.3957	3.9082	3.5919	3.3679	3.2000	3.0687	1.8910
40	7.3141	5.1785	4.3126	3.8283	3.5138	3.2910	3.1238	2.9930	1.8047
45	7.2339	5.1103	4.2492	3.7674	3.4544	3.2325	3.0658	2.9353	1.7374
50	7.1706	5.0566	4.1993	3.7195	3.4077	3.1864	3.0202	2.8900	1.6831
55	7.1194	5.0132	4.1591	3.6809	3.3700	3.1493	2.9834	2.8534	1.6383
60	7.0771	4.9774	4.1259	3.6490	3.3389	3.1187	2.9530	2.8233	1.6006

B.3 Z Table: Standard Normal Distribution

The entries in the table correspond to the area to the left of the particular Z value. The rows correspond to the first two digits of Z, and the columns correspond to the third digit. For example, to find the area to the left of $Z = 1.03$, find the row entry for 1.0 and the column entry for 0.03. The area is 0.8485.

Z	0	0.01	0.02	0.03	0.04	0.05	0.06	0.07	0.08	0.09
0.0	0.5000	0.5040	0.5080	0.5120	0.5160	0.5199	0.5239	0.5279	0.5319	0.5359
0.1	0.5398	0.5438	0.5478	0.5517	0.5557	0.5596	0.5636	0.5675	0.5714	0.5753
0.2	0.5793	0.5832	0.5871	0.5910	0.5948	0.5987	0.6026	0.6064	0.6103	0.6141
0.3	0.6179	0.6217	0.6255	0.6293	0.6331	0.6368	0.6406	0.6443	0.6480	0.6517
0.4	0.6554	0.6591	0.6628	0.6664	0.6700	0.6736	0.6772	0.6808	0.6844	0.6879
0.5	0.6915	0.6950	0.6985	0.7019	0.7054	0.7088	0.7123	0.7157	0.7190	0.7224
0.6	0.7257	0.7291	0.7324	0.7357	0.7389	0.7422	0.7454	0.7486	0.7517	0.7549
0.7	0.7580	0.7611	0.7642	0.7673	0.7704	0.7734	0.7764	0.7794	0.7823	0.7852
0.8	0.7881	0.7910	0.7939	0.7967	0.7995	0.8023	0.8051	0.8078	0.8106	0.8133
0.9	0.8159	0.8186	0.8212	0.8238	0.8264	0.8289	0.8315	0.8340	0.8365	0.8389
1.0	0.8413	0.8438	0.8461	0.8485	0.8508	0.8531	0.8554	0.8577	0.8599	0.8621
1.1	0.8643	0.8665	0.8686	0.8708	0.8729	0.8749	0.8770	0.8790	0.8810	0.8830
1.2	0.8849	0.8869	0.8888	0.8907	0.8925	0.8944	0.8962	0.8980	0.8997	0.9015
1.3	0.9032	0.9049	0.9066	0.9082	0.9099	0.9115	0.9131	0.9147	0.9162	0.9177
1.4	0.9192	0.9207	0.9222	0.9236	0.9251	0.9265	0.9279	0.9292	0.9306	0.9319
1.5	0.9332	0.9345	0.9357	0.9370	0.9382	0.9394	0.9406	0.9418	0.9429	0.9441
1.6	0.9452	0.9463	0.9474	0.9484	0.9495	0.9505	0.9515	0.9525	0.9535	0.9545
1.7	0.9554	0.9564	0.9573	0.9582	0.9591	0.9599	0.9608	0.9616	0.9625	0.9633
1.8	0.9641	0.9649	0.9656	0.9664	0.9671	0.9678	0.9686	0.9693	0.9699	0.9706
1.9	0.9713	0.9719	0.9726	0.9732	0.9738	0.9744	0.9750	0.9756	0.9761	0.9767
2.0	0.9772	0.9778	0.9783	0.9788	0.9793	0.9798	0.9803	0.9808	0.9812	0.9817
2.1	0.9821	0.9826	0.9830	0.9834	0.9838	0.9842	0.9846	0.9850	0.9854	0.9857
2.2	0.9861	0.9864	0.9868	0.9871	0.9875	0.9878	0.9881	0.9884	0.9887	0.9890
2.3	0.9893	0.9896	0.9898	0.9901	0.9904	0.9906	0.9909	0.9911	0.9913	0.9916
2.4	0.9918	0.9920	0.9922	0.9925	0.9927	0.9929	0.9931	0.9932	0.9934	0.9936
2.5	0.9938	0.9940	0.9941	0.9943	0.9945	0.9946	0.9948	0.9949	0.9951	0.9952
2.6	0.9953	0.9955	0.9956	0.9957	0.9959	0.9960	0.9961	0.9962	0.9963	0.9964
2.7	0.9965	0.9966	0.9967	0.9968	0.9969	0.9970	0.9971	0.9972	0.9973	0.9974
2.8	0.9974	0.9975	0.9976	0.9977	0.9977	0.9978	0.9979	0.9979	0.9980	0.9981
2.9	0.9981	0.9982	0.9982	0.9983	0.9984	0.9984	0.9985	0.9985	0.9986	0.9986
3.0	0.9987	0.9987	0.9987	0.9988	0.9988	0.9989	0.9989	0.9989	0.9990	0.9990
3.1	0.9990	0.9991	0.9991	0.9991	0.9992	0.9992	0.9992	0.9992	0.9993	0.9993
3.2	0.9993	0.9993	0.9994	0.9994	0.9994	0.9994	0.9994	0.9995	0.9995	0.9995
3.3	0.9995	0.9995	0.9995	0.9996	0.9996	0.9996	0.9996	0.9996	0.9996	0.9997
3.4	0.9997	0.9997	0.9997	0.9997	0.9997	0.9997	0.9997	0.9997	0.9997	0.9998
3.5	0.9998	0.9998	0.9998	0.9998	0.9998	0.9998	0.9998	0.9998	0.9998	0.9998
3.6	0.9998	0.9998	0.9999	0.9999	0.9999	0.9999	0.9999	0.9999	0.9999	0.9999
3.7	0.9999	0.9999	0.9999	0.9999	0.9999	0.9999	0.9999	0.9999	0.9999	0.9999
3.8	0.9999	0.9999	0.9999	0.9999	0.9999	0.9999	0.9999	0.9999	0.9999	0.9999
3.9	1.0000	1.0000	1.0000	1.0000	1.0000	1.0000	1.0000	1.0000	1.0000	1.0000
4.0	1.0000	1.0000	1.0000	1.0000	1.0000	1.0000	1.0000	1.0000	1.0000	1.0000

B.4 Chi-Squared Table

The entries in the table correspond to the χ^2 value for degrees of freedom (*row*) and a particular α (*column*). For example, the χ^2 value for $\alpha = 5\%$ and 3 degrees of freedom is 7.8147.

	α				
Degrees of freedom	0.25	0.1	0.05	0.025	0.001
1	1.3233	2.7055	3.8415	5.0239	10.8276
2	2.7726	4.6052	5.9915	7.3778	13.8155
3	4.1083	6.2514	7.8147	9.3484	16.2662
4	5.3853	7.7794	9.4877	11.1433	18.4668
5	6.6257	9.2364	11.0705	12.8325	20.5150
6	7.8408	10.6446	12.5916	14.4494	22.4577
7	9.0371	12.0170	14.0671	16.0128	24.3219
8	10.2189	13.3616	15.5073	17.5345	26.1245
9	11.3888	14.6837	16.9190	19.0228	27.8772
10	12.5489	15.9872	18.3070	20.4832	29.5883
11	13.7007	17.2750	19.6751	21.9200	31.2641
12	14.8454	18.5493	21.0261	23.3367	32.9095
13	15.9839	19.8119	22.3620	24.7356	34.5282
14	17.1169	21.0641	23.6848	26.1189	36.1233
15	18.2451	22.3071	24.9958	27.4884	37.6973
16	19.3689	23.5418	26.2962	28.8454	39.2524
17	20.4887	24.7690	27.5871	30.1910	40.7902
18	21.6049	25.9894	28.8693	31.5264	42.3124
19	22.7178	27.2036	30.1435	32.8523	43.8202
20	23.8277	28.4120	31.4104	34.1696	45.3147
21	24.9348	29.6151	32.6706	35.4789	46.7970
22	26.0393	30.8133	33.9244	36.7807	48.2679
23	27.1413	32.0069	35.1725	38.0756	49.7282
24	28.2412	33.1962	36.4150	39.3641	51.1786
25	29.3389	34.3816	37.6525	40.6465	52.6197
26	30.4346	35.5632	38.8851	41.9232	54.0520
27	31.5284	36.7412	40.1133	43.1945	55.4760
28	32.6205	37.9159	41.3371	44.4608	56.8923
29	33.7109	39.0875	42.5570	45.7223	58.3012
30	34.7997	40.2560	43.7730	46.9792	59.7031
35	40.2228	46.0588	49.8018	53.2033	66.6188
40	45.6160	51.8051	55.7585	59.3417	73.4020
45	50.9849	57.5053	61.6562	65.4102	80.0767
50	56.3336	63.1671	67.5048	71.4202	86.6608
55	61.6650	68.7962	73.3115	77.3805	93.1675
60	66.9815	74.3970	79.0819	83.2977	99.6072
65	72.2848	79.9730	84.8206	89.1771	105.9881
70	77.5767	85.5270	90.5312	95.0232	112.3169
80	88.1303	96.5782	101.8795	106.6286	124.8392
100	109.1412	118.4980	124.3421	129.5612	149.4493

B.5 Table of Coefficients for Orthogonal Polynomials

This table presents the coefficients of polynomial comparisons.

k	Polynomial	Values of the coefficients							
3	Linear	−1	0	1					
	Quadratic	1	−2	1					
4	Linear	−3	−1	1	3				
	Quadratic	1	−1	−1	1				
	Cubic	−1	3	−3	1				
5	Linear	−2	−1	0	1	2			
	Quadratic	2	−1	−2	−1	2			
	Cubic	−1	2	0	−2	1			
	Quartic	1	−4	6	−4	1			
6	Linear	−5	−3	−1	1	3	5		
	Quadratic	5	−1	−4	−4	−1	5		
	Cubic	−5	7	4	−4	−7	5		
	Quartic	1	−3	2	2	−3	1		
7	Linear	−3	−2	−1	0	1	2	3	
	Quadratic	5	0	−3	−4	−3	0	5	
	Cubic	−1	1	1	0	−1	−1	1	
	Quartic	3	−7	1	6	1	−7	3	
8	Linear	−7	−5	−3	−1	1	3	5	7
	Quadratic	7	1	−3	−5	−5	−3	1	7
	Cubic	−7	5	7	3	−3	−7	−5	7
	Quartic	7	−13	−3	9	9	−3	−13	7

B.6 Tukey's Studentized Range Table for α = 0.05

The rows refer to the degrees of freedom corresponding to the within-treatment term and the column headings refer to the number of treatments. See Chapter 7 for details.

	Range						
Degrees of freedom	2	3	4	5	6	7	8
2	6.0796	8.3308	9.7990	10.8811	11.7336	12.4346	13.0282
3	4.5007	5.9097	6.8245	7.5017	8.0371	8.4783	8.8525
4	3.9265	5.0402	5.7571	6.2870	6.7064	7.0526	7.3465
5	3.6354	4.6017	5.2183	5.6731	6.0329	6.3299	6.5823
6	3.4605	4.3392	4.8956	5.3049	5.6284	5.8953	6.1222
7	3.3441	4.1649	4.6813	5.0601	5.3591	5.6057	5.8153
8	3.2612	4.0410	4.5288	4.8858	5.1672	5.3991	5.5962
9	3.1992	3.9485	4.4149	4.7554	5.0235	5.2444	5.4319
10	3.1511	3.8768	4.3266	4.6543	4.9120	5.1242	5.3042
11	3.1127	3.8196	4.2561	4.5736	4.8230	5.0281	5.2021
12	3.0813	3.7729	4.1987	4.5077	4.7502	4.9496	5.1187
13	3.0552	3.7341	4.1509	4.4529	4.6897	4.8842	5.0491
14	3.0332	3.7014	4.1105	4.4066	4.6385	4.8290	4.9903
15	3.0143	3.6734	4.0760	4.3670	4.5947	4.7816	4.9399
16	2.9980	3.6491	4.0461	4.3327	4.5568	4.7406	4.8962
17	2.9837	3.6280	4.0200	4.3027	4.5237	4.7048	4.8580
18	2.9712	3.6093	3.9970	4.2763	4.4944	4.6731	4.8243
19	2.9600	3.5927	3.9766	4.2528	4.4685	4.6450	4.7944
20	2.9500	3.5779	3.9583	4.2319	4.4452	4.6199	4.7676
21	2.9410	3.5646	3.9419	4.2130	4.4244	4.5973	4.7435
22	2.9329	3.5526	3.9270	4.1959	4.4055	4.5769	4.7217
23	2.9255	3.5417	3.9136	4.1805	4.3883	4.5583	4.7018
24	2.9188	3.5317	3.9013	4.1663	4.3727	4.5413	4.6838
25	2.9126	3.5226	3.8900	4.1534	4.3583	4.5258	4.6672
26	2.9070	3.5142	3.8796	4.1415	4.3451	4.5115	4.6519
27	2.9017	3.5064	3.8701	4.1305	4.3329	4.4983	4.6378
28	2.8969	3.4993	3.8612	4.1203	4.3217	4.4861	4.6248
29	2.8924	3.4926	3.8530	4.1109	4.3112	4.4747	4.6127
30	2.8882	3.4864	3.8454	4.1021	4.3015	4.4642	4.6014
31	2.8843	3.4806	3.8383	4.0939	4.2924	4.4543	4.5909
32	2.8807	3.4752	3.8316	4.0862	4.2839	4.4451	4.5811
33	2.8772	3.4702	3.8254	4.0790	4.2759	4.4365	4.5718
34	2.8740	3.4654	3.8195	4.0723	4.2684	4.4284	4.5632
35	2.8710	3.4610	3.8140	4.0659	4.2614	4.4207	4.5550
36	2.8682	3.4568	3.8088	4.0600	4.2548	4.4135	4.5473
37	2.8655	3.4528	3.8039	4.0543	4.2485	4.4068	4.5401
38	2.8629	3.4490	3.7992	4.0490	4.2426	4.4003	4.5332
39	2.8605	3.4455	3.7949	4.0439	4.2370	4.3942	4.5267
120	2.8000	3.3561	3.6846	3.9169	4.0960	4.2412	4.3630
inf.	2.7718	3.3145	3.6332	3.8577	4.0301	4.1696	4.2863

References

Agresti, A. (1990). *Categorical data analysis.* New York: Wiley.

Aldrich, J. H., & Nelson, F. D. (1984). *Linear probability, logit, and probit models.* Beverly Hills, CA: Sage.

Anscombe, F. J. (1948). The transformation of Poisson, binomial and negative binomial data. *Biometrika, 35*, 246–254.

Aronson, E., Ellsworth, P., Carlsmith, J., & Gonzales, M. (1990). *Methods of research in social psychology.* New York: McGraw-Hill.

Bartlett, M. S. (1936). Square root transformation in analysis of variance. *Journal of the Royal Statistical Society Supplement, 3*, 68–78.

Berengut, D., & Petkau, A. J. (1979). *On testing the equality of two proportions* (Tech. Rep. No. 267). Stanford, CA: Stanford University.

Bishop, Y. M. M., Fienberg, S. E., & Holland, P. W. (1975). *Discrete multivariate analysis.* Cambridge, MA: MIT Press.

Bonett, D. G. (2008). Confidence intervals for standardized linear contrasts of means. *Psychological Methods, 13*, 99–109.

Box, G. E. P. (1953). Non-normality and tests on variances. *Biometrika, 40*, 318–335.

Box, G. E. P., & Cox, D. R. (1964). An analysis of transformation. *Journal of the Royal Statistical Society, 26*, 211–252.

Brown, M. B., & Forsythe, A. B. (1974). The ANOVA and multiple comparison for data with heterogeneous variances. *Biometrics, 30*, 719–724.

Castellan, N. J. (1965). On the partitioning of contingency tables. *Psychological Bulletin, 64*, 330–338.

Clark, H. H. (1973). The language-as-fixed-effect fallacy: A critique of language statistics in psychological research. *Journal of Verbal Learning and Verbal Behavior, 12*, 335–359.

Cleveland, W. S. (1993). *Visualizing data.* Summit, NJ: Hobart Press.

Cliff, N. (1996). *Ordinal methods for behavioral data analysis.* Hillsdale, NJ: Erlbaum.

Cochran, W. G. (1940). The analysis of variance when experimental errors follow the Poisson or binomial laws. *Annals of Mathematical Statistics, 11*, 335–347.

Cochran, W. G. (1943). Analysis of variance for percentages based on unequal numbers. *Journal of the American Statistical Association, 38*, 287–301.

Cochran, W. G. (1954). Some methods for strengthening the common χ^2 tests. *Biometrics, 10*, 417–451.

Cochran, W. G. (1963). *Sampling techniques* (2nd ed.). New York: Wiley.

Cochran, W. G. (1977). *Sampling techniques.* New York: Wiley.

Cohen, J. (1967). An alternative to Marascuilo's "large sample multiple comparisons" for proportions. *Psychological Bulletin, 67*, 199–201.

Cohen, J. (1987). *Statistical power analysis for the behavioral sciences.* Hillsdale, NJ: Erlbaum.

Collins, L. M., Murphy, S. A., Nair, V. N., & Strecher, V. J. (2005). A strategy for optimizing and evaluating behavioral interventions. *Annals of Behavioral Medicine, 30*, 65–73.

Conover, W. J. (1971). *Practical nonparametric statistics.* New York: Wiley.

Cook, T. D., & Campbell, D. T. (1979). *Quasi-experimentation: Design and analysis issues.* Boston: Houghton Mifflin.

Cox, D. R., & Hinkley, D. V. (1974). *Theoretical statistics.* Cambridge: Chapman and Hall.

Cox, D. R., & Snell, E. J. (1989). *Analysis of binary data.* London: Chapman and Hall.

Cressie, N. (1978). Testing for the equality of two binomial proportions. *Annals of the Institute of Statistical Mathematics, 30*, 421–427.

Curtiss, J. H. (1943). On transformations used in the analysis of variance. *Annals of Mathematical Statistics, 14*, 107–122.

D'Agostino, R. B. (1971). A second look at analysis of variance on dichotomous data. *Journal of Educational Measurement, 8*, 327–333.

D'Agostino, R. B. (1972). Relation between the Chi-square and ANOVA tests for testing the equality of k independent dichotomous populations. *American Statistician, 26*, 30–32.

Donaldson, T. S. (1968). Robustness of the F-test to errors of both kinds and the correlation between the numerator and the denominator of the F-ratio. *Journal of the American Statistical Association, 63*, 660–676.

Edwards, A. L. (1964). *Expected values of discrete random variables and elementary statistics.* New York: Wiley.

Edwards, A. L. (1979). *Multiple regression and the analysis of variance and covariance.* New York: Freeman.

Edwards, A. L. (1985). *Experimental design in psychological research* (5th ed.). New York: Harper & Row.

Emerson, J. D. (1991). Introduction to transformation. In D. C. Hoaglin, F. Mosteller, & J. W. Tukey (Eds.), *Fundamentals of exploratory analysis of variance* (pp. 365–400). New York: Wiley.

Falmagne, J. (1985). *Elements of psychophysical therapy.* Oxford, UK: Clarendon Press.

Fern, E. F., & Monroe, K. B. (1996). Effect-size estimates: Issues and problems in interpretation. *Journal of Consumer Research, 23*, 89-105.

Finney, D. J. (1971). *Probit analysis.* Cambridge, UK: Cambridge University Press.

Fisher, R. A. (1934). Discussion on Dr. Wishart's paper. *Journal of the Royal Statistical Society Supplement, 1*, 51–53.

Fisher, R. A. (1936). *Statistical methods for research workers* (6th ed.). Edinburgh, UK: Oliver & Boyd.

Fisher, R. A. (1942). *The design of experiments* (3rd ed.). Edinburgh, UK: Oliver & Boyd.

Fisher, R. A. (1954). The analysis of variance with various binomial transformations. *Biometrics, 10*, 130–139.

Fisher, R. A., & Yates, F. (1948). *Statistical tables for biological, agricultural and medical research* (3rd ed.). Edinburgh, UK: Oliver & Boyd.

Fleiss, J. L. (1981). *Statistical methods for rates and proportions.* New York: Wiley.

Freeman, M. F., & Tukey, J. W. (1950). Transformations related to the angular and the square root. *Annals of Mathematical Statistics, 21*, 607–611.

Gaito, J. (1965). Unequal intervals and unequal *n* in trend analyses. *Psychological Bulletin, 65*, 125–127.

Gart, J. J., Pettigrew, H. M., & Thomas, D. G. (1985). The effect of bias, variance estimation, skewness and kurtosis of the empirical logit on weighted least squares analyses. *Biometrika, 72*, 179–90.

Ghurye, S. G. (1949). Transformation of a binomial variate for the analysis of variance. *Journal of the Indian Society of Agricultural Statistics, 2*, 94–109.

Gonzalez, R., & Griffin, D. (2001). A statistical framework for modeling homogeneity and interdependence in groups. In M. Clark & G. Fletcher (Eds.), *Blackwell handbook of social psychology* (Vol. 2, pp. 505–534). Malden, MA: Blackwell Publishers.

Goodman, L. A. (1964). Simultaneous confidence intervals for contrasts among multinomial populations. *Annals of Mathematical Statistics, 35*, 716–725.

Grandage, A. (1958). Orthogonal coefficients for unequal intervals. *Biometrics, 14*, 287–289.

Greenhouse, S. W., & Geisser, S. (1959). On methods in the analysis of profile data. *Psychometrika, 24*, 95–112.

Greenwald, A. G. (1976). Within-subjects designs—to use or not to use. *Psychological Bulletin, 83*, 314–320.

Greenwald, A. G., Gonzalez, R., Harris, R. J., & Guthrie, D. (1996). Effect sizes and p values: What should be reported and what should be replicated? *Psychophysiology, 33*, 175–183.

Griffin, D., & Gonzalez, R. (1995). Correlation models for dyad-level models: I. Models for the exchangeable case. *Psychological Bulletin, 118*, 430–439.

Grizzle, J. E. (1961). A new method of testing hypotheses and estimation parameters for the logistic model. *Biometrics, 17*, 372–385.

Grizzle, J. E., Starmer, C. F., & Koch, G. G. (1969). Analysis of categorical data by linear models. *Biometrics, 25*, 489–504.

Groves, R., Fowler, F., Couper, M., Lepkowski, J., Singer, E., & Tourangeau, R. (2004). *Survey methodology.* New York: Wiley.

Guthrie, D. (1981). Analysis of dichotomous variables in repeated measures experiments. *Psychological Bulletin, 90*, 189–195.

Hays, W. L. (1988). *Statistics* (4th ed.). New York: Holt, Rinehart and Winston.

Hoaglin, D. C., Mosteller, F., & Tukey, J. W. (1983). *Understanding robust and exploratory data analysis.* New York: Wiley.

Hochberg, Y., & Tamhane, A. C. (1987). *Multiple comparison procedures.* New York: Wiley.

Hogg, R. V., & Craig, A. T. (1978). *Introduction to mathematical statistics* (4th ed.). New York: Macmillan.

Holland, P. W. (1973). Covariance stabilizing transformations. *Annals of Statistics, 1*, 84–92.

Huynh, H., & Feldt, L. (1970). Conditions under which mean square ratios in repeated measurements designs have exact F-distributions. *Journal of the American Statistical Association, 65*, 1582–1589.

Hyman, R., & Vogt, E. Z. (1967). Water witching: Magical ritual in contemporary United States. *Psychology Today, 1*, 35–42.

Jaynes, E. T. (2003). *Probability theory.* Cambridge, UK: Cambridge University Press.

Judd, C., Smith, E. R., & Kidder, L. H. (1991). *Research methods in social relations* (6th ed.). Orlando, FL: Holt, Rinehart and Winston.

Kastenbaum, M. A. (1960). A note on the additivity partitioning of Chi-square in contingency tables. *Biometrics, 16*, 416–422.

Kenny, D. A., Kashy, D. A., & Cook, W. L. (2006). *Dyadic data analysis.* New York: Guilford Press.

Keselman, H. J., Toothaker, L. E., & Shooter, M. (1975). An evaluation of two unequal *nk* forms of the Tukey multiple comparison statistics. *Journal of the American Statistical Association, 70*, 584–587.

Kirk, R. (1995). *Experimental design: Procedures for the behavioral sciences* (3rd ed.). Pacific Grove, CA: Brooks/Cole.

Knoke, J. D. (1976). Multiple comparisons with dichotomous data. *Journal of the American Statistical Association, 71*, 849–853.

Krantz, D. H., Luce, R. D., Suppes, P., & Tversky, A. (1971). *Foundations of measurement* (Vol. 1). San Diego: Academic Press.

Langer, E. J., & Abelson, R. P. (1972). The semantics of asking a favor: How to succeed in getting help without really dying. *Journal of Personality and Social Psychology, 24*, 26–32.

Levene, H. (1960). Robust test for equality of variance. In I. Olkin (Ed.), *Contributions to probability and statistics.* Stanford, CA: Stanford University Press.

Levin, J. R., Serlin, R. C., & Seaman, M. A. (1994). A controlled, powerful multiple-comparison strategy for several situations. *Psychological Bulletin, 115*, 153–159.

Levy, K. J. (1975). Large-sample many–one comparisons involving correlations, proportions, or variances. *Psychological Bulletin, 82*, 177–179.

Little, R., & Rubin, D. (2002). *Statistical analysis with missing data* (2nd ed.). New York: Wiley.

Loftus, G., & Masson, M. (1994). Using confidence intervals in within-subjects designs. *Psychonomic Bulletin and Review, 1*, 476–490.

Lunney, G. H. (1970). Using analysis of variance with a dichotomous dependent variable: An empirical study. *Journal of Educational Measurement, 7*, 263–269.

Marascuilo, L. A. (1966). Large-sample multiple comparisons. *Psychological Bulletin, 1966*, 280–290.

Maxwell, S. E., & Delaney, H. D. (1990). *Designing experiments and analyzing data: A model comparison perspective.* Belmont, CA: Wadsworth.

Maxwell, S. E., & Delaney, H. D. (1993). Bivariate median splits and spurious statistical significance. *Psychological Bulletin, 113*, 181–190.

McCullagh, P., & Nelder, J. A. (1989). *Generalized linear models* (2nd ed.). New York: Chapman and Hall.

Mitchell, J. (1990). *An introduction to the logic of psychological measurement.* Hillsdale, NJ: Erlbaum.

Mosteller, F., & Tukey, J. (1949). The uses and usefulness of binomial probability paper. *Journal of the American Statistical Association, 44*, 174–212.

Narens, L. (2001). *Theories of meaningfulness.* Hillsdale, NJ: Erlbaum.

Neale, J. M., & Liebert, R. M. (1986). *Science and behavior* (3rd ed.). Englewood Cliffs, NJ: Prentice-Hall.

Neter, J., Wasserman, W., & Kutner, M. H. (1996). *Applied linear statistical models.* Boston: McGraw-Hill.

O'Brien, R. G. (1981). A simple test of variance effects in experimental designs. *Psychological Bulletin, 89*, 570–574.

Peugh, J. L., & Enders, C. K. (2005). Using the SPSS MIXED procedure to fit cross-sectional and longitudinal multilevel models. *Educational and Psychological Measurement, 65*, 717–741.

Pfungst, O. (1911). *Clever Hans.* New York: Holt, Rinehart and Winston.

Prentice, D. A., & Miller, D. T. (1992). When small effects are impressive. *Psychological Bulletin, 112*, 160–164.

Ratcliff, R. (1993). Methods for dealing with reaction time outliers. *Psychological Bulletin, 114*, 510–532.

Raudenbush, S. W. (1993). Hierarchical linear models and experimental design. In L. K. Edwards (Ed.), *Applied analysis of variance in behavioral science* (pp. 365–400). New York: Marcel Dekker.

Raudenbush, S. W., & Bryk, A. S. (2002). *Hierarchical linear models: Applications and data analysis methods.* Thousand Oaks, CA: Sage.

Rindskopf, D. (1990). Nonstandard log-linear models. *Psychological Bulletin, 108*, 150–162.

Robbins, H. (1977). A fundamental question of practical statistics. *American Statistician, 31*, 97.

Roberts, F. S. (1979). *Measurement theory.* Reading, MA: Addison-Wesley.

Rodger, R. S. (1969). Linear hypotheses in 2 × 2 frequency tables. *British Journal of Mathematical and Statistical Psychology, 22*, 30–48.

Rosenthal, R. (1966). *Experimenter effects in behavioral research.* Englewood Cliffs, NJ: Prentice-Hall.

Rosenthal, R. (1967). Covert communication in the psychological experiment. *Psychological Bulletin, 67*, 356–367.

Rosenthal, R., & Rosnow, R. (1985). *Contrast analysis.* Cambridge, UK: Cambridge University Press.

Rubin, D. B. (1997). Estimating causal effects from large data sets using propensity scores. *Annals of Internal Medicine, 127*, 757–763.

Rubin, D. B. (2005). Causal inferences using potential outcomes: Design, modeling, decisions. *Journal of the American Statistical Association, 100*, 322–331.

Satterthwaite, F. E. (1946). An approximate distribution of estimates of variance components. *Biometrics, 2*, 110–114.

Scheffé, H. (1953). A method for judging all contrasts in the analysis of variance. *Biometrika, 40*, 87–104.

Steiger, J. H. (1980). Comparison of two methods for testing linear hypotheses in tables of proportions. *Psychological Bulletin, 88*, 772–775.

Stevens, S. S. (1946). On the theory of scales of measurement. *Science, 103*, 677–680.

Tabachnik, B. G., & Fidell, L. S. (1983). *Using multivariate statistics.* New York: Harper & Row.

Tiku, M. L. (1971). Power function of the F test under non-normal situations. *Journal of the American Statistical Association, 66*, 913–916.

Tukey, J. (1977). *Exploratory data analysis.* Reading, MA: Addison-Wesley.

Tukey, J. (1994). *The collected works of John W. Tukey: Multiple comparisons, Volume VIII.* New York: Chapman & Hall.

Underwood, B. J. (1975). Individual differences as a crucible in theory construction. *American Psychologist, 30*, 128–134.

Underwood, B. J., & Shaughnessy, J. J. (1975). *Experimentation in psychology.* New York: Wiley.

Upton, G. J. G. (1982). A comparison of alternative tests for the 2 × 2 comparative trial. *Journal of the Royal Statistical Society (A), 145*, 86–105.

Wald, A. (1943). Tests of statistical hypotheses concerning several parameters when the number of observations is large. *Transactions of the American Mathematical Society, 54*, 426–482.

Walker, H. M. (1940). Degrees of freedom. *Journal of Educational Psychology, 31*, 253–269.

Welch, B. L. (1938). The significance of the difference between two means when the population variances are unequal. *Biometrika, 29*, 350–362.

Welch, B. L. (1947). The generalization of "Student's" problem when several different population variances are involved. *Biometrika, 34*, 28–35.

Wickens, T. D. (1989). *Multiway contingency tables analysis for the social sciences.* Hillsdale, NJ: Erlbaum.

Wilcox, R. R. (1993). Comparing one-step M-estimators of location when there are more than two groups. *Psychometrika, 58*, 71–78.

Wilcox, R. R. (1995). ANOVA: A paradigm for low power and misleading measures of effect size? *Review of Educational Research, 65*, 51–77.

Winer, B. J. (1971). *Statistical principles in experimental design.* New York: McGraw-Hill.

Wolter, K. M. (1985). *Introduction to variance estimation.* New York: Springer.

Woodward, J. A., Bonett, D. G., & Brecht, M. L. (1990). *Introduction to linear models and experimental design.* San Diego, CA: Harcourt Brace Javonovich.

Wu, C., & Hamada, M. (2000). *Experiments: Planning, analysis, and parameter design optimization.* New York: Wiley.

Yule, G. U. (1911). *An introduction to the theory of statistics.* London: Griffin.

Index